ETHYLENE IN PLANT BIOLOGY

ETHYLENE IN PLANT BIOLOGY

FREDERICK B. ABELES

U. S. Department of Agriculture
Agricultural Research Service
National Agricultural Research Center
Beltsville, Maryland

ACADEMIC PRESS New York and London 1973

A Subsidiary of Harcourt Brace Jovanovich, Publishers

ACADEMIC PRESS, INC.
111 Fifth Avenue, New York, New York 10003

United Kingdom Edition published by
ACADEMIC PRESS, INC. (LONDON) LTD.
24/28 Oval Road, London NW1

Library of Congress Cataloging in Publication Data

Abeles, Frederick.
Ethylene in plant biology.

Includes bibliographies.
1. Plants, Effect of ethylene on. I. Title.
[DNLM: 1. Ethylene. 2. Plants–Physiology.
QK 898.E8 A141e 1973]
QK753.E8A2 581.1'927 72–9319
ISBN 0–12–041450–3

PRINTED IN THE UNITED STATES OF AMERICA

I dedicate this book
to my parents

OSCAR ABELES

META HESS ABELES

Contents

5. Stress Ethylene

6. Growth and Developmental Effects of Ethylene

7. Phytogerontological Effects of Ethylene

8. Regulation of Metabolic and Physiological Systems by Ethylene

9. Ethylene Analogs and Antagonists

10. Mechanism of Ethylene Action

11. Air Pollution and Ethylene Cycle

Preface

This book should prove useful to both undergraduate students and professional workers. A background in plant anatomy, plant physiology, or biochemistry is helpful but not essential.

Photographs of living material are used to illustrate many of the basic phenomena observed by ethylene physiologists. Much of hormone physiology is descriptive. We expose plants to growth regulators and try to explain what we see. Although the explanations are interesting, they are usually incomplete. The phenomena, however, are reproducible. The use of illustrative material provides a base for our explanations.

In an effort to be complete, I found that some observations had to be examined from different points of view. It is difficult to discuss effects of ethylene on respiration without discussing fruit ripening, and, similarly, it is impossible to cover fruit ripening without including respiration. Thus, there is overlap in some portions of the text.

I have researched the existing literature, and have attempted to include all the relevant data available. Hopefully this will ease the task of future researchers.

I would like to thank Mary D. Nelson for editing and typing the manuscript. This book could not have been completed without her aid. I am also indebted to the many colleagues mentioned in the text for supplying photographs and other materials from their research. A wise husband knows that his wife is responsible for providing the atmosphere needed to get a job done. I thank my wife Ann for her aid and comfort. Finally,

I would like to acknowledge an old debt to Allan H. Brown, my advisor and friend, who supplied the early support and understanding essential to a scientist's development and training.

FREDERICK B. ABELES*

* *Present address:* U.S. Army Medical Research Institute for Infectious Diseases, Frederick, Maryland 21701. The author's affiliation when he started writing the book was: Vegetation Control Laboratory, U.S. Army, Fort Detrick, Frederick, Maryland.

Chapter 1

Introduction

The trails of fire ants consist of a substance that leaves the body from the extruded sting. This organ touches the ground intermittently like a moving pen dispensing ink. Once the trail has been laid, groups of workers follow it from nest to food site. When the pheromone dose is massive, whole colonies follow it. The trails laid by a worker usually diffuse to below threshold concentrations within 2 minutes. Old useless trails do not linger to confuse food seekers, and the odor's intensity indicates the amount of food remaining. Individual workers moving to and from a food site lay a returning trail only when rewarded. Thus the fading trail informs the ants that their food supply is dwindling (1).

Harvester ants produce alarm substances as an aid to colony defense. Released as a vapor when the colony is threatened, the pheromone diffuses to a sphere of about 6 cm in diameter and the alarm signal lasts about 35 seconds. A single pheromone can serve a number of functions. The same 9-keto-2-decenoic acid that prevents honeybees from constructing queen cells also acts as the virgin queen's sex attractant on her mating flight. Once the drones detect the pheromone odor, they immediately head for the queen. The pheromone acts as a releaser by altering flight direction, as an aphrodisiac when the drone has caught up with the queen, as an inhibitor of ovarian development in adult workers, and as an attractant to keep swarm members together in flight.

Atkins (1) has also described the role of gaseous regulators in mammalian systems. Black-tail deer release a substance from the metatarsal gland that serves as a territorial marker. Male mice produce a substance that causes abortion in pregnant female mice to which they are not mated. The odor of a strange male blocks the pregnancy of a newly impregnated female by suppressing prolactin secretion so that the corpus luteum fails to develop and estrus is restored. These compounds may be important in the regulation of population density. In humans there are known sexual differences in the ability to smell certain substances. Immature girls and men are insensitive to some of the synthetic lactones that mature females readily perceive during ovulation, and men become more sensitive to the odor after estrogen administration.

Fungal spore germination can be regulated by long-chain alcohols and aldehydes such as pelargonaldehyde (9). Sporangiophores of phycomycetes release a volatile repellent that guides the growth of the hyphal filament through fibrous substrates. The release of this repellent appears to be important in determining the spacial organization of the sporangiophores. As the sporangiophore comes in close proximity with an obstacle or another sporangiophore, the buildup of repellent concentrations guides the growth of the filament in the opposite direction (2).

Plants also use a gaseous substance as a regulator of growth and development. This substance is ethylene, a simple C_2 unsaturated gas. The history of its discovery can be traced back to Girardin (11). He reported that leaking illuminating gas from gas mains defoliated shade trees in a number of German cities. As the use of gas for street illumination became extensive, similar reports by other workers appeared. An example of tree damage due to leaking illuminating gas is shown in Fig. 11-4 p. 261. Similarly, a good deal of the early literature on ethylene was associated with the use of smoke to accelerate fruit ripening (8), alter the sex of cucumbers (15), and promote flowering in pineapples (21).

The first worker to show that ethylene was the active component of illuminating gas was Neljubov (Fig. 1-1) in 1901 (17). While employed as a graduate student in the laboratory and greenhouse of the Botanical Institute of St. Petersburg University, he noticed that pea seedlings germinated in the laboratory grew in a horizontal direction. This was followed by a number of experiments that showed that the observed effects were not due to cultural conditions such as light and temperature but rather to the composition of the air in the laboratory. He observed that plants grown in the presence of air drawn from the outside of the laboratory grew normally in a vertical fashion. By adding illuminating gas to outside air he found that he obtained the same growth phenomenon observed with laboratory air. He suspected that the hydrocarbon content of illuminating gas was the

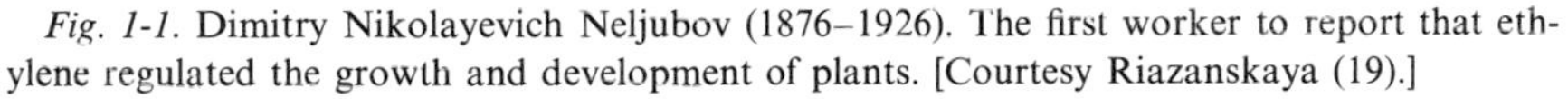

Fig. 1-1. Dimitry Nikolayevich Neljubov (1876–1926). The first worker to report that ethylene regulated the growth and development of plants. [Courtesy Riazanskaya (19).]

active factor because laboratory air passed over a CuO ignition tube lost its ability to alter the growth of pea seedlings (18) (see Fig. 1-2). Neljubov systematically examined a number of constituents of coal gas for their effects on plants. While SO_2, CS_2, benzol, xylol, and naphthalene caused injury, inhibited growth, and caused stem thickening, none of them caused the characteristic horizontal growth associated with illuminating gas.

However, acetylene and ethylene both caused horizontal growth. Ethylene was the more active of the pair since as little as 0.06 ppm caused horizontal growth. Application of ethylene during the course of vertical growth caused horizontal growth, and removal of the gas caused a return to normal vertical growth. A biographical article describing these and other research efforts of this little-known Russian plant physiologist has appeared (19).

American workers such as Crocker and Knight (6, 14) and Harvey (12) subsequently performed experiments similar to those reported by Neljubov. They confirmed the fact that the active component of illuminating gas and smoke was ethylene and in addition to causing growth effects on seedlings also accelerated the senescence of carnations.

The initial suggestion that plants produced ethylene stemmed from the report of Cousins in 1910 (5). He reported that oranges produced a gas that

Fig. 1-2. Neljubov's demonstration that the physiologically active component of illuminating gas could be oxidized by hot copper oxide. 1, Laboratory air; 2, laboratory air passed over heated CuO; 3, laboratory air passed over unheated CuO. [Courtesy Neljubov (18).]

promoted the ripening of bananas. Shortly afterward, Sievers and True (20) in California described the use of kerosene fumes to degreen lemons. Denny in 1924 (7) demonstrated that ethylene was the active component of combustion fumes from these stoves and described the use of ethylene as a ripening agent. By the middle 1930's most of the physiological effects of ethylene had been described, and in an outstanding series of papers, the Boyce Thompson workers (Fig. 1-3) characterized a great deal of what is currently known about the physiological effects of ethylene. They also demonstrated that vegetative tissue produced ethylene using etiolated pea seedling (triple response) and leaf epinasty bioassays. In 1934 Gane in England (10) was the first to prove chemically that plants produced ethylene, thereby providing the remaining evidence that ethylene was a plant hormone. A few years later, fungi were also shown to produce ethylene (3). The use of the term hormone for ethylene is an accepted practice although a careful consideration of the definition of the word leads one to think that its use in this context is not completely accurate. Hormone was a term coined by animal workers to describe endogenous chemicals active in low concentration produced in one site of the organism and transported to

Fig. 1-3. The Boyce Thompson Institute investigators. A, W. Crocker; B, F. E. Denny; C, A. E. Hitchcock; D, F. Wilcoxon; E, P. W. Zimmerman. (Photographs courtesy of Boyce Thompson Institute for Plant Research, Inc.)

A

B

C

D

E

another to bring about its physiological effect. Generally ethylene is produced by the cell or tissue that responds to the gas and for all practical purposes transport is not an important aspect of ethylene physiology. However, there is little point in originating a new name for gaseous plant regulators since ethylene is the only one known thus far and the usage of the word hormone is adequate as long as the distinction concerning translocation is remembered. The physiological significance or evolutionary advantage of a gaseous hormone is hard to assess. One obvious advantage is that detoxification or degradation mechanisms are not essential, offering an advantage in economy in the regulation of hormone levels. Unlike other known plant hormones, ethylene is removed from plants by venting to the external atmosphere. Why plants evolved ethylene as a regulator is hard to understand. As this book demonstrates, this simple C_2 unsaturated molecule plays a profound role in plant growth and regulation. The forces in time that led to use of ethylene as a hormone are unknown.

The discovery of another hormone, auxin, took place in much the same time period as that for ethylene. Early work by Charles Darwin in the 1890's demonstrated that the tip of grass coleoptiles was the site of light perception in phototropism. The Hungarian scientist Paal found that the tip provided the material required for growth phenomena in the subtending portion of the coleoptile. In the middle of the 1910's he reported that removal of the coleoptile tip stopped elongation of the remaining portion of the coleoptile. Replacing it caused the coleoptile to resume growth, while placing it on one side of the decapitated coleoptile caused growth in that direction. At about the same time, the Danish worker P. Boysen-Jensen reported that phototropism did not occur if a strip of mica was inserted between the illuminated tip and the rest of the coleoptile. Replacement of the impervious mica with a block of gelatin caused the plant to regain phototropic sensitivity, demonstrating that the tip provided a growth promoter which caused stem curvature. Chemical identification of the growth substance stemmed from the work of Fritz Went. In 1926 he reported that tips of oat coleoptiles secreted a substance into agar blocks that replaced the physiological effectiveness of intact coleoptile apices. The substance was referred to as auxin. The chemical structure of auxin was not determined directly from these plant extracts but rather from urine. Kögle and his associates in 1934 isolated, purified, and identified indoleacetic acid (IAA) as a substance in urine that mimicked the physiological effectiveness of coleoptile tips. In 1935 K. V. Thimann subsequently identified IAA as a component of plant tissues.

Research with other plant hormones was active in the years subsequent to the discovery of auxin and occupied the major interest and effort of hormone physiologists. Research with gibberellins stems from the work of

E. Kurosawa in Japan in 1926. He reported that an exudate from the fungus *Gibberella fujikuroi* caused bizarre growth promotions when applied to rice seedlings. The relationship between the fungus and perturbations of normal growth had been reported earlier by Japanese workers and had been referred to as bakanae (silly seedling) disease. The active ingredient called gibberellin was isolated in the 1930's by T. Yabuta and T. Hayashi. The second World War disrupted communications between the Japanese and the Western world, and the structure was not elucidated until the 1950's; this was accomplished by English and American workers, P. W. Brian, C. A. West, and B. O. Phinney. Gibberellins, of which thirty or more are now known, have a wide variety of physiological effects including growth promotion, breaking of dormancy, overcoming the need for cold to induce flowering, and the induction of parthenocarpic fruit.

Cytokinins were discovered in the 1950's as a result of work with promoters of tissue culture cell division. Aged or autoclaved DNA from herring sperm was found to promote cell division in tobacco pith culture. The active ingredient was called kinetin because it was able to promote cell division. The active compound was isolated and subsequently shown to be a purine derivative. A number of these purine derivatives have been discovered and as a class they are referred to as cytokinins. Kinetin itself was never found to be a product of plants, although another purine derivative discovered by D. S. Letham and C. Miller and subsequently called zeatin was isolated from corn. In addition to promoting cell division in tissue culture, cytokinins act as juvenility substances or aging retardants and promote lateral bud development.

An abscission-stimulating substance from cotton was isolated and identified as abscisic acid by F. T. Addicott, K. Ohkuma, and J. Lyon in 1965. This substance was found to be identical to a dormancy-inducing substance being studied simultaneously in England by P. F. Wareing and co-workers. This compound has a number of interesting effects, including promotion of senescence, inhibition of seed germination, induction of dormancy, and regulation of transpiration by controlling the shape of guard cells.

For reasons not entirely clear, research in ethylene physiology slowed in the late 1930's through the late 1950's although a number of workers such as W. C. Hall and H. K. Pratt actively contributed to this field. A number of possible reasons for this lack of interest exists. One reason may have been that the discovery of new hormones occupied the forefront of scientific interest and displaced contemporary research with ethylene. Second, it was difficult using bioassays and other crude chemical techniques to measure ethylene production by plants and unless some estimation of rate of production was possible its role as a controlling factor was difficult to

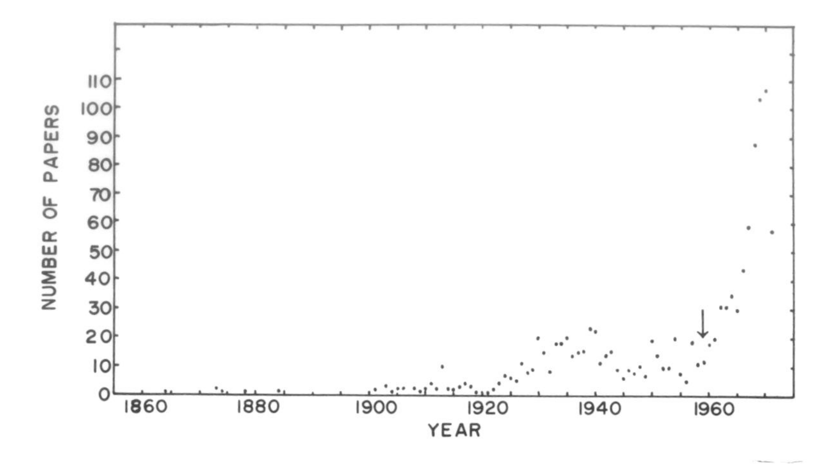

Fig. 1-4. Number of papers on ethylene physiology which have been published from 1864 to 1971. By July of 1972, twenty-seven papers had appeared. Arrow indicates the introduction of gas chromatography as an analytic tool for ethylene analysis.

determine. All of this changed with the introduction of gas chromatography as an analytic technique in 1959 by Burg and Stolwijk (4) in America and Huelin and Kennett (13) in England. The introduction of gas chromatography had a profound effect on this area of research as evidenced by the increase in the number of papers published in the last few years (see Fig. 1-4). The increased use of Ethrel as a convenient means of treating plants with ethylene in the field also gave an added push to the number of papers published as horticultural workers explored practical and commercial applications of ethylene.

Another more subtle factor may have played a role in the declining interest in ethylene physiology. The Boyce Thompson workers (23) discovered and called attention to the fact that the then newly discovered hormone auxin promoted ethylene production and that some of the effects of auxin were due indirectly to increased ethylene production. To the investigators committed to auxin research, this statement represented a challenge to the idea that auxin was the prime regulator of growth and development, and as a result Michener (16) wrote a paper indicating that ethylene did not replace auxin as a growth (elongation) promoting hormone and felt somehow that this discredited the significance of a hormonal role for ethylene. Went and Thimann, then recognized leaders in hormone research, echoed this point of view in a book (22) and suggested that the action of ethylene on plants was indirectly due to a change in auxin content and distribution. This attitude was picked up and repeated by other leading investigators of the time and a generation of students found little or no discussion of ethylene in their courses or textbooks. Except for workers directly concerned with fruits and postharvest physiology, there was little reason to be concerned with the varying and profound effects of a substance whose origins could be traced back to why air pollution damaged plants.

The remainder of this book is intended to summarize our knowledge on the role of ethylene in plant physiology.

References

1. Atkins, H. (1971). *The Sciences* **11,** 14.
2. Bergman, K., Burke, P. V., Cerda-Olmedo, E., David, C. N., Delbreck, M., Foster, K. W., Goodell, E., Heisenber, M., Meisser, G., Zalokar, M., Dennison, B. F., and Shropshire, D. (1969). *Bacteriol. Rev.* **69,** 99.
3. Biale, J. B. (1940). *Calif. Citrogr.* **25,** 186.
4. Burg, S. P., and Stolwijk, J. A. J. (1959). *J. Biochem. Microbiol. Technol. Eng.* **1,** 245.
5. Cousins, H. H. (1910). *Annu. Rep. Dep. Agr. Jamaica.*
6. Crocker, W. and Knight, L. I. (1908). *Bot. Gaz.* **46,** 259.
7. Denny, F. E. (1924). *J. Agr. Res.* **27,** 757.
8. Eaks, I. L. (1967). *Proc. Amer. Soc. Hort. Sci.* **91,** 868.
9. French, R. C., and Weintraub, R. L. (1957). *Arch. Biochem. Biophys.* **72,** 235.
10. Gane, R. (1934). *Nature (London)* **134,** 1008.
11. Girardin, J. P. L. (1864). *Jahresber. Agrikult.-Chem. Versuchssta., Berlin* **7,** 199.
12. Harvey, E. M. (1915). *Bot. Gaz.* **60,** 193.
13. Huelin, F. E., and Kennett, B. H. (1959). *Nature (London)* **184,** 996.
14. Knight, L. I., and Crocker, W. (1913). *Bot. Gaz.* **55,** 337.
15. Mekhanik, F. J. (1958). *Dokl. Vseso. Akad. Sel'skokhoz. Nauk* **21,** 20.
16. Michener, H. D. (1938). *Amer. J. Bot.* **25,** 711.
17. Neljubov, D. (1901). *Beih. Bot. Zentralbl.* **10,** 128.
18. Neljubov, D. (1913). *Imp. Acad. Sci. (St. Petersburg)* **31,** No. 4, 1.
19. Riazanskaya, K. V. (1958). *Proc. Inst. Hist. Nat. Sci. Tech., Acad. Sci. USSR* **24** (*Hist. Biol. Sci.* **5**), p. 85.
20. Sievers, A. F., and True, R. H. (1912). *U. S., Dep. Agr., Bur. Plant Ind., Bull.* **232.**
21. Traub, H. P., Cooper, W. C., and Reece, P. C. (1940). *Proc. Amer. Soc. Hort. Sci.* **37,** 521.
22. Went, F. W., and Thimann, K. V. (1937). "Phytohormones." Macmillan, New York.
23. Zimmerman, P. W., and Wilcoxon, F. (1935). *Contrib. Boyce Thompson Inst.* **7,** 209.

Chapter 2

Chemistry and Analysis of Ethylene

I. Chemistry*

A. Physical Properties

Ethylene (ethene) is a C_2 hydrocarbon gas with a double bond and a molecular weight of 28.05. It freezes at $-181°C$, melts at $-169°C$, and boils at $-103°C$. It is a colorless gas with a sweet etherlike odor. It is flammable with explosion limits of 2.75–28.60% in air. Because of its double bond it absorbs UV light at 161, 166, and 175 nm with extinction coefficients of 3.94, 3.80, and 3.70, respectively.

Ethylene is about five times as soluble in water as is oxygen. At 0°C its absorption coefficient (α) is 0.266 and at 25°C, 0.108. The absorption coefficient is the volume of gas, reduced to standard conditions (0°C and 760 mm Hg), dissolved in one volume of water where the pressure of the gas is 760 mm. At a concentration of 1 ppm in the gas phase and a temperature at 0°C, the molarity of ethylene in water is 10.1×10^{-9}, at 25°C, 4.43×10^{-9}.

B. Preparation

Ethylene is readily synthesized from a variety of compounds including alcohol in the presence of H_2SO_4 or aluminum oxide.

*See reference (82).

$$CH_3CH_2OH + H_2SO_4 \xrightarrow{(-H_2O)} CH_3CH_2OSO_3H \xrightarrow{150°C} CH_2CH_2 + H_2SO_4$$

$$CH_3CH_2OH \xrightarrow[375°C]{Al_2O_3} CH_2CH_2 + H_2O$$

At lower temperatures the production of diethyl ether is favored. Smith and Hoskins (90) reported that significant amounts of CO were also formed when H_2SO_4 was used as a catalyst. This fact was uncovered when they were preparing ethylene for inhalation experiments with mice. They found that enough CO was produced to kill mice when high levels of ethylene were tested. Interestingly enough, they reported that an atmosphere of 75% ethylene and 25% oxygen caused intoxification of a mouse. Had they increased levels a little higher to 85%, they would have discovered the anesthetic qualities of ethylene. This discovery was made 20 years later and resulted in ethylene becoming an important anesthetic until the introduction of less flammable compounds.

Triethylamine oxide also forms ethylene when heated.

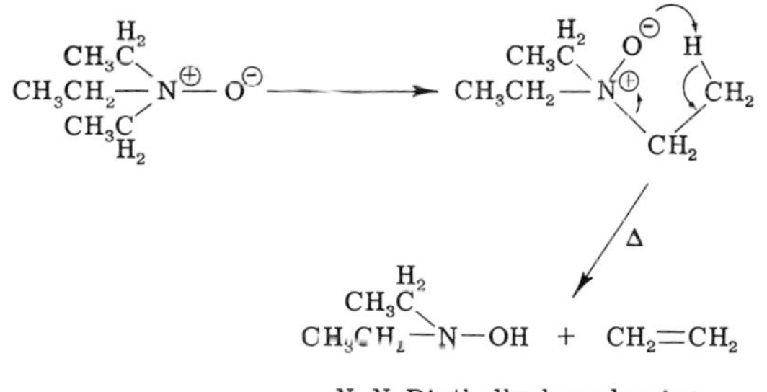

N, N-Diethylhydroxylamine

In some reactions light can serve as a catalyst. Methyl-*n*-propylketone and methionine (111) dissociate in the presence of light to form ethylene.

$$CH_3-\overset{O}{\overset{\|}{C}}-CH_2-CH_2-CH_3 \xrightarrow[\text{vapor phase}]{UV,\ h\nu} CH_3-\overset{O}{\overset{\|}{C}}-CH_3 + CH_2=CH_2$$

$$\underset{5}{CH_3-S}\ \underset{4}{CH_2}-\underset{3}{CH_2}-\underset{2}{CHNH_2}-\underset{1}{COOH} \xrightarrow[FMN]{h\nu} \underset{5}{CH_3SH} + \underset{4,\,3}{C_2H_4} + NH_3 + \underset{2}{HCOOH} + \underset{1}{CO_2}$$

Ethylene is also readily formed from β-hydroxyethylhydrazine and 2-chloroethylphosphonic acid.

$$HO{-}CH_2{-}CH_2{-}NH{-}NH_2 \longrightarrow C_2H_4 + HONH{-}NH_2$$

β-Hydroxyethyl-
hydrazine

$$Cl{-}CH_2{-}CH_2{-}P(=O)(O^{\ominus}){-}O^{\ominus} \xrightarrow{OH^{\ominus}} C_2H_4 + Cl^{\ominus} + {}^{\ominus}O{-}P(=O)(OH){-}O^{\ominus}$$

2-Chloroethyl-
phosphonic acid

The latter reaction was initially described by Maynard and Swan (59) and takes place with increasing speed as the pH is raised above 4.5. 2-Chloroethylphosphonic acid, otherwise known as Ethrel or Ethephon, is an important plant growth regulator.

It should be pointed out that practically any compound will form some ethylene when heated, for example, monoethyl phosphate, diethyl phosphate, butyric acid, ethyl iodide, methional (53), methanol, 1-propanol, 1-butanol, acetaldehyde, propanal, butanal, pentanal, hexanal, heptanal, acetic acid, propionic acid, butyric acid, ethyl ether, propyl ether, butyl ether, acrolein, acrylic acid, acetone, and pyruvate (54). A list of methods used to prepare ethylene from a variety of hydrocarbon precursors has been described by Faraday and Freeborn (24).

C. Chemical Properties

Ethylene is a reactive chemical and combines with halogens, hydrogen halides, hypohalous acids (HOCl and HOBr), water, and sulfuric acid to form, respectively, the dihalogens (i.e., 1,2-dibromoethane), ethylene chloride, ethylene bromide, 2-chloro- or 2-bromoethanol, ethanol, and ethyl sulfate. Ethylene can be oxidized with O_2, O_3, and $KMnO_4$ to form ethylene glycol or formaldehyde.

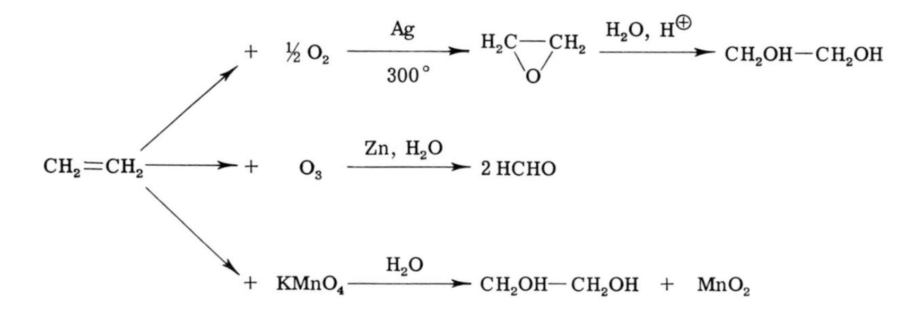

The double bond makes ethylene especially susceptible to high energy irradiation. Direct γ irradiation of ethylene results in its destruction and the formation of acetylene, butane, ethane, hydrogen, and methane (58). Ethylene labeled with ^{14}C undergoes self-irradiation to form a number of products. For each 100 eV absorbed, 10–2000 molecules are converted into polyethylene, 20 into other compounds, and 3.8 into H_2 and methane. For comparison, 100 eV of energy will convert 6.5 molecules of ethane into H_2, 1 into methane, 0.34 into propane, and 6.6 into other compounds (103).

D. Coordination Compounds with Metals

The binding of olefins to metals takes place through a coordination bond. The reaction of ethylene with various metals has been reviewed by Douglas (22). Ethylene and other olefins react with platinum but the stability decreases as the chain length or size of the molecule increases. Coordination compounds with NO, CO, and styrene are reported to be more stable than ethylene. Ethylene can be recovered by reacting Zeise's salt ($PtCl_2 \cdot CH_2{=}CH_2$) with KCN or pyridine. In water the ethylene is converted to acetaldehyde as it is released.

Ethylene reacts with palladium to form colored unstable complexes of the formula $PdCl_2 \cdot CH_2{=}CH_2$. Palladium complexes are less stable than platinum complexes and the order of stability is cyclohexene, ethylene, styrene, butylene, pinene, and camphorene. Olefin-metal complexes can also occur with Ag^+, Al^{3+}, Cu^+, Hg^+, and Zn^{2+}. Metals reported not to complex with olefins include: Cd^{2+}, Co^{2+}, Cr^{3+}, Cu^{2+}, Fe^{2+}, Fe^{3+}, Ir^{2+}, Pb^{2+}, and Ti^+. Muhs and Weiss (64) determined the equilibrium constants for a number of olefins and silver. In the case of the straight-chain compounds, the relative order closely approximates their biological activity: ethylene, 22.3; propene, 9.1; butene, 7.7; pentene, 4.9; and allene, 0.8. However, compounds not associated with biological activity can have equal or greater binding efficiency. For example, cyclooctatetraene, 91; 2-norbornene, 62; 1,5-hexadiene, 28.8; 2-methyl-1,5-hexadiene, 22.1; and 4-vinyl-1-cyclohexene, 11.2.

The mechanism of bonding is unknown although data have shown that the double bond remains intact and there is no cis-trans isomerism. In the Zeise's salt the C-C axis is approximately perpendicular to the plane of the $PtCl_3$ group and probably symmetrically arranged with respect to the platinum atom. The nature of the bonding in complexes of unsaturated compounds with Ag and other transition metals with nearly filled *d* orbitals has been pictured to involve a σ bond formed by the overlap of the filled π orbital of the olefin with the free *s* orbital and a π bond formed by overlap of the vacant antibonding π orbitals of the olefin with filled *d* orbitals of the silver.

II. Ethylene Analysis

A. Sampling Techniques

Most ethylene determinations on biological materials are performed by enclosing the tissue in a container fitted with a septum that can be punctured by a syringe needle. A sample of gas is withdrawn and transferred to a gas chromatograph for analysis. This technique is used when rates of production are of interest. A number of investigators have tried to estimate internal levels of ethylene by extracting intercellular gases by one technique or another. Earlier investigators have removed gases from tissues by boiling them out (66, 81). More commonly, tissue is placed under water and the system is subjected to a vacuum. Bubbles of gas arising from the tissue are then collected and sampled (8, 13, 14, 81). A novel technique utilizing balloons has been described (9). The technique entails placing plant material in a balloon and transferring the gas collected after extension under vacuum to a gas chromatograph. The efficiency of this technique is about 70–80%. With bulky tissues such as fruits, the gas phase can be measured directly by inserting a syringe needle into the cavity, placing the fruit under water, and withdrawing a sample (10).

B. Qualitative Analysis

The first investigator to demonstrate chemically that ethylene was produced by plants was Gane in 1935 (33). He noted that apples produced a gas that caused leaf epinasty and that was removed by bromine, ozone, fuming nitric acid, and fuming sulfuric acid. The reaction with bromine yields an oil, ethylene dibromide, which was further identified by the formation of the aniline derivative, diphenylethylenediamine.

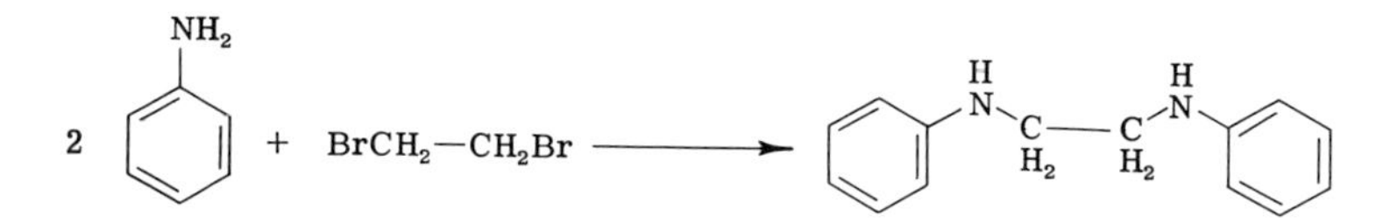

Pratt *et al.* (80) identified ethylene production from avocados in a similar fashion and in addition formed a further derivative of phenylisothiocyanate.

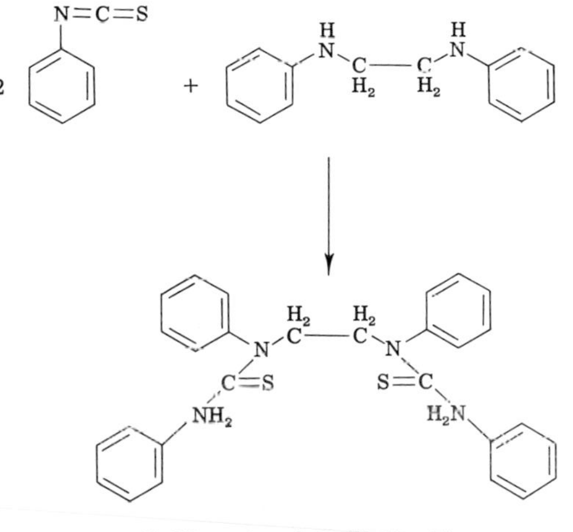

N, *N'*-Diphenyl-*N*, *N'*-dianiline-thioformylethylenediamine

This technique was also used by Mukerjee to identify ethylene production from mangos (65).

Niederl and co-workers (70, 71) identified ethylene by converting the dibromoethylene into silver acetylide by the following method.

$$BrCH_2CH_2Br + KOH + CH_3CH_2OH \longrightarrow CH{\equiv}CH$$

$$CH{\equiv}CH + CuCl_2 \longrightarrow CuC{\equiv}CCu$$

$$CuC{\equiv}CCu + 2AgNO_3 \longrightarrow AgC{\equiv}CAg + 2CuNO_3$$

Gibson and Crane (37) developed a paper chromatographic technique for qualitative analysis of ethylene–mercuric acetate addition compounds. The spots were identified with a mercury stain. This technique was useful for the identification of ethylene and other olefin compounds.

C. Quantitative Analysis

1. *Microbromination*

Christensen and Hansen (14, 47) developed a microbromination technique for ethylene analysis in the range of 25–1000 ppm in a volume of 40 ml. Ethylene was removed from the plant by subjecting it to a vacuum and transferring the gas to a flask containing potassium bromate, sulfuric acid, and potassium bromide. Subsequently, potassium oxide was introduced, and the iodine liberated was titrated with sodium thiosulfate.

$$BrO_3^- + 5\,Br^- + 6\,H^+ \longrightarrow 3\,Br_2 + 3\,H_2O$$
$$CH_2{=}CH_2 + Br_2 \longrightarrow BrCH_2CH_2Br$$
$$2\,KI + Br_2 \longrightarrow 2\,Br^- + I_2$$
$$2\,S_2O_3^{2-} + I_2 \longrightarrow S_4O_4^{2-} + 2\,I^-$$

2. Gravimetric Analysis

Christensen *et al.* (15) also developed another analytic method involving gravimetric analysis of ethylene. The technique involved absorbing CO_2, aldehydes, and other gases from a gas stream passed over fruit and then combusting the remaining hydrocarbons to CO_2. CO_2 was determined by weighing the amount absorbed on alkali. Ethylene was determined by the difference obtained by insertion of a bromine trap in the gas line. Sensitivity of this technique was in the order of 100–1000 ppm.

3. Ethylsulfuric Acid

Walls (109) developed an analytic technique involving the formation of ethylsulfuric acid. Ethylene was absorbed in a concentrated solution of silver sulfate in sulfuric acid. The ethylsulfuric acid was converted into alcohol, and chromic acid was used to oxidize the alcohol. Ethylene was measured by determining the amount of acetic acid formed.

$$H_2SO_4 + CH_2{=}CH_2 \longrightarrow CH_3CH_2SHO_4$$
$$CH_3CH_2SHO_4 + H_2O \longrightarrow H_2SO_4 + CH_3CH_2OH$$
$$CH_3CH_2OH + O_2 \xrightarrow[K_2Cr_2O_7]{} H_2O + CH_3COOH$$

4. $KMnO_4$ Oxidation

Nelson and other workers (42, 46, 66–68) have used oxidation by potassium permanganate as an analytic tool. Ethylene is absorbed in an excess of potassium permanganate to form ethylene glycol and manganese dioxide. The unused potassium permanganate was reacted with excess ferrous sulfate and the unoxidized ferrous ions were titrated with permanganate. Sensitivity of this technique was in the order of 100 ppm for volumes of 100 ml or more.

5. Perchloratocerate Oxidation

Gross (43) developed an analytic technique utilizing oxidation with perchloratocerate acid. The unreacted excess perchloratocerate was determined by titration with sodium oxalate. The method will determine 0.3 mg ethylene or 0.5 ppm ethylene in a 1 cubic foot per minute gas stream run for 12 hours.

6. Manometric Technique

Young *et al.* (112) developed a sensitive manometric technique for ethylene determinations. It is a simple technique and has been used subsequently by others (35, 73, 74, 77, 78, 112). Mercuric perchlorate used to trap ethylene from an air stream was placed in Warburg vessels and the trapped ethylene released by the addition of HCl or LiCl was measured manometrically. The technique measured quantities as small as 50 μl with an accuracy of 10%.

7. Bromocoulometric Method

A bromocoulometric method was developed by Nicksic and Rostenbach (69) for measuring ethylene levels in auto exhaust. The device utilized the reaction between bromine and ethylene to use up bromine from a solution. After the reaction between the olefin and bromine was complete, the bromine was regenerated and the amount of current utilized to regenerate the bromine taken as an indication of the amount of ethylene present. Sensitivity of the device permitted measurement of ethylene levels as low as 0.5 ppm.

8. Colorimetric Assays

Reaction of ethylene with metals has been the basis of a number of colorimetric assays for ethylene. Tomkins (104) described the use of 2% palladium chloride adsorbed on filter paper to measure ethylene in a concentration of 1% or more. The assay is not very selective, as palladium will also react with CO, H_2S, and H_2.

Stitt and co-workers (98–100) developed a rapid test for ethylene based on its reaction with mercuric oxide to form mercury vapor. Sufficient quantities of ethylene for the assay were collected on cold silica gel (−78°C). The silica gel was then heated to 150°C and the ethylene was passed over granular mercuric oxide. The mercury vapor was passed over strips of paper coated with red selenium; the length of darkening was proportional to the quantity of ethylene present.

$$CH_2{=}CH_2 + 6\,HgO \longrightarrow 2\,CO_2 + 2\,H_2O + 6\,Hg$$

An instrument was developed that could measure as little as 5 ppm ethylene in a 38-ml sample.

A more sophisticated analytic method based on the reaction of ethylene with palladium chloride has been developed by a number of investigators. Polis *et al.* (76) described a technique that permitted the determination of ethylene levels of 100 ppm in 50-ml samples. The methods were based on the reaction of ethylene with palladium chloride, phosphomolybdic acid, and acetone. Shepherd (87) developed an indicating gel for the rapid colorimetric determination of CO in air at levels of 2 ppm. The material also measured ethylene. A yellow silicomolybdate complex, the heteropoly acid $Hg[Si(Mo_2O_7)_6]$, was formed between the reaction of silica gel with ammonium molybdate. The palladium served to catalyze its reduction and the reaction product appeared to be a mixture of oxides, predominantly the blue trioctoxide.

A specialized colorimetric reagent for olefins was developed by MacPhee (57). The reagent consisted of sodium molybdate, palladous sulfate, acetic acid, and sulfuric acid. The yellow color turned green-blue in the presence of reducing agents such as ethylene, other olefins, and CO. The composition of the reagent is 6.1 g sodium molybdate, 25 ml water, 0.35 ml of a 10% palladous sulfate solution, 40 ml glacial acetic acid, and 40 ml concentrated H_2SO_4, mixed in that order. The reagent can be stored several weeks in the dark. The absorption maximum of the reduced molybdate was 685 nm. A 2-liter flask containing 45 ppm ethylene will give an absorbance of 0.80 with 20 ml reagent after incubation periods of 1.5 hours at 60°C. This reagent can also be used after absorption on silica gel. Kobayashi (51) developed similar detecting tubes that could measure ethylene levels as low as 0.01 ppm. These inexpensive detectors are presently marketed by a number of companies* as Kitagawa gas analyzers. The tubes can measure 0.1–100 ppm ethylene in a 500-ml gas sample. A hand-held pump is required to pass air over the gel at a predetermined rate.

9. Gas Chromatography

Gas chromatography is the simplest, most accurate, and most sensitive method for measuring ethylene. Standard commercial instruments are capable of measuring 5 ppb ethylene (peak height twice base line signal) in a 2-ml sample. The analysis time is normally in the order of 1 minute. This sensitivity is sufficient to measure ethylene in most samples of ambient air. In cases where greater sensitivity is required, samples of air can be passed over silica gel held at Dry Ice temperatures to absorb the ethylene. Other gases such as CO_2 and water can be removed with Ascarite and cold traps. The ethylene is then released by heating the silica gel to 150°C and sweeping the ethylene into a gas chromatograph. Some investigators have been able to improve the sensitivity of their gas chromatographs and have reported the ability to measure quantities as low as 1 ppb (7, 41).

The earliest reference to gas chromatographic analysis of ethylene was that of Turner (108) in 1943. He developed an instrument designed to measure the hydrocarbon content of natural gas. Natural gas samples

*Unico Environmental Instruments, Fall River, Massachusetts.

were absorbed on activated charcoal and eluted by raising the temperature. The column effluent was surveyed by a thermal conductivity device. Although ethylene was present in standards used to calibrate his machine, no ethylene was detected in the samples of natural gas he studied.

Sixteen years later a number of reports appeared describing the use of gas chromatography to measure ethylene evolution by plants. Burg and Stolwijk (12) and Huelin and Kennett (49) measured ethylene production from apples with gas chromatographs equipped with thermal conductivity detectors. The sensitivity of these machines was in the order of 10–100 ppm. The use of a flame ionization detector by Meigh (61) increased sensitivity of the technique by a factor of 10,000. Practically all gas chromatographs currently in use to measure ethylene utilize the flame ionization detector.

Table 2-1 presents a summary of ten different operating parameters used to measure ethylene. Not all the variation described in the literature is described here, but rather a selective sample to indicate the variety of parameters used. In my opinion, the easiest and most reliable column to use is number 2, an 18-inch piece of tubing. I prefer ¼-inch copper, filled with aluminum oxide and operated at near room temperature. Such columns have been in steady use for years without any significant change in performance. Separation of deuterated isomers of ethylene can be achieved by using column number 1. Automatic repetitive sampling of ethylene can be performed with commercially available equipment (56).

10. Other Techniques

Infrared spectrophotometers have also been used to measure ethylene. A 40-meter cell set at a wavelength of 10.5 μ was capable of measuring concentrations as low as 41 ppm in a 100-ml gas sample. A mass spectrometer was used by Davis and Squires (19) to measure hydrocarbon production from cow dung. By freezing out ethylene they were able to measure quantities as low as 0.05 ppm.

D. Dilution Techniques

The introduction of small quantities of ethylene into air streams has been an essential feature of a number of experimental programs. Different techniques employed include a system of Mariotte bottles and permeation tubes. The Mariotte bottle (79) will maintain a slow constant flow of liquid under constant pressure; hence it can be used to displace a small flow of gas from a reservoir through a flow meter into a larger airflow. Pratt and co-workers have described the assembly, a system that will maintain a concentration of ethylene from 0.003 to 1000 ppm. The system described

Table 2-1. Operating Parameters for Gas Chromatography of Ethylene

No.	Length	Diameter	Tubing	Liquid phase	Support[a]	Ratio liquid phase to support	Carrier gas	Flow (ml/minute)	Temp. (°C)	Ref.
1	50 feet	—	Nylon	$AgNO_3$ + ethylene glycol	45–60 Chromosorb P	20:80	N_2	20–40	0	52
2	18 inches	¼ inch	Tygon	—	Al_2O_3	—	He	27	23	12
3	5 feet	¼ inch	Stainless steel	—	80–100 Porapak S	—	Ar	40	50	32
4	5 feet	⅛ inch	Stainless steel	—	30–60 silica gel	—	N_2	25	65	45
5	1 m	6 mm	—	Tricresyl phosphate	40–60 C-22 firebrick	20:80	N_2	40	20	49
6	150 feet	0.01 inch	—	Golay column R polypropylene glycol	—	—	—	—	50	55
7	10 feet	½ inch	—	—	60–80 silica gel	—	He	75	180	56
8	152 cm	0.32 cm	—	—	60–72 Al_2O_3	—	N_2	50	55	60
9	5.5 m	5 mm	—	Silicone oil 550	60–80 C-22 firebrick	30:100	N_2	—	22	62
10	152 cm	2 mm	Stainless steel	*b*	80–100 Chromosorb W	15:85	N_2	15	24	97

[a] Numbers refer to mesh size.
[b] Diethyleneglycol succinate on hexamethyldiisosilazone-treated Chromosorb.

is useful if limited facilities are a problem and the duration of the experiments is not excessively long.

O'Keeffe and Ortman (72) described a metering system based on the permeation of gases through sections of plastic tubing. The permeation rates of gases enclosed in sections of plastic tubing permit the dispensing of nanogram quantities of gases. Following an initial period of a few hours to several weeks, permeation proceeds at a constant rate until the enclosed gas is exhausted. The rate of permeation is temperature dependent, but is independent of normal changes in pressure and composition of the atmosphere. Permeation of the gas was almost independent of the wall thickness of the Teflon tube, usually 0.01–0.03 inch thick. Tube diameter varied from 0.1875–0.062 inch. The permeation rate in nanograms per centimeter length per minute at 20°C for SO_2 was 203; for NO_2, 1110; for propane, 27; for acetylene, 30; for propylene, 290; and for methane, 400.

Although higher initial cost is involved, the simplest and most reliable technique is to purchase dilutions of ethylene from commercial suppliers of compressed gas and use these mixtures directly or dilute them further in an air stream.

E. Ethylene Traps

A number of methods have been used to trap and subsequently release ethylene from air streams or larger volumes of air. Liquid nitrogen (BP, −195.8°C) can be used to freeze ethylene from an air stream (ethylene FP, −181°C).

A second technique is to pass air over silica gel at alcohol Dry Ice temperature (−72°C). The gas stream should be prepurified by passing over alkali to remove CO_2 and a cold trap to remove water. Ethylene is removed from the silica gel by heating to 150°C. This technique is not specific for ethylene and other hydrocarbons will be trapped and released at the same time. A system based on the reaction of ethylene and brominated activated charcoal and subsequent release of dibromoethane has been described. The dibromoethane can then be measured by infrared analysis or gas chromatography (106).

Mercuric perchlorate is a specific trap for olefins and operates effectively at room temperature. The mercuric perchlorate is prepared by pouring 40 ml water and 210 ml 60% perchloric acid in a glass mortar. Red mercuric oxide (54.2 g) is added slowly and ground with a glass pestle to prevent caking. If the reagent is not clear it can be filtered through asbestos or a sintered glass funnel. The mixture is then placed in a glass container and diluted to 1 liter. The solution is stable and can be stored at room temperature for a year or so. Caution: perchloric acid should be stored by itself in a fireproof location and all contaminated surfaces flushed with water. An equal volume of NaCl or LiCl (4 *M*) is used to destroy the ethylene mercury complex. Solutions containing the ethylene complex should be stored at 5°C until used. Ethylene can also be absorbed on mercuric nitrate in nitric acid and released with HCl (48).

F. Radiochemical Techniques

Various biochemical pathways of ethylene formation have been proposed and tested by feeding plants and fungi with labeled precursors and assaying the radioactive ethylene produced. The most common technique is to trap the [^{14}C]ethylene in 0.25 *M* mercuric perchlorate–2 *M* perchloric acid solution. The ethylene is then released by additional volumes of 4 *M* LiCl and subsequently retrapped in 0.1 *M* mercuric acetate in methanol (35, 36, 83, 88). While the technique is specific for olefins, Burg and Burg have warned that alcohol is also absorbed by mercuric perchlorate and can be converted into ethylene (11). According to Das (18, cf. 37) the reaction between ethylene and mercuric acetate in alcohol proceeds as follows. The reaction also takes place in water but is much slower.

$$Hg(CH_3COO^-)_2 + CH_3OH \longrightarrow Hg(OCH_3)(O{-}CO{-}CH_3) + CH_3COOH$$

$$Hg(OCH_3)(O{-}CO{-}CH_3) + CH_2{=}CH_2 \longrightarrow CH_3{-}O{-}CH_2{-}CH_2{-}Hg{-}O{-}CO{-}CH_3$$

The mercuric acetate is prepared by adding 20 g mercuric acetate to 500 ml methanol plus 1 ml glacial acetic acid and filtering. This technique results in a 90% or better recovery of ethylene with no contamination with $^{14}CO_2$. The mercuric acetate is soluble in standard liquid scintillation preparations as opposed to the mercuric perchlorate complex which precipitates out. However, Thompson and Spencer (102) assayed [^{14}C]ethylene–mercuric perchlorate complex directly by utilizing a special scintillation cocktail. Earlier, Spencer (96) reported that a perchlorate–ethylene glycol mixture could be counted with a gas flow counter.

Radioactive ethylene can be assayed directly by dissolving it in scintillation fluid. Wang *et al.* (110) trapped ethylene on silica gel in a Dry Ice trap. The ethylene was released, gas chromatographed, and collected from the column effluent with a liquid nitrogen trap. Scintillation fluid was added directly to the liquified ethylene and a count was made. Labeled ethylene can also be assayed directly by transferring it to a Cary ionization counting chamber (11, 38, 50).

Because ethylene has caused significant economic loss in fruit storage and greenhouses, investigators have made a number of attempts to develop systems that would remove ethylene from the air. Most of the work has been aimed at oxidation with permanganate and ozone, reaction with bromine to form dibromoethane, and absorption with heavy metal ions.

G. Air Purification

1. Brominated Activated Carbon (BAC)

Smith and Gane (89) originally used bromine water to remove ethylene from the air. In an attempt to improve the efficiency of this system, Smock and Southwick adsorbed the bromine on charcoal and passed air through filters made of this material (91–95). Because of its small size, ethylene has little affinity for activated carbon. However, as the ethylene reacts with the bromine, its molecular weight increases from 28 to 188, and the larger dibromoethane becomes firmly bound to the carbon. A number of reports of the use of BAC in air purification in flower storage areas (28, 29), greenhouses (17), and apple storage facilities (91) have appeared. A commercial device called "Flower Saver" based on this principle has been described (3). In spite of its wide use, little data on performance characteristics of this material have appeared. Antoniani *et al.* (5) reported that BAC reduced ethylene levels from 400 ppm to 30 ppm at a flow rate of 900 liters/hour (30 cubic feet/hour) in apple storage areas.

The ratio of bromine to charcoal used varies from 5% (89) to 24% (26). The average value is 10%. The manner of preparation determines the capacity of BAC to react with ethylene. Morrow *et al.* (63) found that slow isothermal bromination was better than rapid bromination which gives off heat. Rapid bromination followed by heating to 550°C produced unreactive BAC. The isothermal bromination was achieved by pouring bromine in sintered glass funnels placed upside down so that the fumes could escape and bind to the carbon which was arranged in a layer two or three granules thick (106). At saturation pressure, capillary condensation of bromine occurs and charcoals with larger pore diameters desorb their bromine more rapidly and to a greater extent than do carbons with small pore diameters.

BAC filters are relatively efficient in that 90% of the bromine will react with ethylene (107). However, the reaction of bromine with water and its release from the carbon has imposed limitations on the practical use of this material. Turk and Van Doren (107) reported that bromine reacted with water in the following manner to release HBr.

$$Br_2 + H_2O \longrightarrow HBr + HOBr$$

$$HOBr \xrightarrow[\text{activated carbon catalyst}]{} HBr + \tfrac{1}{2}\,O_2$$

The release of HBr and Br_2 has been reported to corrode equipment in contact with BAC including that made from Monel and stainless steel. However, plastic and tinware were found to be resistant (26). The Br_2 and HBr released can also cause damage to plants. Flowers were damaged when BAC was used to purify air in flower storage areas (2). Insertion of a second activated carbon filter upstream from the BAC was required to reduce the levels of escaping HBr and Br_2. Another problem associated with BAC is the fact that it is hygroscopic and becomes wet in humid atmospheres (105).

2. $KMnO_4$

As the importance of ethylene in determining the storage life of fruits, vegetables, and flowers became apparent, investigators turned to substitutes for BAC as an air purifier. Apeland (6) found that ventilation, ozone, and permanganate were equally effective in reducing the yellowing of cucumbers in storage and extending their storage life. Dostal and Hoff (21) found that permanganate solutions were more effective ethylene absorbants than solutions of chromic acid supplemented with $SnCl_2$, $FeSO_4$, and U_2O_5. Forsyth *et al.* (30) reported that $KMnO_4$ and Celite (diatomaceous earth) was the most effective absorbant compared with permanganate and charcoal, BAC, or $AgNO_3$ in concentrated H_2SO_4.

Since that time a number of investigators have shown that permanganate adsorbed on Celite, vermiculite, silica gel, or alumina pellets is an effective ethylene absorbant.

Permanganate-Celite was prepared by spreading 500 g $KMnO_4$ over 375 g Celite in a Pyrex dish and adding 600 ml water so that the permanganate would permeate the Celite (30, 31).

Permanganate-vermiculite was prepared by mixing coarse-grade vermiculite with a saturated solution of $KMnO_4$ (6.4% at 20°C). Three and one-half grams of vermiculite will support 2.8 g of saturated permanganate solution (101).

Permanganate-silica gel was prepared by mixing 120 ml 0.1 *M* $KMnO_4$ with 100 g of 16-mesh silica gel and drying the slurry at 110°C for 16 hours (1).

Permanganate-alumina is a commercial preparation called Purafil and is available from H. E. Burroughs and Assoc., 3550 Broad Street, Chamblee, Georgia 30341.

These materials have proved to be reasonably effective under a variety of conditions. Permanganate-Celite reduced ethylene levels from 2500 ppm to 10 ppm in apple storage areas and extended the shelf life 29 days (4, 85). In the case of Chinese gooseberries, ethylene levels dropped to 0.5 ppm compared with 7 ppm for controls and extended the useful storage life for 4 weeks (101). A permanganate-Celite mixture inserted in an air-exchange unit protected carnations from ethylene air pollution (23). Purafil was used to protect bean plants from polluted air containing ethylene (1). Permanganate scrubbers are noncorrosive and in addition are effective against other air pollutants such as H_2S, SO_2, NH_3, NO, and O_3. The useful life remaining in the permanganate scrubber can be determined by observing the color change from purple to brown as the MnO_4^- is reduced to MnO_2.

3. Ozone

A number of investigators (6, 16, 27, 34, 44, 92) have examined the possibility that a reaction between ozone and ethylene could be used as a basis for a system for air purification. The idea is to mix air containing ethylene with ozone, either added separately, or formed by direct UV irradiation of the contaminated air, for a sufficient length of time to result in the destruction of ethylene. The formaldehyde and formic acid formed from this reaction as well as the unreacted ozone would be removed by subsequent passage through activated carbon. This idea has been tested most extensively by Colbert (16). According to his findings, 0.25 g ozone was capable of reacting with 15.4 mg ethylene in an air stream moving at 60 cubic feet per minute. However, this system was not totally effective as the air stream leaving the carbon filter still contained 10 ppm ethylene. A number of investigators using this system, or one like it, have reported both beneficial effects (6, 44) or unsatisfactory results (92). While the reaction between ozone and ethylene may take place with reasonable speed at high ethylene concentrations, it apparently does not do so at low levels encountered in polluted air to be a useful system for ambient air purification. With a 30-second reaction time, levels of ozone up to 10 ppm did not reduce levels of ethylene from 10 ppb to 500 ppb.*

Scott *et al.* (86) have reported that a UV lamp emitting light primarily at 253.7 nm was effective in destroying ethylene and other hydrocarbon gases.

*Unpublished results.

The mechanism involved was not determined, although ozone was thought to be involved since the lamp also emitted light at 184.9 nm. They also suggested that production of molecular oxygen might be an important part of the mechanism. Radiation from this lamp reduced ethylene levels from 250 ppb to 80 ppb in 10 minutes, and to less than 2.5 ppb in 2 hours.

4. *X Rays*

Maxie *et al.* (58) reported that 35 krad of gamma irradiation reduced ethylene levels from 1.1 ppm to 0 ppm. However, the length of time required to achieve this result was not described.

5. *Metal Binding*

The ability of ethylene to bind to mercury ions and other metals has been known for a long time (40, 84). This fact has been employed in a number of systems designed to measure ethylene both quantitatively and qualitatively (18, 20, 47, 48, 75, 81, 109). A number of workers have attempted to use this reaction as a basis for traps for air purification (21, 25, 30). However, these systems have been impractical because the reaction is too slow and the liquids too corrosive. Recently, Gillot and Delafosse (39) reported that oxidation of ethylene by oxygen was catalyzed by magnesium chromate. This suggests that catalytic oxidation of ethylene should be more fully explored.

References

1. Abeles, F. B., Forrence, L. E., and Leather, G. R. (1971). *Plant Physiol.* **48,** 504.
2. Akamine, E. K., and Sakamoto, H. I. (1951). *Amer. Orchid. Soc., Bull.* **20,** 149.
3. Anonymous. (1954). *Refrig. Eng.* **63,** 75.
4. Anonymous. (1968). *Agr. Gaz. N. S. W.* **79,** 52.
5. Antoniani, C., Monzini, A., and Kaderavek, G. P. (1959). *Ann. Sper. Agr.* **13,** 945.
6. Apeland, J. (1961). *Buli. Inst. Intern. Froid, Annexe* **4,** 45.
7. Bellar, J. T., Sigsby, J. E., Clemons, C. A., and Altshuller, A. P. (1962). *Anal. Chem.* **34,** 763.
8. Beyer, E. M., and Morgan, P. W. (1970). *Plant Physiol.* **46,** 352.
9. Blanpied, G. D. (1971). *HortScience* **6,** 133.
10. Burg, S. P., and Burg, E. A. (1962). *Plant Physiol.* **37,** 179.
11. Burg, S. P., and Burg, E. A. (1964). *Nature* (*London*) **203,** 869.
12. Burg, S. P., and Stolwijk, J. A. J. (1959). *J. Biochem. Microbiol. Technol. Eng.* **1,** 245.
13. Bussel, J., and Maxie, E. C. (1966). *Proc. Amer. Soc. Hort. Sci.* **88,** 151.
14. Christensen, B. E., Hansen, E., and Cheldelin, V. H. (1939). *Ind. Eng. Chem., Anal. Ed.* **11,** 114.
15. Christensen, B. E., Hansen, E., Cheldelin, V. H., and Stark, J. B. (1939). *Science* **89,** 319.
16. Colbert, J. W. (1952). *Refrig. Eng.* **60,** 265.
17. Darley, E. F. (1961). *Calif. Agr.* **15,** 4.

18. Das, M. N. (1954). *Anal. Chem.* **26,** 1086.
19. Davis, J. B., and Squires, R. M. (1954). *Science* **119,** 381.
20. Denny, F. E. (1938). *Contrib. Boyce Thompson Inst.* **9,** 431.
21. Dostal, H. C., and Hoff, J. E. (1968). *HortScience* **3,** 46.
22. Douglas, B. E. (1956). *In* "The Chemistry of the Coordination Compounds" (J. C. Bailar, ed.). Van Nostrand-Reinhold, Princeton, New Jersey.
23. Eaves, C. A., and Forsyth, F. R. (1969). *Florists' Rev.* **145,** 61.
24. Faraday, J. E., and Freeborn, A. S. (1957). "Faraday's Encyclopedia of Hydrocarbon Compounds" (Ethylene sheet). Chemindex, London.
25. Fidler, J. C. (1948). *J. Hort. Sci.* **24,** 178.
26. Fidler, J. C. (1950). *J. Hort. Sci.* **25,** 81.
27. Fidler, J. C. (1955). *J. Sci. Food Agr.* **6,** 293.
28. Fischer, C. W. (1949). *N. Y. State Flower Growers, Bull.* **52,** 5.
29. Fischer, C. W. (1950). *N. Y. State Flower Growers, Bull.* **61,** 1.
30. Forsyth, F. R., Eaves, C. A., and Lockhart, C. L. (1967). *Can. J. Plant Sci.* **47,** 717.
31. Forsyth, F. R., Eaves, C. A., and Lightfoot, H. J. (1969). *Can. J. Plant Sci.* **49,** 567.
32. Galliard, T., and Grey, T. C. (1969). *J. Chromatogr.* **41,** 441.
33. Gane, R. (1935). *J. Pomol. Hort. Sci.* **13,** 351.
34. Gane, R. (1937). *New Phytol.* **36,** 170.
35. Gibson, M. S. (1963). Ph.D. Thesis, Purdue University, Lafayette, Indiana.
36. Gibson, M. S. (1964). *Arch. Biochem. Biophys.* **106,** 312.
37. Gibson, M. S., and Crane, F. L. (1963). *Plant Physiol.* **38,** 729.
38. Gibson, M. S., and Young, R. E. (1966). *Nature* (*London*) **210,** 529.
39. Gillot, B., and Delafosse, D. (1970). *C. R. Acad. Sci.* **271,** 1152.
40. Gluud, W., and Schneider, G. (1924). *Ber. Deut. Chem. Ges.* **57,** 254.
41. Goeschl, J., Pratt, H. K., and Bonner, B. A. (1967). *Plant Physiol.* **42,** 1077.
42. Gronsberg, Ye. Sh. (1970). *Khim. Prom.* **10,** 755.
43. Gross, C. R. (1951). *Cornell Univ. Sta., Mimeo* **s-313.**
44. Hall, E. G. (1955). *Food Preserv. Quart.* **15,** 66.
45. Hall, I. V., and Forsyth, F. R. (1967). *Can. J. Bot.* **45,** 1163.
46. Hall, W. C. (1951). *Bot. Gaz.* **113,** 55.
47. Hansen, E., and Christensen, B. E. (1939). *Bot. Gaz.* **101,** 403.
48. Hansen, E., and Hartman, H. (1935). *Oreg., Agr. Exp. Sta., Bull.* **342.**
49. Huelin, F. E., and Kennett, B. H. (1959). *Nature* (*London*) **184,** 996.
50. Ketring, D. L., Young, R. E., and Biale, J. B. (1968). *Plant Cell Physiol.* **9,** 617.
51. Kobayashi, Y. (1957). *J. Soc. Org. Syn. Chem., Tokyo* **14,** 137.
52. Krummer, W. A., Pitts, J. N., and Steer, R. P., (1971). *Environ. Sci. Technol.* **5,** 1045.
53. Kumamoto, J., Dollwet, H. H. A., and Lyons, J. M. (1969). *J. Amer. Chem. Soc.* **91,** 1207.
54. Lieberman, M., and Kunishi, A. T. (1967). *Science* **158,** 938.
55. Lockard, J. D., and Kneebone, L. R. (1962). *Proc. Int. Conf. Sci. Aspects Mushroom Growing, 5th* p. 281.
56. Lougheed, E. C., Franklin, E. W., and Smith, R. B. (1969). *Can. J. Plant Sci.* **49,** 386.
57. MacPhee, R. D. (1954). *Anal. Chem.* **26,** 221.
58. Maxie, E. C., Rae, H. L., Eaks, I. L., and Sommer, N. F. (1966). *Radiat. Bot.* **6,** 445.
59. Maynard, J. A., and Swan, J. M. (1963). *Aust. J. Chem.* **16,** 596.
60. McGlasson, W. B. (1969). *Aust. J. Biol. Sci.* **22,** 489.
61. Meigh, D. F. (1959). *Nature* (*London*) **184,** 1072.
62. Meigh, D. F., Norris, K. H., Craft, C. C., and Lieberman, M. (1960). *Nature* (*London*) **186,** 902.

63. Morrow, J. I., Turk, A., and Davis, S. (1970). *Atmos. Environ.* **4,** 87.
64. Muhs, M. A., and Weiss, F. T. (1962). *J. Amer. Chem. Soc.* **84,** 4697.
65. Mukerjee, P. K. (1960). *Hort. Advan.* **4,** 138.
66. Nelson, R. C. (1937). *Plant Physiol.* **12,** 1004.
67. Nelson, R. C. (1939). *Food Res.* **4,** 173.
68. Nickerson, W. J. (1948). *Arch. Biochem.* **17,** 225.
69. Nicksic, S. W., and Rostenbach, R. E. (1961). *J. Air. Pollut. Contr. Ass.* **11,** 417.
70. Niederl, J. B., and Brenner, M. W. (1938). *Mikrochemie* **24,** 134.
71. Niederl, J. B., Brenner, M. W., and Kelly, J. N. (1938). *Amer. J. Bot.* **25,** 357.
72. O'Keeffe, A. E., and Ortman, G. C. (1966). *Anal. Chem.* **38,** 760.
73. Phan, C. T. (1962). *Sacc. Rev. Gen. Bot.* **69,** 505.
74. Phan, C. T. (1962). *Advan. Hort. Sci. Appl., Proc. Int. Hort. Congr., 15th, 1958.* Vol. 2, p. 238.
75. Phan, C. T. (1965). *Phytochemistry* **4,** 353.
76. Polis, B. D., Berger, L. B., and Schrenk, H. H. (1944). *U. S., Bur. Mines, Rep. Invest.* **RI-3785.**
77. Pratt, H. K. (1954). *Plant Physiol.* **29,** 16.
78. Pratt, H. K., and Greiner, C. W. (1957). *Anal. Chem.* **29,** 862.
79. Pratt, H. K., Workman, M., Martin, F. W., and Lyons, J. M. (1960). *Plant Physiol.* **35,** 609.
80. Pratt, H. K., Young, R. E., and Biale, J. B. (1948). *Plant Physiol.* **23,** 526.
81. Rakitin, Iu. V., and Povolotskaya, K. L. (1960). *Sov. Plant Physiol.* **7,** 303.
82. Roberts, J. D., and Caserio, M. C. (1967). "Modern Organic Chemistry." Benjamin, New York.
83. Sakai, S., and Imaseki, H. (1970). *Agr. Biol. Chem.* **34,** 1584.
84. Schoeller, A., Schrauth, W., and Essers, W. (1913). *Ber. Deut. Chem. Ges.* **46,** 2864.
85. Scott, K. J., McGlasson, W. B., and Roberts, E. A. (1970). *Aust. J. Exp. Agr. Anim. Husb.* **10,** 237.
86. Scott, K. J., Wills, R. B. H., and Patterson, B. D. (1971). *J. Sci. Food Agr.* **22,** 496.
87. Shepherd, M. (1947). *Anal. Chem.* **19,** 77.
88. Shimokawa, K., and Kasai, Z. (1965). *Radioisotopes* **14,** 137.
89. Smith, A. J. M., and Gane, R. (1932). *Gt. Brit., Food Invest. B., Rep.* p. 156.
90. Smith, J. L., and Hoskins, A. P. (1901). *J. Hyg.* **1,** 123.
91. Smock, R. M. (1943). *N. Y., Agr. Exp. Sta., Ithaca, Bull.* **799.**
92. Smock, R. M. (1944). *Proc. Amer. Soc. Hort. Sci.* **44,** 134.
93. Smock, R. M., and Southwick, F. W. (1943). *N. Y., Agr. Exp. Sta., Ithaca, Bull.* **813.**
94. Southwick, F. W. (1945). *J. Agr. Res.* **71,** 297.
95. Southwick, F. W., and Smock, R. M. (1943). *Plant Physiol.* **18,** 716.
96. Spencer, M. S. (1958). *Can. J. Biochem. Physiol.* **36,** 595.
97. Stahmann, M. A., Clare, B. G., and Woodbury, W. (1966). *Plant Physiol.* **41,** 1505.
98. Stitt, F., Tjensvold, A. H., and Tomimatsu, Y. (1951). *Anal. Chem.* **23,** 1138.
99. Stitt, F., and Tomimatsu, Y. (1951). *Anal. Chem.* **23,** 1098.
100. Stitt, F., and Tomimatsu, Y. (1953). *Anal. Chem.* **25,** 181.
101. Strachen, G. (1970). *Orchardist, N. Z.* **43,** 32.
102. Thompson, J. E., and Spencer, M. S. (1966). *Nature* (*London*) **210,** 595.
103. Tolbert, B. M., and Lemmon, R. M. (1955). *Radiat. Res.* **3,** 52.
104. Tomkins, S. S. (1933). *Amer. Gas. Ass. Mon.* **15,** 511.
105. Turk, A., and Messer, P. J. (1953). *J. Agr. Food Chem.* **1,** 264.
106. Turk, A., Morrow, J., Levy, P. F., and Weissman, P. (1961). *Int. J. Air Water Pollut.* **5,** 14.

107. Turk, A., and Van Doren, A. (1953). *J. Agr. Food Chem.* **1,** 145.
108. Turner, N. C. (1943). *Petrol. Refiner* **22,** 140.
109. Walls, L. P. (1942). *J. Pomol. Hort. Sci.* **20,** 59.
110. Wang, C. H., Persyn, A., and Krackov, J. (1962). *Nature* (*London*) **195,** 1306.
111. Yang, S. F., Ku, H. S., and Pratt, H. K. (1966). *Biochem. Biophys. Res. Commun.* **24,** 739.
112. Young, R. E., Pratt, H. K., and Biale, J. B. (1948). *Amer. J. Bot.* **35,** 814.

Chapter 3

Physiology and Biochemistry of Ethylene Production

I. Variety of Organisms and Rates of Production

A. Role of Ethylene in Animal Physiology

1. Ethylene Production

In spite of a few reports to the contrary, ethylene is not a normal metabolite of animals. Kokonov (80) reported that healthy rats produced ethylene at a rate of 17–20 μl/kg/day. The rate of production increased following inhalation of ethylene or subcutaneous injection of aluminum oxide. Subcutaneous abscesses or exposure to UV light reduced the rate of ethylene production. Rats with M_1 sarcomas also produced more ethylene than controls. The maximum rate of ethylene production corresponded with intensive growth of the tumor, and as the tumors regressed, the rate of ethylene production decreased again. The only other report of ethylene production by an intact animal was that of Ram Chandra and Spencer (128). They reported that air exhaled by a human subject contained between 7 and 25 ppb ethylene.

A number of workers reported that subcellular preparations from beef (54) and rats (88, 119, 129) produced ethylene. However, ethylene can be produced from practically any organic substance subjected to heat or oxidation, and the physiological significance of these observations remains in doubt.

It is known that a variety of fungi produce ethylene, though the role of gas in the development of the organism is unknown. Nickerson (118) reported that human pathogens causing lung diseases in man, *Blastomyces dermatitides, Blastomyces brasiliensis,* and *Histoplasma capsulatum,* also produced ethylene.

2. *Effects of Ethylene*

Except for its anesthetic qualities, ethylene has no effect on animals. Harvey (64) reported that guinea pigs exposed to 1000 ppm ethylene for 4 days showed no growth effects and did not develop any lesions. Rats exposed to 1% ethylene in air for prolonged periods of time also appeared normal (67). The development of nematodes was also found to be unaffected by an ethylene–releasing chemical, Ethrel (120). However, Friede and Ebert (47) found that ethylene reduced the rate of growth of animal tissue cultures. Ethylene also reduced the growth of tumor cultures and altered their capacity for implantation in healthy animals. Ethylene also affected previously developed tumors. Approximately 15% of the diseased animals treated with ethylene showed a regression and disappearance of tumorous tissue, while other animals demonstrated a reduction in the intensity of tumor growth.

Early experiments on the effect of ethylene on animals were hampered by the fact that the investigators were dealing with impure preparations resulting in death due to the presence of CO. For example, Smith and Hoskins (139a) studying the toxic effect of illuminating gas reported that ethylene made from H_2SO_4 and alcohol contained enough CO to kill mice. CO-free ethylene made from ethylene dibromide applied as a 72% ethylene–28% oxygen mixture to mice caused intoxication and sleepiness. When the animals were returned to air they regained a normal appearance.

According to Luckhardt and Lewis (99), the first successful use of ethylene to cause sleep was by Luessem in 1885 using a 75% ethylene–25% oxygen mixture. However, the anesthetic qualities of ethylene were not utilized until the work of Luckhardt in 1923 (98, 99). The concentration required for anesthesia in human patients averaged about 82% ethylene. In some cases, however, the concentration had to be raised to 90 and 95% to effect anesthesia. The effect of ethylene was rapid; only 15 seconds were required to induce sleep. Patients normally regained consciousness 3 minutes after the gas was removed. Guinea pigs have been ethylenized for 6 hours and recovered immediately with no apparent aftereffects (23). Besides inducing sleep, ethylene had few other effects on the patients. Some workers reported a reduction in rectal temperature (22, 23) and sweating (99).

Other gases such as ether, acetylene, N_2O, and cyclopropane are also anesthetic and seem to share with ethylene a planar molecular structure. The mode of action of these gases in animals is different from that observed in plants because of the high concentrations required for an effect and the relative ineffectiveness of ether, N_2O, and cyclopropane as metabolic regulators in plants. In addition, CO_2 has a promotive or synergistic effect on ethylene anesthesia as opposed to its competitive action in plants (79).

Ethylene lost favor as an anesthetic as other less flammable gases came into use. As of 1933 there had been 1,005,375 ethylene anesthesias administered with twenty explosions, resulting in five deaths and one injury. In 1939 there were twenty-six explosions, and three patients and one anesthetist died. The hazard due to explosions could be reduced by keeping the relative humidity above 56% to reduce ignition by static electricity (59).

B. Bacteria

Little is known concerning the role of ethylene in bacterial physiology. Certain strains of *Pseudomonas solanacearum* associated with prematurely ripening bananas were found to produce ethylene (46). Rate of ethylene production followed the growth of the cells as well as the rate of respiration. When cells were removed from culture they continued to produce ethylene in a medium of basal salts. Ethylene production was also associated with bacterial rot of cauliflower. However, in this case the pathogen, *Erwinia carotovora,* did not produce ethylene (100). Enzymes released by the bacteria caused the host tissue to produce the ethylene observed. Ethylene production by *Streptomycetes* has also been reported (68).

Ethylene consumption by various strains of hydrocarbon-utilizing bacteria has been reported (147, 161). Zobell (161) pointed out that hydrocarbon bacteria require oxygen and convert 80–90% of the assimilated carbon into CO_2. Uptake of ethylene and other hydrocarbons by the soil has been demonstrated (7). The mechanism was probably biological since heat and anaerobiosis inhibited the removal of ethylene from the air.

Nitrogen-fixing bacteria are able to reduce acetylene to ethylene by means of a pathway associated with the reduction of nitrogen into ammonia. This technique has been used to assay for nitrogenase as well as the presence of nitrogen-fixing bacteria in the soil (41).

C. Fungi

Ethylene production from fungi was first suggested by a report by Gane in 1935 (52). He found that yeast produced a substance causing abnormal growth of seedlings. The active component could be removed by bromine water and ozone. While the substance was probably ethylene, its origin was somewhat in doubt because compressed gas was the source of oxygen for his experiments. When the experiments were repeated with air the yeast no longer produced ethylene. A possible explanation for these observations is the fact that commercial oxygen is often contaminated with levels of ethylene equal to or greater than 0.05 ppm.

The original observation by Biale (17) and Miller *et al.* (116) in 1940 that *Penicillium digitatum* produces ethylene probably represents the first correct description of ethylene production by fungi. Since that time many other workers have reported ethylene production by a variety of fungi. A list of effective ethylene-producing fungi is presented in Table 3-1. However, not all fungi tested produce ethylene, and a number of inactive species have been described (17, 18, 68). Ilag and Curtis (68) reported that only 25% of the 228 species they tested produced ethylene.

Table 3-1

Summary of Fungi Known to Produce Ethylene (68)

Species	ppm Ethylene accumulated after 24 hours
Alternaria solani	0.32
Ascochyta imperfecti	9.93
Aspergillus clavatus	514.0
Aspergillus flavus	12.8
Aspergillus ustus	0.86
Aspergillus variecolor	0.25
Botrytis spectabilis	0.25
Cephalosporium gramineum	12.4
Chaetomium chlamaloides	0.41
Dematium pullulans	0.61
Hansenula subpelticulosa	3.21
Myrothecium roridum	0.90
Neurospora crassa	0.90
Penicillium corylophilum	10.7
Penicillium luteum	0.18
Penicillium patulum	2.41
Schizophyllum commune	3.64
Sclerotinia laxa	0.48
Scopulariopsis brevicaulis	0.32
Thamnidium elegans	15.5
Thielavia alata	0.25

The biochemical pathway for ethylene formation in fungi is not the same as that employed by higher plants. For example, Owens *et al.* (121) found that the methionine analog rhizobitoxin blocked ethylene production by sorghum seedlings but not *Penicillium.* As described more fully later in this chapter, methionine is thought to be the precursor of ethylene in higher plants, while ethanol may be the precursor in fungi. However, the pathway in both higher plants and fungi is the same to the extent that oxygen is required for ethylene production.

The rate of ethylene production generally follows the growth curve of the fungus (44, 53, 122, 140). See Figs. 3-1 and 3-2. However, Meheriuk and Spencer (110) reported that the greatest rates of ethylene production from *Penicillium* were associated with spore germination and the rates fell off as the mycelium matured. Rates of ethylene production are extremely variable and are known to be regulated by growth conditions, media composition, and different isolates from a single culture. Spalding and Lieberman (140) found that twelve single spore cultures of *Penicillium* after 7 days' growth in a closed system varied in ethylene production from virtually none (0.1 μg) to 36 μg.

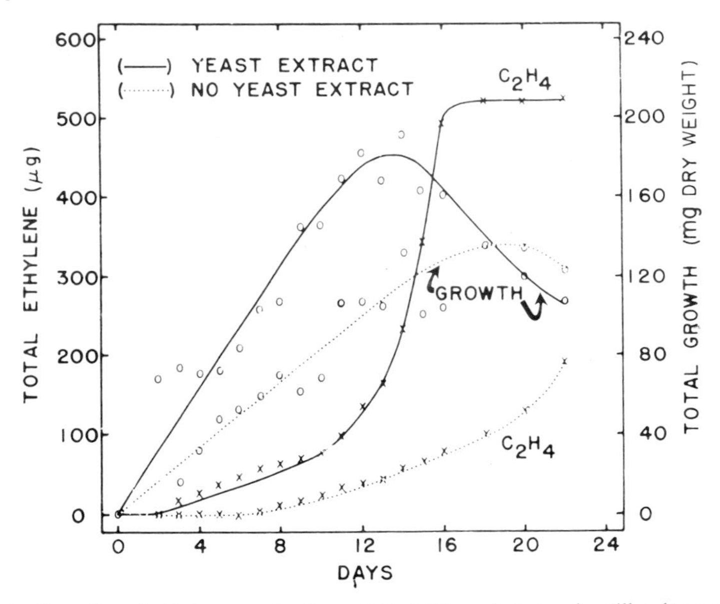

Fig. 3-1. Growth and ethylene production of *Penicillium digitatum* in still culture at 25°C for 22 days in an open system containing modified Pratt's medium, with and without yeast extract. The values obtained for ethylene were converted to production on a 24-hour basis, and the daily values were added to give a cumulative figure for total ethylene. [Courtesy Spalding and Lieberman (140).]

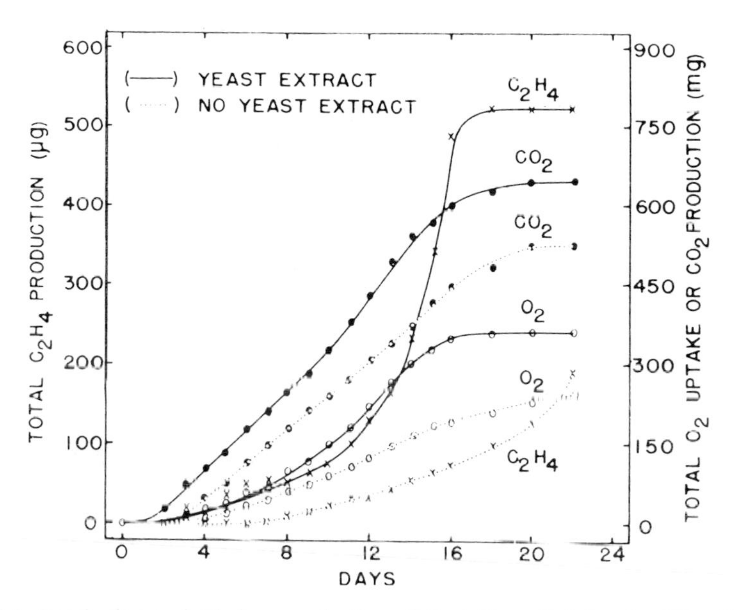

Fig. 3-2. Respiration and ethylene production of *Penicillium digitatum* in still culture at 25°C in an open system containing modified Pratt's medium, with and without yeast extract. [Courtesy Spalding and Lieberman (140).]

The biological significance of fungal ethylene is not known. The gas has been shown to regulate spore germination and the rate of mycelial growth. It may also serve as a means of hastening senescence in host tissue, thereby facilitating the growth of the fungus by increased availability of substrates for growth. See Chap. 6 for a more detailed review of the role of ethylene in fungal physiology.

D. Higher Plants

Rates of ethylene production during the development of higher plants vary from organ to organ and time of development. Differences in ethylene production by various parts of the etiolated pea seedling are shown in Fig. 3-3. The highest rates of ethylene production were associated with meristematic and nodal tissue, while lower rates were observed in the internodal regions. Blanpied (21) followed changes in ethylene production by apple leaves and flowers during development. He observed that ethylene production was highest in dormant buds and decreased slowly as the leaves and flowers expanded (Table 3-2). Ethylene production increased again during the senescence and abscission of floral and leaf tissue (Tables 3-2 and 3-3)

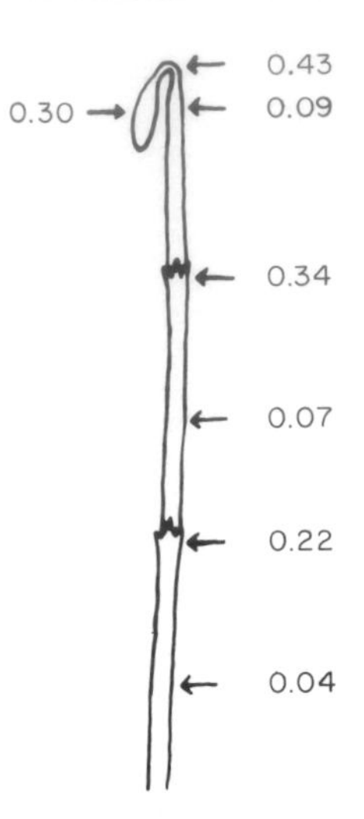

Fig. 3-3. Distribution of ethylene production along the etiolated pea stem. Ethylene production is expressed as μl/kg fresh wt/hr. [Courtesy Burg and Burg (28).]

Table 3-2

EXTRACTABLE ETHYLENE CONTENT OF GOLDEN DELICIOUS APPLE TISSUE DURING VARIOUS STAGES OF FLOWER BUD DEVELOPMENT AND BLOOM (21)

		μl Ethylene/kg fresh wt		
Stage of flower (bud) development	Days[a]	Vegetative buds	Flower buds	Flowers
Dormant	0	1.09	0.99	—
Dormant	14	0.86	0.73	—
Dormant	31	0.75	0.68	—
Green tip	40	—	0.41	—
Half-inch green	45	0.59	0.28	—
Tight cluster	48	0.53	0.22	0.03
Pink	58	—	—	0.05
Full pink	59	0.14	—	0.06
Full bloom	66	—	—	0.05
Petal fall[b]	67	—	—	0.29
Flower abscission				
Abscising flowers	79	—	—	0.28
Developing fruits	79	—	—	0.11

[a] Starting March 13.
[b] Flowers without petals.

Table 3-3

EXTRACTABLE ETHYLENE CONTENT OF LEAVES AND FRUITS OF SEVERAL APPLE CULTIVARS DURING A GROWING SEASON[a]

	μl Ethylene/kg fresh wt					
	Red Astrachan		McIntosh		Golden Delicious	
Month	Leaves	Fruits	Leaves	Fruits	Leaves	Fruits
June	0.15	0.05	0.19	0.03	0.18	0.03
July	0.13	0.01	0.10	0.01	0.17	0.01
August	0.24	15.00	0.20	0.01	0.26	0.01
September	0.21	—	0.14	0.80	0.12	0.02
October	0.40	—	0.22	6.30	0.34	1.11

[a] Data from Blanpied (21).

indicating that in some varieties ethylene played an active role in regulating leaf and flower drop. Note, however, that in McIntosh leaves only a slight increase in ethylene production was observed.

Ethylene production increases during fruit development. Figure 3-4 shows a typical increase in ethylene production and respiration during the development of a climacteric fruit, the tomato. Immature 17- and 25-day-old fruit, while capable of responding to ethylene in terms of increased coloration and respiration, were not sufficiently mature to produce ethylene normally associated with fruit maturation and ripening. As more mature fruit were studied the increase in ethylene production came sooner after picking. Not all fruits demonstrate accelerated respiration during ripening even though rates of ethylene production increase. Forsyth and Hall (45) observed an increase in ethylene production and maturation in developing cranberries but no increase in respiration (see Fig. 3-5).

E. RELATIONSHIP BETWEEN INTRACELLULAR ETHYLENE ACCUMULATION AND THE RATE OF ETHYLENE PRODUCTION

The diffusion gradient in tissue is determined by the surface-to-volume ratio, the resistance to diffusion, and the rate of ethylene production. The amount of ethylene present outside the plant has no effect on the gradient but changes the internal level. For example, if the gradient between internal and external levels of ethylene is 1 ppm in ethylene-free air, then the addition of 0.1 ppm to the external gas phase will cause the internal levels to rise to 1.1 ppm. The gradient, however, remains the same.

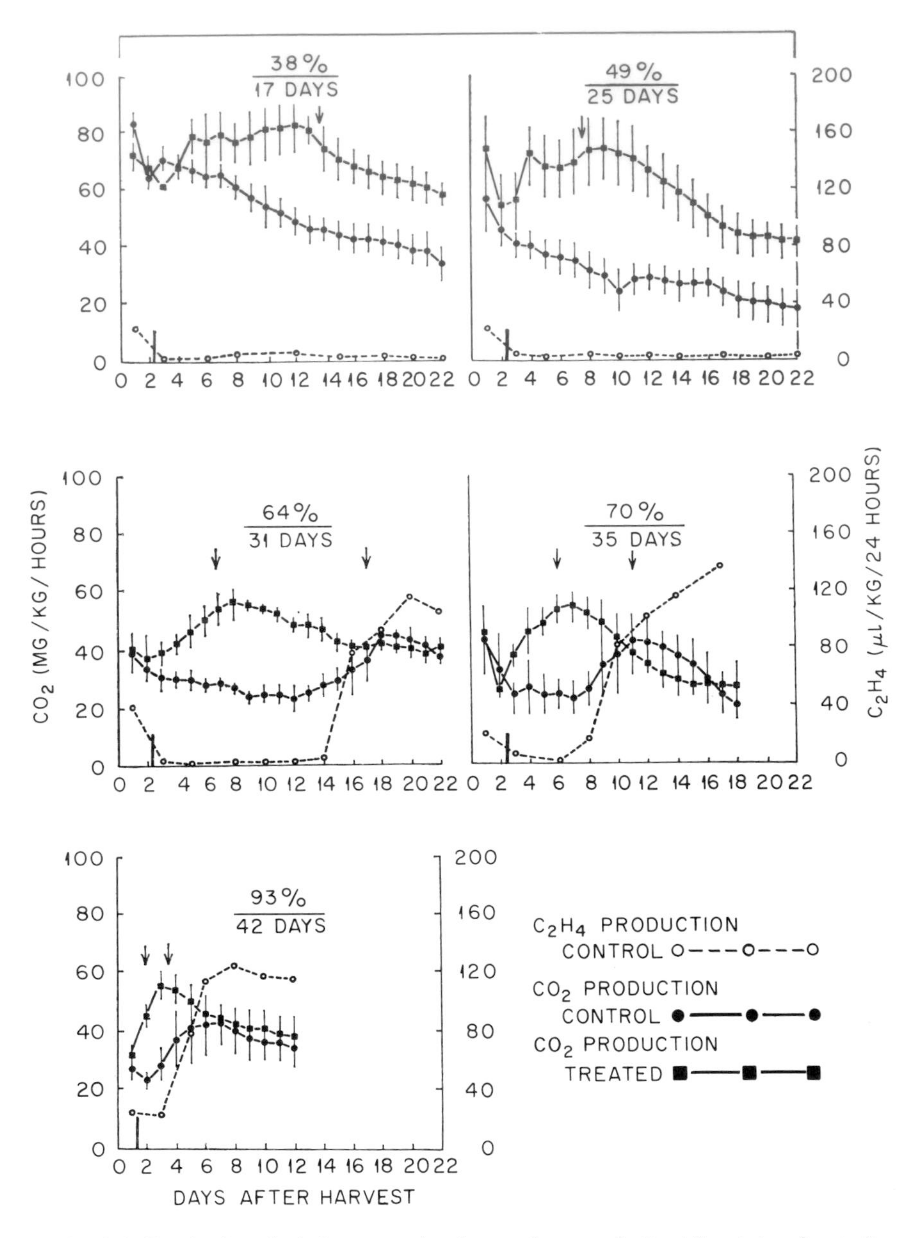

Fig. 3-4. Respiration of ethylene-treated and control tomato fruits at five states of maturity. Also shown is ethylene production by control fruits. The arrows above the curves indicate the time at which the fruits showed the first detectable red color (breaker stage), and the vertical line in the lower left-hand corner indicates the time when the ethylene treatment commenced. [Courtesy Lyons and Pratt (102).]

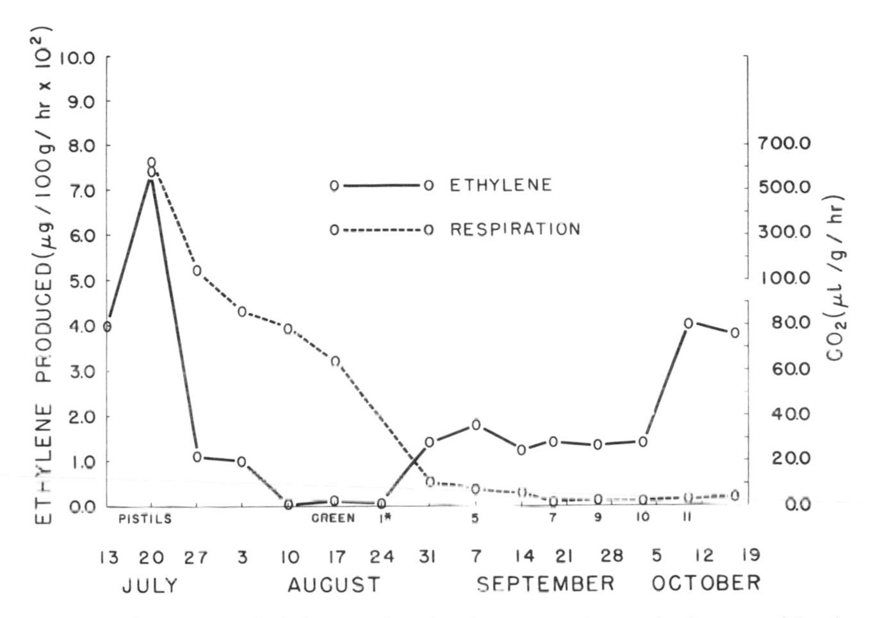

Fig. 3-5. The amount of ethylene produced and accompanying respiration rate of developing flowers and fruit of *Vaccinium macrocarpon*. Numbers along the abscissa indicate increasing redness of the cranberries. A rating of 1 indicates the first trace of red coloration. [Courtesy Forsyth and Hall (45).]

A number of workers have compared the rate of ethylene production with internal levels of ethylene. The ratio between internal levels of ethylene and the rate of production can be expressed as a conversion constant which is equal to: ppm ethylene in the tissue per μl ethylene produced per kg fresh weight per hour. The data in Table 3-4 show that the conversion constant for fruit is about 10 times greater than that observed for vegetative tissue. The difference between surface-to-volume ratios of bulky fruit tissue as opposed to vegetative tissue is one reason for the differences observed. The surface-to-volume ratio for fruit has been calculated as 1 cm^2/cm^3, while similar values for leaves were 82, for stems 78, and roots 121 (13). In addition to surface-to-volume relationships, the difference in epidermal anatomy between fruit and leaves is an important factor in regulating gas movement. As assimilatory structures, leaves have greater porosity compared to fruits, which are modified to minimize water loss. Ben-Yehoshua and Eaks (14) calculated that the surface of an orange has fourteen stomates per square millimeter, with an average size of 13 $\mu m \times 5$ μm. The citrus leaf has 468 stomates per square millimeter, with an average size of 23 $\mu m \times 17$ μm.

The conversion constant is remarkably uniform from one fruit to another

Table 3-4

RELATIONSHIP BETWEEN THE RATE OF ETHYLENE PRODUCTION BY TISSUE AND INTERNAL LEVELS

Tissue	Conversion constant (ppm per 1 μl ethylene · kg · hr)	Internal levels (ppm)	Ref.
Fruit			
Orange (Navel)	0.5–1	0.1–1.0	14, 15
Apple (McIntosh and Baldwin)	2.0–13	25.0–2500	25
Avocado (Cloquette)	2.2	29.0–74.0	25
Banana (Gros Michel)	1.9	0.05–2.1	25
Lemon	1.8	0.11–0.17	25
Lime	3.4	0.3–1.96	25
Mango	3.8	0.04–3.0	25
Nectarine	13.0	3.6–602	25
Orange (Valencia)	4.0	0.13–0.32	25
Passion fruit	9.0	466–530	25
Peach (Elberta)	5.2	0.9–20.7	25
Pear (Bosc)	5.0	80	25
Pineapple	2.0–7.0	0.16–0.40	25
Plum	3.9	0.14–0.23	25
Summer squash	1.9	0.04–2.1	25
Tomato	3.2–4.4	3.6–30	25
Pineapple	3000	80–1140	42
Cantaloupe	2.5	1.0–16	101
Tomato	3.0	—[a]	102
Banana	1.0	0.01	151
Leaf			
Huisache	0.024	—	13
Cotton	0.25–0.34	0.25–0.75	16
Roots			
Pea	0.3	2.0	35
Pea	0.16	—	36
Stem			
Pea	0.13	—	58
Bean	0.42	0.8	75

[a] Data not available.

and over a wide range of ethylene production. As Burg and Burg (25) pointed out, the range in internal levels varied by a factor of 1000 between different fruits, while the conversion constant varied only by a factor of 10. The conversion constant is not specific for ethylene. Any other gas produced by plants, such as CO_2, subject to the same physical limitation as ethylene, will yield a similar constant. Kang and Ray (75) reported that ethylene and CO_2 both had conversion constants of 0.42 ppm/μl/kg/hr. Leonard and Wardlaw (86) presented CO_2 production data which yielded a conversion constant of 0.5 for bananas.

Zimmerman *et al.* (160) reported that ethylene transport occurred across dead as well as living tissue, and that fumigating one leaf caused ethylene symptoms throughout the plant. The diffusion path in plants depends on the structure of tissue. Movement in wood is 100 times faster longitudinally than horizontally (43). In some fruits, the major exit for ethylene is the point of stem attachment. In pepper and tomato, 60% of the ethylene was reported to diffuse through the pedicel (27). Similarly, the pedicel was found to be the primary route for ethylene movement in citrus (9). In apple, however, the primary site of diffusion was the skin (27).

II. Environmental Control

A. Temperature

The optimum temperature for ethylene production by apples is 30°C (31, 63, 123). As the temperature was raised the rate of production fell until ethylene production ceased at 40°C. Burg and Thimann (31) reported that Q_{10} between 10°C and 25°C was 2.8. Over the same temperature range, the Q_{10} values for oxygen uptake and CO_2 production were found to be 2.75 and 2.5, respectively. The temperature inhibition at 40°C was reversible. After 5 hours at room temperature, about 50% of the temperature-induced inhibition usually disappears. A similar effect had been reported for pears (62). Temperature-regulated ethylene production has been shown to play a role in seed germination and senescence.

Ketring and Morgan (77) reported that heat shock (40°–45°C) promoted the germination of dormant Virginia-type peanuts and also increased the rate of ethylene production by fivefold. Ethylene itself was shown to promote germination.

The ability of plums to ripen at elevated temperatures was found to be correlated with the rate of ethylene production (150). Plums that ripened normally at 32°C produced measurable amounts of ethylene at that temperature, but production was suppressed in varieties that did not ripen normally at high temperatures. At 32°C, the Beauty variety evolved large amounts of ethylene and ripened normally. Santa Rosa produced an intermediate amount of ethylene and the fruit ripened but without full color. Durate and Kelsey at 32°C produced no ethylene and did not ripen normally. Ethylene treatment prior to holding at high temperature resulted in an improved ripening response and increased the production of ethylene by the fruit. At 32°C, all varieties ripened normally after ethylene treatment, indicating that the high temperature effect was directed at an inhibition of ethylene production as opposed to an inhibition of ripening.

Nichols (117) demonstrated that the keeping qualities of carnations were extended by storing the plants under refrigeration, and the wilting of the flowers was closely correlated with the rate of ethylene production. For example, at 21°C, wilting and increased ethylene production occurred after 6 days; at 10°C, 16 days; and at 2°C the flowers retained water for over 30 days, and no increase in ethylene production was observed.

A physiological disease called "premature ripening" of Bartlett pears has caused a 5 million dollar loss in Washington, Oregon, and California (152). No disease or insect was associated with the disease. Drop was correlated with low temperatures during the growing season and was prevalent in pears grown at higher elevations. Cold days consisted of 17°C days and 8°C nights, while control temperatures were 24°C days and 15°C nights. Approximately 17 days after the start of the experiment ethylene production increased in the cold-treated pears followed by accelerated respiration and ripening.

B. Carbon Dioxide

Depending on the tissue, CO_2 can inhibit, promote, or have no effect on ethylene production. Ethylene production by apple fruit was inhibited with CO_2 concentrations between 10% (127) and 80% (31). The effect may have been due to the inhibition of ripening by CO_2 and a subsequent slowdown of ethylene production associated with the climacteric.

The observation that removal of CO_2 by KOH reduced ethylene production by sweet potato roots suggested to Imaseki *et al.* (70) that CO_2 acted as a stimulator of ethylene production. However, CO_2 was found to have no effect on ethylene production by beans (5) and citrus fruits (15, 131).

C. Oxygen

Except for rice (83), low concentrations of oxygen invariably inhibit ethylene production. Rice may represent an unusual system since these plants are adapted to growth under low oxygen tensions. Apparently respiration operates at sufficient efficiency (under low oxygen levels) to meet the energy needs of these plants.

The inhibition of ethylene production under anaerobic conditions has been observed by many workers for a variety of tissues (12, 31, 34, 39, 40, 60, 96, 97, 106, 127). Figure 3-6 shows the effect of varying oxygen concentrations on ethylene production and respiration by apple tissue. The data show that the effect of oxygen on ethylene production is similar to its effect on respiration and suggest that ethylene production is dependent

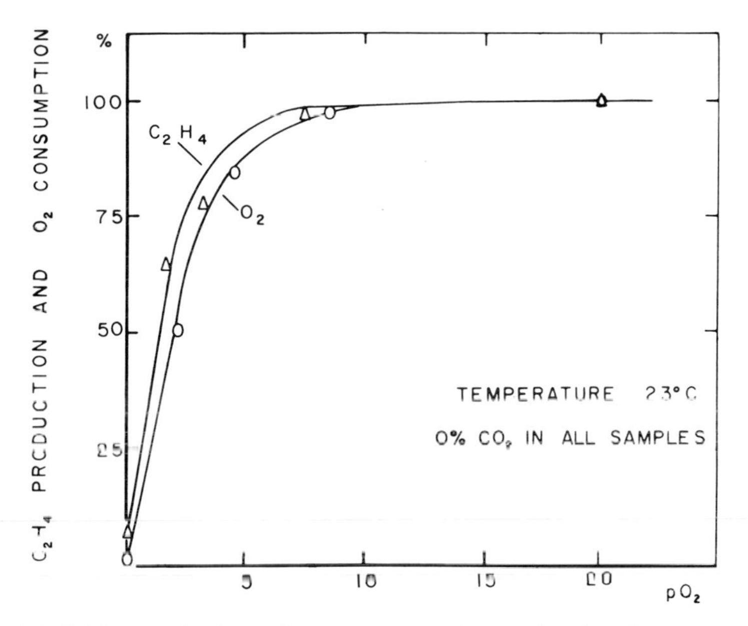

Fig. 3-6. Ethylene production and oxygen consumption as a function of oxygen concentrations (pO_2). Measurements were made over a 1-hour period after the gas mixture had been vacuum-impregnated into McIntosh apple tissue sections. All mixtures contained only O_2 and N_2. The rates of both ethylene production and oxygen consumption of the controls are represented at 100%. [Courtesy Burg and Thimann (31).]

on aerobic production of energy. Since apple tissue shows a Pasteur effect, it appears that ethylene production can not occur with the energy supplied by fermentation. Removal of oxygen results in a rapid decrease in the rate of ethylene production. Figure 3-7 shows that apple tissue stopped making ethylene soon after they were placed in nitrogen (curve A). When the sections were returned to air (curve B) ethylene production started immediately and at a rate greater than controls. Presumably, an ethylene precursor accumulated in the tissue under anaerobic conditions which could be converted rapidly into ethylene in the presence of oxygen.

Low concentration of oxygen delays fruit ripening. The effect appears to be due to an inhibition of ethylene production as opposed to a lack of respiratory energy for ripening or a lack of oxygen to permit ethylene action.

A number of workers (65, 106) have shown that low oxygen levels (0.5–5%) delay ripening of bananas. The addition of ethylene to the air initiated normal ripening changes in terms of respiration and coloration. Apparently the limiting factor under these conditions was the availability of ethylene and not oxygen.

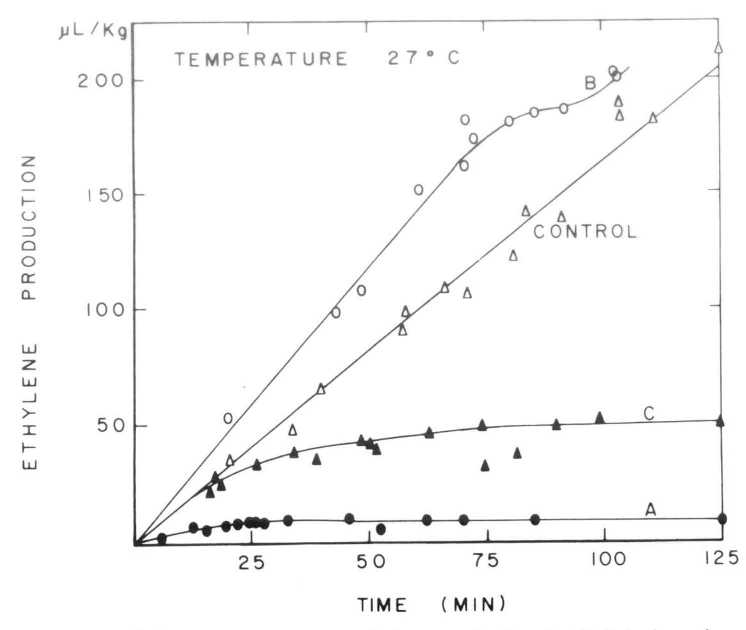

Fig. 3-7. Effects of nitrogen treatment on ethylene production by McIntosh apple sections. Control curve, production of ethylene by sections kept under atmospheric conditions for 4 hours before experiment. Curve A, ethylene production immediately after tissue was placed under nitrogen. Curve B, ethylene production by tissue previously stored under nitrogen for 4 hours, and then replaced in air. Curve C, in these experiments, the tissue had been stored in nitrogen for 4 hours, placed in air for 5 minutes, reevacuated, and returned to nitrogen, after which the ethylene production was recorded. [Courtesy Burg and Thimann (31).]

D. Light

Ethylene production can be regulated by light. Depending on the tissue involved, light can increase or decrease the rate of ethylene production, and in some cases it appears as if the change in the rate of ethylene production mediates the effect formerly attributed to light.

An increase in the rate of ethylene production following illumination has been observed in lettuce seeds (8), cranberries (37), sorghum seedlings (38), rose tissue cultures (85), and oat seedlings (110). In all cases the effect was small; only a doubling in ethylene production was observed. The action spectrum has not been established. The effect of light on lettuce seeds was shown not to be due to a red far-red system, eliminating phytochrome as the mediating pigment. Unpublished work of L. E. Craker has shown that the blue end of the spectrum was most efficient in promoting ethylene production from sorghum. While this is the same region most effective in phototropic effects, it has not been established that the two pigment systems are related. The effect of light on cranberries was found to be twofold. Not only did it promote ethylene production (35), but it was also required for ethylene-induced anthocyanin formation.

The decrease in ethylene production associated with hook opening and other light-mediated processes in seedling development has been shown to be regulated by phytochrome. Goeschl *et al.* (57) found that ethylene production decreases after etiolated pea seedlings were exposed to a single dose of red light. Far-red irradiation following the red treatment reversed the effect of red light. Figure 3-8 shows the change in the rate of ethylene

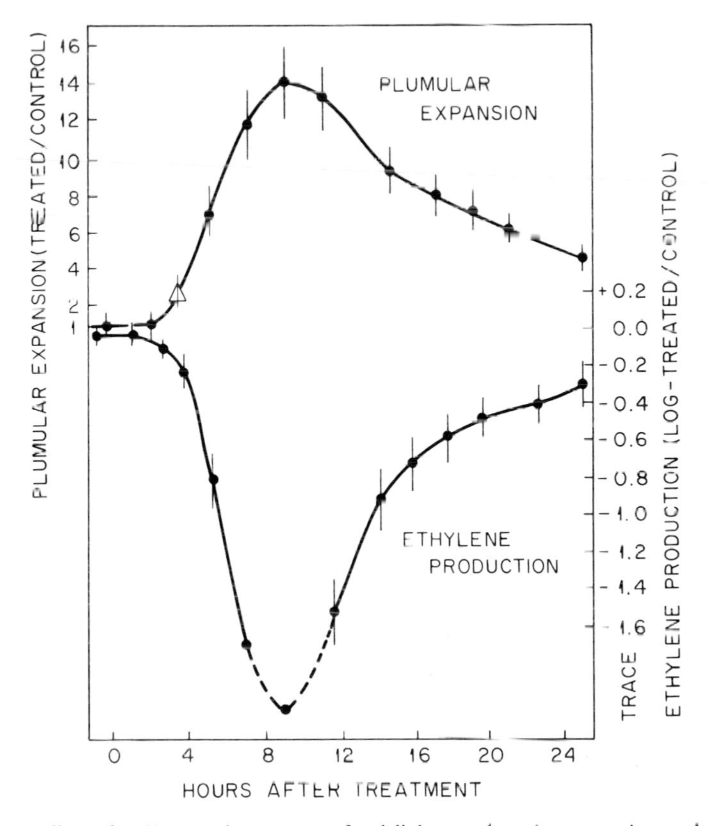

Fig. 3-8. Effect of a 60-second exposure of red light on plumular expansion and ethylene production in etiolated Alaska pea seedlings. The curve of relative plumular expansion was based on the relative increase in weight of treated tissue compared to control tissue. Ethylene production was determined from the net concentration in the effluent air stream from chambers enclosing the upper part of the epicotyls. The relative ethylene production rate is presented on a log scale, since preliminary experiments showed leaf expansion to be inversely proportional to the log of the concentration of ethylene treatments. [Courtesy Goeschl *et al.* (57).]

production following illumination with red light. Goeschl *et al.* (57) observed that red light also caused the plumular hook to expand (Fig. 3-8) and suggested that ethylene intervened as a regulator in the phytochrome control of plumular expansion. A number of subsequent workers (29, 56) confirmed this idea and extended the observations to other systems. However, the inhibition of ethylene production by red light does not account for all of the red light effect. For example, the sensitivity of pea hooks to ethylene decreases following irradiation. In the light, the saturating dose for hook closure was 0.64 ppm, and the half-maximum concentration was 0.05 ppm. In the dark, the values decreased to 0.04 and 0.009 ppm, respectively. In addition Kang *et al.* (76) have shown that red light increased the production of CO_2 which would also result in increased hook opening since CO_2 is a competitive inhibitor of ethylene action. The fact that ethylene is only a part of phytochrome-mediated development in seedlings is discussed further in Chap. 6.

Other examples of physiological processes under the partial control of phytochrome-mediated ethylene production have been described. Red light inhibits ethylene production in rice coleoptiles, and the effect can be reversed by far-red light (69). In addition, red light increased the elongation of intact rice coleoptiles. As discussed further in Chap. 6, ethylene promotes the growth of rice seedlings. Red light was also shown to promote the opening of bean hooks and reduce the rate of ethylene production (75). Similar to the pea, the opening of bean hooks occurred following the reduction of ethylene levels and the application of CO_2. Accumulation of carotene in the shoot apex of peas is also blocked by ethylene (73). The exposure of this tissue to red light resulted in the formation of carotene as well as a reduction in ethylene production. Kang and Burg (73) suggested that red light-induced carotenogenesis was partially or wholly caused by phytochrome-mediated inhibition of ethylene biosynthesis. In addition, they suggested that this effect may also account for the increase in cell division, the differentiation of metaxylem, and the induction of anthocyanin in *Sinapsis alba*.

The regulation of ethylene production by light appears to be confined in the light-responsive part of the seedling. Light had no effect on the straight portion of the bean hook (76).

While phytochrome appears to have a regulator effect on ethylene production, the reverse is not true. Ethylene was found to have no effect on the ability of bean hooks to respond to red or far-red light.* Suge *et al.* (145) similarly demonstrated that ethylene had no effect on the relative shape of the action spectrum regulating the growth of rice seedlings.

* D. Peterson, unpublished results, 1968.

E. Mineral Nutrition

Little is known concerning the effect of mineral nutrition on the rate of ethylene production. Hewitt (66) observed that nitrogen application to pears had no effect on ethylene production by the fruit.

III. Biochemistry of Ethylene Production

A. Fungi

Even though Gane established the fact in 1935 that plants produced ethylene (50, 51), biochemical studies on the pathway of ethylene did not start until the late 1950's. Most research has centered around work with higher plants, although ethylene production by fungi (primarily *Penicillium*) has also been studied. After Miller *et al.* (116) and Biale (18) demonstrated that *Penicillium* produced ethylene, early experiments consisted of growing fungi on different carbon media and determining the relative rates of ethylene production. Fergus (44) found that the best carbon sources were D-mannitol and D-mannose, followed by D-xylose, D-galactose, D-fructose, and other sugars. For the most part, the rate of ethylene production followed growth curves of the fungus, and the greatest rates of gas production were associated with the time mycelium growth was complete (see Figs. 3-1 and 3-2).

The high rate of ethylene production, and the ease of introducing proposed intermediates, offered distinct advantages to investigators working with fungi. However, in spite of a considerable amount of effort by a large number of workers, the biochemistry of ethylene formation in fungi remains unknown. The pathway with the greatest amount of experimental evidence in its favor is also the simplest, namely, the dehydration of ethanol.

$$CH_3{-}CH_2OH \longrightarrow CH_2{=}CH_2 + H_2O \qquad (3\text{-}1)$$

There are a number of observations that support this pathway of ethylene formation. Feeding ethanol to *Penicillium* resulted in a greater production of ethylene (53, 61, 122), and feeding labeled ethanol to the fungus resulted in the formation of labeled ethylene (53, 122). Unpublished inhibitor studies by some former colleagues (J. Lonski and H. E. Gahagan, 1967) demonstrated that dimedone (5,5-dimethyl-1,3-cyclohexanedione), which is an aldehyde-fixing agent, had no effect on fungal growth but completely blocked ethylene formation. The action of dimedone was thought to prevent the normal conversion of acetaldehyde to ethanol and subsequently the formation of ethylene. However, precursor studies have shown that pyruvate (55) and subsequently ethanol form ethylene in an asymmetrical fashion. If ethanol were the immediate precursor of ethylene, then both carbons should have been converted into ethylene with equal efficiency. For example, carbon 1 of ethanol was converted into ethylene with a percent conversion efficiency of 0.35, while carbon 2 was converted with an efficiency of 1.55. If the conversion of ethanol to ethylene was due to a simple dehydration mechanism, then both carbons should have been converted into ethylene with equal efficiency.

The possibility that other compounds may be precursors of ethylene has been extensively examined. The role of these substances as precursors of ethylene has been reviewed (6), but the available evidence fails to support the functioning of any known pathway other than ethanol to form ethylene. Many of the compounds studied were readily interconvertible, and the conversion efficiencies into ethylene were so dissimilar from one compound to the next that a logical pattern was not readily apparent. Unlike the pathway followed in higher plants, methionine does not appear to be an effective precursor (71, 78). Incontrovertible evidence in favor of any one pathway for ethylene production in fungi remains to be established.

B. Higher Plants

A number of substances have been proposed as precursors of ethylene in higher plants. These include linolenic acid (2, 10, 11, 103, 114, 115, 132, 133, 153, 157), acetate (24, 26, 89, 134, 137) glycerol (24, 26, 31, 124), ethanol (19, 26, 61, 85, 89, 124), carbohydrates (24, 26, 33, 61, 72, 124, 134, 135), propanal (10, 89), acrylic acid (137–139, 148, 149), and other compounds (20, 26, 33, 87, 124, 136). None of these compounds has continued to receive serious consideration as a potential intermediate, and for one reason or another all have been shown not to be direct precursors. It is now known that methionine is the most probable precursor of ethylene, and a review (6) of the work on the other substances has been published.

The first report that methionine may serve as a source of ethylene was that of Lieberman and Mapson in 1964 (95). During the course of an investigation in which ethylene production from a reaction mixture consisting of copper, ascorbate, and linolenic acid was being studied, they observed that methionine would also form ethylene. No other amino acid would function in this system and, unlike other reaction mixtures, ethylene was the only hydrocarbon produced. Support for the role of methionine in ethylene production was obtained when Lieberman and Mapson, and subsequently other investigators, reported that tissue fed with methionine produced more ethylene than controls (4, 48, 90, 94, 95, 108). However, some investigators have failed to observe a promotion of ethylene production when methionine was fed to tissue (40, 85).

While increased ethylene production by tissues fed with methionine would normally be considered as evidence in favor of a biochemical pathway from methionine to ethylene, a number of factors complicate the interpretation. First of all, increased ethylene production from plants can be due to stress and a wide range of chemicals, including amino acids, inorganic substances, and herbicides (see Chap. 5). Second, ethylene production can occur nonenzymatically by means of Cu^{2+} + ascorbate, Cu^{+}, or $Fe^{2+} + H_2O_2$ catalyzed reactions (93). This means that ethylene production from methionine might arise if the tissue or homogenate contained a significant amount of heavy metal ions and a system capable of generating H_2O_2. Third, it has been shown that peroxidase is capable of generating ethylene when supplied with methionine derivatives such as methional (107, 109) and 2-keto-*S*-methylthiobutyric acid (KMBA) (84). However, Yang (155) has found that methionine failed to form ethylene when incubated with peroxidase. Finally, though of little significance in tissue studies, methionine can form ethylene in the presence of flavin mononucleotide in the light (158, 159).

Additional support for the idea that methionine served as a precursor to ethylene was obtained when Lieberman *et al.* (93) added [^{14}C]methionine labeled in positions 1, 2, 3, 4, or in the methyl group to apple slices. Only methionine labeled in positions 3 and 4 produced significant amounts of ethylene. In addition, the percent conversion efficiency was high (60%). These findings were subsequently confirmed by Burg and Clagett (30). They reported that etiolated pea stem tissue fed with [U-^{14}C]methionine produced [^{14}C]ethylene when treated with a concentration of IAA capable of increasing ethylene evolution over the normal basal rate. They also demonstrated that the carboxylic acid group of methionine was converted into CO_2 and that carbons 3 and 4 were converted into ethylene. The second carbon was thought to be converted into formate and the CH_3—S— group into nonvolatile components including methionine. See Eq. (3-2).

$$\overset{5}{C}H_3\text{—S—}\overset{4}{C}H_2\text{—}\overset{3}{C}H_2\text{—}\overset{2}{C}HNH_2\text{—}\overset{1}{C}OOH \longrightarrow \overset{5}{C}H_3\text{—S—R} + \overset{4}{C}H_2\text{=}\overset{3}{C}H_2 + NH_3 + H\overset{2}{C}OOH + \overset{1}{C}O_2 \qquad (3\text{-}2)$$

The enzymatic reaction with methionine in tissue appears to favor utilization of the natural L-isomer. Baur and Yang (11) reported that unlabeled L-methionine would reduce incorporation of radioactivity from DL-methionine into ethylene better than unlabeled D-methionine.

A number of what would appear to be logical intermediates between methionine and ethylene have been examined. However, Baur and Yang (11) concluded that in apple slices L-methionine was the most effective substrate, followed by KMBA, DL-homoserine, L-methionine sulfoxide, methional, and β-methyl thiopropylamine.

Mapson *et al.* (104) have suggested that KMBA is an intermediate in the conversion of methionine into ethylene. This idea is based upon the observation that KMBA promoted ethylene production when added to cauliflower tissue and was converted more efficiently into ethylene than methionine. In addition, unlabeled methionine increased incorporation of [^{14}C]KMBA into ethylene, while in the reverse experiment unlabeled KMBA decreased the incorporation of [^{14}C]methionine into ethylene. This idea was questioned by Lieberman and Kunishi (91). They found that cauliflower sections would leak peroxidase into buffer solution surrounding the tissue and suggested that the ethylene production observed was due in fact to an ethylene-forming peroxidase system located outside of the cell.

Peroxidase-catalyzed ethylene production is different from the ethylene production pathway in tissue. The peroxidase system will utilize methional as a substrate (105, 107, 146), while in intact tissue methional is a poor precursor (11). Second, the peroxidase system will utilize KMBA 100 times more effectively than methionine (81), while plant tissue will utilize KMBA only twice as fast (104) or slower than methionine (11). Third, it is known that monophenols or *m*-diphenols promote ethylene production from the peroxidase system while *o*-diphenols are active inhibitors (154). In contrast, ethylene production from intact tissue was not greatly influenced by the addition of either class of phenolic substances (49).

Finally, Kang *et al.* (74) reported that there was no correlation between the amount of peroxidase in pea stem tissue and the rate of ethylene evolution. They found that ethylene production from etiolated pea plants was greater in the apical part of the seedling compared with the subapical portion. However, they found twice as much peroxidase in the subapical part as in the apical part of the seedlings. They also reported that a high dose of IAA caused a burst of ethylene production from pea tissue. However, during the course of the dramatic increase in ethylene production there was a slight decrease in peroxidase content of the tissue.

C. Enhancement of Ethylene Production

A number of workers have found that conversion of methionine to ethylene occurs only during periods of accelerated ethylene production. Burg and Clagett (30) reported no conversion of methionine to ethylene unless the pea stem tissue sections were first treated with IAA. Similarly, Baur *et al.* (12) observed conversion of methionine to ethylene in the climacteric avocado but not in preclimacteric fruit. Baur *et al.* (12) have shown that the increase in ethylene production was not due to changes in the level of methionine in tissue. They found similar amounts of methionine were present in preclimacteric and climacteric tissue even though the rate of ethylene production had increased 3000-fold. During the climacteric peak, levels of methionine had dropped by about 60%. They concluded that, at high rates of ethylene production, methionine must be actively turned over since only a 3-hour supply of methionine was available in the tissue. We have found, however, that [U-^{14}C]-DL-methionine was converted into ethylene in normal primary leaves of bean (*Phaseolus vulgaris*) plants. In addition, when ethylene production from the leaves was enhanced by treatment with NAA, $CuSO_4$, Endothal (3,6-endoxohexahydrophthalic acid), or ozone, the additional ethylene was also produced from methionine (1).

The increase in ethylene production by auxin is thought to represent the synthesis of enzymes required for the conversion of methionine to ethylene (3). This idea is based on the observation that inhibition of protein synthesis prevents or retards the development of auxin-induced ethylene production. Kang *et al.* (74) have shown that the half-life of the biosynthetic pathway is short once auxin is removed (about 2 hours), and that the continued presence of auxin is required for a high rate of ethylene production.

D. Inhibitor Studies

Inhibitor studies with dinitrophenol and other compounds that block the formation of ATP during respiration have shown that oxidative metabolism is required for ethylene formation (32, 94, 135, 141). However, a number of workers have found that cyanide had no effect on ethylene production (32, 135). Similarly, Lieberman *et al.* (94) claimed that CO had no effect on ethylene production by apple slices. The role of oxidative metabolism in ethylene formation was first demonstrated by Hansen in 1942 (62). He found that ethylene production did not occur when oxygen was removed from pears. As discussed earlier in this chapter, other investigators have reported similar findings. Hansen also noticed that there appeared to be a surge of ethylene production following the return of pear tissue to air. A similar phenomenon in apple tissue was also observed by Burg and Thimann (31) (see Fig. 3-7). It is not clear whether oxygen participates directly in the conversion of methionine to ethylene or causes energy-rich metabolites to be made available as a result of respiration.

Baur *et al.* (12) have shown that the oxygen requirement occurs at some point between methionine and ethylene. A nitrogen gas phase inhibited conversion of [^{14}C]methionine to ethylene. Returning the tissue to air caused an increase in ethylene production, which again suggests the formation of some oxygen-dependent intermediate. If molecular oxygen is required for ethylene biogenesis, then there appears to be no recognition of this fact in present proposed biochemical pathways.

The formation of ethylene from methionine also appears to be sensitive to rhizobitoxin, a phytotoxin produced by certain strains of the bacterium *Rhizobium japonicum* (121). The precise structure remains to be elucidated; however, it is known to be a basic sulfur-containing amino acid which yields a derivative of homoserine upon desulfurization. Rhizobitoxin inhibits the enzyme β-cystathionase, which plays a role in methionine biosynthesis in bacteria. It also inactivates β-cystathionase isolated from spinach leaves. Owens *et al.* (121) applied rhizobitoxin to sorghum seedlings to test the idea that methionine biosynthesis was important in ethylene formation. Rhizobitoxin reduced ethylene production, and the addition of methionine partially overcame the effect of the inhibitor. The lack of complete reversal was seen as due to a second effect of rhizobitoxin on the conversion of methionine to ethylene. They found that conversion of radioactivity from [^{14}C]methionine into ethylene was blocked to the same extent as ethylene production.

E. Subcellular Localization

The site of ethylene production is, as yet, unknown. A number of investigators have experimented with the possibility that subcellular organelles contain enzymes required for ethylene biosynthesis. This idea was encouraged by the observation that ethylene production was associated with some tissue component sensitive to disruption and osmotic shock. Subdividing apples into smaller slices decreases ethylene production (32, 92). Tomatoes, on the other hand, produced more ethylene as they were sectioned (92). However, ethylene produced from both fruits stopped when they were homogenized. Burg and Thimann (32) found that the molarity of the bathing solution in which apple tissue was suspended had a profound effect on the rate of gas production. Soaking apple tissue in water reduced ethylene production by 50%, while solutions of glycerol, KCl, and other compounds prevented the reduction in ethylene evolution. Nichols (117) reported that ethylene production from flowers was halted when they were exposed to freezing temperatures. In addition to inhibiting ethylene production, these low temperatures caused a collapse of the petals and a rapid loss of water.

A number of investigators have examined the idea that subcellular fractions of cells might be the site of ethylene production. Ethylene production from chloroplasts (125, 126) and mitochondria (111–113, 130, 142–144) has been described. However, there are a number of reasons for questioning the significance of these findings. Spencer and Meheriuk (143) reported that exposing mitochondria to 0° and 100°C failed to prevent ethylene production. Even though animals are not generally recognized as being capable of producing ethylene, mitochondria and other particulate preparations from beef heart (53, 54) and rats (88, 124) produced ethylene. Ku and Pratt (82) and Stinson and Spencer (144) failed to observe ethylene production by purified preparations of mitochondria but reported that other subcellular fractions did produce ethylene.

Systems consisting of soluble enzymes have also been studied. We (3) found that preparations from etiolated pea seedlings would produce ethylene from substrates later shown to be methionine (155) when flavin mononucleotide or Fe^{2+} ions were added. Mapson *et al.* (107, 109) and Yang (156) have elucidated a system involving peroxidase as the enzyme responsible for ethylene production. However, for reasons discussed earlier, it is unlikely that natural ethylene production is mediated by peroxidase.

References

1. Abeles, A. L., and Abeles, F. B. (1972). *Plant Physiol.* **50,** 496.
2. Abeles, F. B. (1966). *Nature (London* **210,** 23.
3. Abeles, F. B. (1966). *Plant Physiol.* **41,** 585.
4. Abeles, F. B. (1967). *Physiol. Plant.* **20,** 442.
5. Abeles, F. B. (1968). *Weed Sci.* **16,** 498.
6. Abeles, F. B. (1972). *Annu. Rev. Plant Physiol.* **23,** 259.
7. Abeles, F. B., Craker, L. E., Forrence, L. E., and Leather, G. R. (1971). *Science* **173,** 914.
8. Abeles, F. B., and Lonski, J. (1969). *Plant Physiol.* **44,** 277.
9. Barmore, C. R., and Biggs, R. H. (1970). *HortScience* **5,** Sect. 2, 67th Meet. Abstr., 358.
10. Baur, A. H., and Yang, S. F. (1969). *Plant Physiol.* **44,** 189.
11. Baur, A. H., and Yang, S. F. (1969). *Plant Physiol.* **44,** 1347.
12. Baur, A. H., Yang, S. F., Pratt, H. K., and Biale, J. B. (1971). *Plant Physiol.* **47,** 696.
13. Baur, J. R., and Morgan, P. W. (1969). *Plant Physiol.* **44,** 831.
14. Ben-Yehoshua, S., and Eaks, I. L. (1969). *J. Amer. Soc. Hort. Sci.* **94,** 292.
15. Ben-Yehoshua, S., and Eaks, I. L. (1970). *Bot. Gaz.* **13,** 144.
16. Beyer, E. M., and Morgan, P. W. (1971). *Plant Physiol.* **48,** 208.
17. Biale, J. B. (1940). *Calif. Citrogr.* **25,** 186.
18. Biale, J. B. (1940). *Science* **91,** 458.
19. Biale, J. B. (1960). *Advan. Food Res.* **10,** 293.
20. Bitancourt, A. A. (1968). *Cienc. Cult. (Sao Paulo)* **20,** 400.

21. Blanpied, G. D. (1972). *Plant Physiol.* **49,** 627.
22. Bouckaert, J. J. (1924). *C. R. Soc. Biol.* **91,** 907.
23. Bouckaert, J. J. (1926). *Arch. Int. Pharmacodyn. Ther.* **31,** 159.
24. Burg, S. P. (1959). *Arch. Biochem. Biophys.* **84,** 543.
25. Burg, S. P., and Burg, E. A. (1962). *Plant Physiol.* **37,** 179.
26. Burg, S. P., and Burg, E. A. (1964). *Nature (London)* **203,** 869.
27. Burg, S. P., and Burg, E. A. (1965). *Physiol. Plant* **18,** 870.
28. Burg, S. P., and Burg, E. A. (1968). *Plant Physiol.* **43,** 1069.
29. Burg, S. P., and Burg, E. A. (1968). *In* "Biochemistry and Physiology of Plant Growth Substances" (F. Wrightman and G. Setterfield, eds.), p. 1275. Runge Press, Ottawa.
30. Burg, S. P., and Clagett, C. O. (1967). *Biochem. Biophys. Res. Commun.* **27,** 125.
31. Burg, S. P., and Thimann, K. V. (1959). *Proc. Nat. Acad. Sci. U.S.* **45,** 335.
32. Burg, S. P., and Thimann, K. V. (1960). *Plant Physiol.* **35,** 24.
33. Burg, S. P., and Thimann, K. V. (1961). *Arch. Biochem. Biophys.* **91,** 450.
34. Bussel, J., and Maxie, E. C. (1966). *Proc. Amer. Soc. Hort. Sci.* **88,** 151.
35. Chadwick, A. V., and Burg, S. P. (1967). *Plant Physiol.* **42,** 415.
36. Chadwick, A. V., and Burg, S. P. (1970). *Plant Physiol.* **45,** 192.
37. Craker, L. E. (1971). *HortScience* **6,** 137.
38. Craker, L. E., Standley, L. A., and Starbuck, M. J. (1971). *Plant Physiol.* **48,** 349.
39. Curtis, R. W. (1969). *Plant Physiol.* **44,** 1368.
40. Curtis, R. W. (1969). *Plant Cell Physiol.* **10,** 909.
41. Dilworth, M. J. (1966). *Biochim. Biophys. Acta* **127,** 285.
42. Dull, G. G., Young, R. E., and Biale, J. B. (1967). *Physiol. Plant.* **20,** 1059.
43. Elers, T. L. (1965). *Forest Prod. J.* **15,** 134.
44. Fergus, C. L. (1954). *Mycologia* **46,** 543.
45. Forsyth, F. R., and Hall, I. V. (1969). *Natur. Can.* **96,** 257.
46. Freebairn, H. T., and Buddenhagen, I. W. (1964). *Nature (London)* **202,** 313.
47. Friede, K. A., and Ebert, M. K. (1934). *Z. Krebsforsch.* **40,** 431.
48. Fuchs, Y., and Lieberman, M. (1968). *Plant Physiol.* **43,** 2029.
49. Gahagan, H. E., Holm, R. E., and Abeles, F. B. (1968). *Physiol. Plant.* **21,** 1270.
50. Gane, R. (1934). *Nature (London)* **134,** 1008.
51. Gane, R. (1935). *Gt. Brit., Food Invest. Bd., Rep.* p. 122.
52. Gane, R. (1935). *Gt. Brit., Food Invest. Bd., Rep.* p. 130.
53. Gibson, M. S. (1963). Ph.D. Thesis, Purdue University, Lafayette, Indiana.
54. Gibson, M. S. (1963). *Biochim. Biophys. Acta (London)* **78,** 528.
55. Gibson, M. S., and Young, R. E. (1966). *Nature (London)* **210,** 529.
56. Goeschl, J. D., and Pratt, H. K. (1968). *In* "Biochemistry and Physiology of Plant Growth Substances" (F. Wrightman and G. Setterfield, eds.), p. 1229. Runge Press, Ottawa.
57. Goeschl, J. D., Pratt, H. K., and Bonner, B. A. (1967). *Plant Physiol.* **42,** 1077.
58. Goeschl, J. D., Rappaport, L., and Pratt, H. K. (1966). *Plant Physiol.* **41,** 877.
59. Guthrie, D., and Woodhouse, K. W. (1940). *J. Amer. Med. Ass.* **114,** 1846.
60. Haber, E. S. (1926). *Proc. Amer. Soc. Hort. Sci.* **23,** 201.
61. Hall, W. C. (1951). *Bot. Gaz.* **113,** 55.
62. Hansen, E. (1942). *Bot. Gaz.* **103,** 543.
63. Hansen, E. (1945). *Plant Physiol.* **20,** 631.
64. Harvey, R. B. (1928). *Minn., Agr. Exp. Sta., Bull.* **247.**
65. Hesselman, C. W., and Freebairn, H. T. (1969). *J. Amer. Soc. Hort. Sci.* **94,** 635.
66. Hewitt, A. A. (1967). *Proc. Amer. Soc. Hort. Sci.* **91,** 90.
67. Hirschfelder, A. D., and Ceder, E. T. (1930). *Amer. J. Physiol.* **91,** 624.

68. Ilag, L., and Curtis, R. W. (1968). *Science* **159,** 1357.
69. Imaseki, H., Pjon, C. J., and Furuya, M. (1971). *Plant Physiol.* **48,** 241.
70. Imaseki, H., Teranishi, T., and Uritani, I. (1968). *Plant Cell Physiol.* **9,** 769.
71. Jacobsen, D. W., and Wang, C. H. (1968). *Plant Physiol.* **43,** 1959.
72. Kaltaler, R. E. L., and Boodley, J. W. (1970). *HortScience* **5,** Sect. 2, 67th Meet. Abstr., 355.
73. Kang, B. G., and Burg, S. P. (1972). *Plant Physiol.* **49,** 631.
74. Kang, B. G., Newcomb, W., and Burg, S. P. (1971). *Plant Physiol.* **47,** 504.
75. Kang, B. G., and Ray, P. M. (1969). *Planta* **87,** 206.
76. Kang, B. G., Yocum, C. S., Burg, S. P., and Ray, P. M. (1967). *Science* **156,** 958.
77. Ketring, D. L., and Morgan, P. W. (1969). *Plant Physiol.* **44,** 326.
78. Ketring, D. L., Young, R. E., and Biale, J. B. (1968). *Plant Cell Physiol.* **9,** 617.
79. Kleindorfer, G. B. (1931). *J. Pharmacol. Exp. Thera.* **43,** 445.
80. Kokonov, M. T. (1960). *Vop. Med. Khim.* **6,** 158.
81. Ku, H. S., and Leopold, A. C. (1970). *Biochem. Biophys. Res. Commun.* **41,** 1155.
82. Ku, H. S., and Pratt, H. K. (1968). *Plant Physiol.* **43,** 999.
83. Ku, H. S., Suge, H., Rappaport, L., and Pratt, H. K. (1969). *Planta* **90,** 333.
84. Ku, H. S., Yang, S. F., and Pratt, H. K. (1969). *Phytochemistry* **8,** 567.
85. LaRue, T. A. G., and Gamborg, O. L. (1971). *Plant Physiol.* **48,** 394.
86. Leonard, E. R., and Wardlaw, C. W. (1941). *Ann. Bot.* (*London*) **5,** 379.
87. Lieberman, M., and Craft, C. C. (1961). *Nature* (*London*) **189,** 243.
88. Lieberman, M., and Hochstein, P. (1966). *Science* **152,** 213.
89. Lieberman, M., and Kunishi, A. T. (1967). *Science* **158,** 938.
90. Lieberman, M., and Kunishi, A. T. (1968). *In* "Biochemical Regulation in Diseased Plants or Injury," p. 165. Phytopathol. Soc. Jap.
91. Lieberman, M., and Kunishi, A. T. (1971). *Plant Physiol.* **47,** 576.
92. Lieberman, M., and Kunishi, A. T. (1971). *HortScience* **6,** 355.
93. Lieberman, M., Kunishi, A. T., Mapson, L. W., and Wardale, D. A. (1965). *Biochem. J.* **97,** 449.
94. Lieberman, M., Kunishi, A. T., Mapson, L. W., and Wardale, D. A. (1966). *Plant Physiol.* **41,** 376.
95. Lieberman, M., and Mapson, L. W. (1964). *Nature* (*London*) **204,** 343.
96. Lieberman, M., and Spurr, R. A. (1955). *Proc. Amer. Soc. Hort. Sci.* **65,** 381.
97. Lougheed, E. C., and Franklin, E. W. (1970). *Can. J. Plant Sci.* **50,** 586.
98. Luckhardt, A. B., and Carter, J. B. (1923). *J. Amer. Med. Ass.* **80,** 1440.
99. Luckhardt, A. B., and Lewis, D. (1923). *J. Amer. Med. Ass.* **81,** 1851.
100. Lund, B. M., and Mapson, L. W. (1970). *Biochem. J.* **119,** 251.
101. Lyons, J. M., McGlasson, W. B., and Pratt, H. K. (1962). *Plant Physiol.* **37,** 31.
102. Lyons, J. M., and Pratt, H. K. (1964). *Proc. Amer. Soc. Hort. Sci.* **84,** 491.
103. Mapson, L. W., March, J. F., Rhodes, M. J. C., and Wooltorton, L. S. C. (1970). *Biochem. J.* **117,** 473.
104. Mapson, L. W., March, J. F., and Wardale, D. A. (1969). *Biochem. J.* **115,** 653.
105. Mapson, L. W., and Mead, A. (1968). *Biochem. J.* **108,** 875.
106. Mapson, L. W., and Robinson, J. (1966). *J. Food Technol.* **1,** 215.
107. Mapson, L. W., Self, R., and Wardale, D. A. (1969). *Biochem. J.* **111,** 413.
108. Mapson, L. W., and Wardale, D. A. (1967). *Biochem. J.* **102,** 574.
109. Mapson, L. W., and Wardale, D. A. (1971). *Phytochemistry* **10,** 29.
110. Meheriuk, M., and Spencer, M. (1964). *Can. J. Bot.* **42,** 337.
111. Meheriuk, M., and Spencer, M. (1964). *Nature* (*London*) **204,** 43.

112. Meheriuk, M., and Spencer, M. (1967). *Phytochemistry* **6,** 535.
113. Meheriuk, M., and Spencer, M. (1967). *Phytochemistry* **6,** 545.
114. Meigh, D. F. (1962). *Nature (London)* **196,** 345.
115. Meigh, D. F., Jones, J. D., and Hulme, A. C. (1967). *Phytochemistry* **6,** 1507.
116. Miller, E. V., Winston, J. R., and Fisher, D. F. (1940). *J. Agr. Res.* **60,** 269.
117. Nichols, R. (1966). *J. Hort. Sci.* **41,** 279.
118. Nickerson, W. J. (1948). *Arch. Biochem.* **17,** 225.
119. Olson, A. O., and Spencer, M. (1968). *In* "Biochemistry and Physiology of Plant Growth Substances" (F. Wrightman and G. Setterfield, eds.), p. 1243. Runge Press, Ottawa.
120. Orion, D., and Minz, G. (1969). *Nematologica* **15,** 608.
121. Owens, L. D., Lieberman, M., and Kunishi, A. T. (1971). *Plant Physiol.* **48,** 1.
122. Phan, C. T. (1962). *Rev. Gen. Bot.* **69,** 505.
123. Phan, C. T. (1963). *Rev. Gen. Bot.* **70,** 679.
124. Phan C. T. (1970). *Physiol. Plant.* **23,** 981.
125. Porutskii, G. V., Luchko, A. S., and Matkovskii, K. I. (1962). *Sov. Plant Physiol.* **9,** 382.
126. Porutskii, G. V., and Matkovskii, K. I. (1963). *Izv. Inst. Biol. Metod. Popov, Bulg. Akad. Nauk* **13,** 147.
127. Potter, N. A., and Griffiths, D. G. (1947). *J. Pomol. Hort. Sci.* **23,** 171.
128. Ram Chandra, G., and Spencer, M. (1963). *Biochim. Biophys. Acta* **69,** 423.
129. Ram Chandra, G., and Spencer, M. (1963). *Nature London)* **197,** 366.
130. Ram Chandra, G., Spencer, M., and Meheriuk, M. (1963). *Nature (London)* **199,** 767.
131. Rasmussen, G. K., and Jones, J. D. (1969). *HortScience* **4,** 60.
132. Rhodes, M. J. C., Wooltorton, L. S. C., Galliard, T., and Hulme, A. C. (1970). *J. Exp. Bot.* **21,** 40.
133. Saad, F. A., Crane, J. C., and Maxie, E. C. (1969). *J. Amer. Soc. Hort. Sci.* **94,** 335.
134. Sakai, S., Imaseki, H., and Uritani, I. (1970). *Plant Cell Physiol.* **11,** 737.
135. Shimokawa, K., and Kasai, Z. (1966). *Plant Cell Physiol.* **7,** 1.
136. Shimokawa, K., and Kasai, Z. (1967). *Science* **156,** 1362.
137. Shimokawa, K., and Kasai, Z. (1970). *Agr. Biol. Chem.* **34,** 1633.
138. Shimokawa, K., and Kasai, Z. (1970). *Agr. Biol. Chem.* **34,** 1640.
139. Shimokawa, K., and Kasai, Z. (1970). *Agr. Biol. Chem.* **34,** 1646.
139a. Smith, J. L., and Hoskins, A. P. (1901). *J. Hyg.* **1,** 123.
140. Spalding, D. H., and Lieberman, M. (1965). *Plant Physiol.* **40,** 645.
141. Spencer, M. S. (1959). *Can. J. Biochem. Physiol.* **37,** 53.
142. Spencer, M. S. (1959). *Nature (London)* **184,** 1231.
143. Spencer, M., and Meheriuk, M. (1963). *Nature (London)* **199,** 1077.
144. Stinson, R. A., and Spencer, M. (1970). *Can. J. Biochem.* **48,** 541.
145. Suge, H., Katsura, N., Inada, K. (1971). *Planta* **101,** 365.
146. Takeo, T., and Lieberman, M. (1969). *Biochim. Biophys. Acta* **178,** 235.
147. Tausz, J., and Donath, P. (1930). *Hoppe-Seyler's Z. Physiol. Chem.* **190,** 141.
148. Thompson, J. E., and Spencer, M. (1966). *Nature (London)* **210,** 595.
149. Thompson, J. E., and Spencer, M. (1967). *Can. J. Biochem.* **45,** 563.
150. Uota, M. (1955). *Proc. Amer. Soc. Hort. Sci.* **65,** 231.
151. Vendrell, M. (1970). *Aust. J. Biol. Sci.* **23,** 1133.
152. Wang, C. Y., Mellenthin, W. M., and Hansen, E. (1971). *J. Amer. Soc. Hort. Sci.* **96,** 122.
153. Wooltorton, L. S. C., Jones, J. D., and Hulme, A. C. (1965). *Nature (London)* **207,** 999.
154. Yang, S. F. (1967). *Arch. Biochem. Biophys.* **122,** 481.
155. Yang, S. F. (1968). *In* "Biochemistry and Physiology of Plant Growth Substances" (F. Wrightman and G. Setterfield, eds.), p. 1217. Runge Press, Ottawa.

156. Yang. S. F. (1969). *J. Biol. Chem.* **244,** 4360.
157. Yang, S. F., and Baur, A. H. (1969). *Qual. Plant. Mater. Veg.* **19,** 201.
158. Yang, S. F., Ku, H. S., and Pratt, H. K. (1966). *Biochem. Biophys. Res. Commun.* **24,** 739.
159. Yang, S. F., Ku, H. S., and Pratt, H. K. (1967). *J. Biol. Chem.* **242,** 5274.
160. Zimmerman, P. W., Hitchcock, A. E., and Crocker, W. (1931). *Contrib. Boyce Thompson Inst.* **3,** 313.
161. Zobell, C. E. (1950). *Advan. Enzymol.* **10,** 443.

Chapter 4

Hormonal and Herbicidal Regulation of Ethylene Production

I. Auxin

A. Background

The initial discovery that auxin regulated ethylene production was made in 1935 by Crocker, Hitchcock, Wilcoxon, and Zimmerman (35, 148). They observed that IAA and NAA (α-naphthaleneacetic acid) produced effects on plants similar to those caused by ethylene, including epinasty, inhibition of growth, root induction, tissue swelling, and anesthesia of *Mimosa pudica* (sensitive plant) pulvinal tissue. Using *Chenopodium album* and marigold as bioassay plants, they showed that tomato plants treated with IAA in lanolin produced an emanation that caused epinasty. They wrote that "some of the effects attributed to so-called growth substances might be due indirectly to the unsaturated hydrocarbon gas produced in the tissues" (148). The only contemporary scientist who investigated this phenomenon further was Michener (97, 98). Michener attempted, but failed to show that the growth promotive effect of auxin was due to ethylene, and concluded that no significant relationship between the two compounds existed. In spite of the fact that the Boyce Thompson scientists clearly stated that *some* and not all of auxin's effects were due to increased ethylene production, Michener's findings were accepted by contemporary scientists as showing no significant role for ethylene in growth and development and the importance of the original discovery remained unappreciated or ignored until 29 years later. Abeles and Rubinstein (6) and Morgan and Hall (103) confirmed the fact that auxin increased ethylene production by using a more precise analytic technique, gas chromatography. Actually Morgan and Hall (102) had demonstrated 2 years earlier that 2,4-D, which can be considered an auxin-like substance, also promoted ethylene production. Since that time, interest and research in this area has grown, and we now know, as Crocker and his colleagues originally suggested, that ethylene acts as an intermediate in a number of phenomena regulated by auxin. To some extent it is possible to think of ethylene as an intermediate hormone, in much the same way as cyclic AMP acts as an intermediate in the action of many animal hormones.

B. Biochemistry of Auxin-Induced Ethylene Production

The promotion of ethylene production by auxin is widespread. The phenomenon has been observed in all cases it has been looked for except in seeds (74, 130), cultures of cell suspensions, such as horseradish and sweet clover (53), and mature fruit such as apples, pears, and tomatoes (6). The lack of an effect on mature fruit may be due to the fact that the potential for ethylene production in these tissues is already fully expressed and limited by factors not under hormonal control, for example, low levels of the precursor, methionine. Abeles and Rubinstein (6) found that the dose-response curve for auxin promotion of ethylene production was similar for that for IAA and the synthetic homolog NAA: no effect at 10^{-7} *M* and below, a threshold effect at 10^{-6} *M*, and an increase in the rate of ethylene production up to 10^{-3} *M*. Weak auxin analogs with longer side chains such as indole propionic acid and indole butyric acid showed a corresponding decrease in activity. Coumarin is another compound that exhibits weak auxin-like effects (5, 83). A number of workers have shown that coumarin is capable of promoting ethylene production from seeds (75), bean hypocotyl hooks (105), and fruit (83). Dose-response curves have been published by others (17, 70, 121, 124) using different tissues and techniques. Nevertheless, the results they obtained were similar and most workers observed a threshold effect at 10^{-6} *M* IAA (see Fig. 4-4, p. 65).

The lag between the application of auxin and increase in ethylene production is about an hour, but varies somewhat according to the tissue under examination. Chadwick and Burg (26, 27) have shown that roots respond more rapidly to auxin. Ethylene production from pea roots increased 15–30 minutes after the application of auxin. Burg and Burg (21) also pointed out that the length of the lag period depended on the concentration of IAA used. They reported an immediate increase in ethylene production following the application of low concentrations of auxin and a lag in production following the application of high concentrations of auxin. Shingo and Imaseki (124) have found no lag in ethylene production following the application of auxin if the tissue were first pretreated with small amounts (10^{-5} *M*) of auxin. These data suggest that auxin may have two effects: one, overcoming the lag phase, and two, promoting ethylene production. Removal of auxin results in a decline in ethylene production after an hour lag period (69, 124). These observations suggest that elevated rates of ethylene production are closely regulated by internal levels of auxin and that the rate of ethylene production can give an approximate estimation of auxin levels in plant tissue. This idea was supported by observations on the rates of ethylene production by tissue treated or manipulated to alter its internal auxin content.

The apical tissue of seedlings is thought to be an important site of auxin synthesis. The hormone is translocated from the apex of stems to the tissue below where it promotes cell elongation. A number of investigators have shown that ethylene production in seedlings is greatest in the apex or immediately below it (19, 20, 121, 124). Removal of the apex by excision or by ringing the stem with an inhibitor of auxin transport, triiodobenzoic acid, reduced the rate of ethylene in the subtending tissue (6, 19). It was possible to restore the original rate of ethylene production by applying IAA to the decapitated stem tissue.

According to Cholodny and Went, tropistic responses are due to an asymmetric distribution of auxin across stimulated tissue. Since auxin controls ethylene production, the auxin gradient caused by a tropistic response should cause a parallel gradient of ethylene production. For example, a geotropically stimulated bean seedling will have more auxin on the lower than on the upper side. If endogenous auxin controls ethylene production, then the lower side should produce more ethylene than the upper side. The same argument would apply to a phototropic response. In this case the dark side should produce more ethylene than the light side. Abeles and Rubinstein (6) found this to be true. Ethylene production from vertical plants was the same from both sides of the stems, but it was higher from the lower compared with the upper side of horizontal stems. And similarly, ethylene production was the same from both sides of uniformly illuminated plants but it was higher from the dark side compared with the light side.

Kang *et al.* (69) examined the relationship between internal levels of auxin and the rate of ethylene production from apical and subapical portions of pea seedlings. They observed that IAA caused a rapid rise in ethylene production followed by a gradual decline and return to pretreatment levels after 24 hours. In the same tissue, 2,4-D, which is more stable, caused ethylene production to remain high during the course of the experiment (Fig. 4-1). These results suggested that ethylene production required the continuous presence of auxin in a physiologically active concentration. Figure 4-2 shows that the level of [^{14}C]IAA in subapical tissue declined rapidly compared with apical sections where the level of [^{14}C]IAA was constant for a longer period of time. In the subapex, the decrease in free IAA was paralleled by an increase in IAA conjugation and decarboxylation. However, the internal IAA did not depend on external supply of IAA since the rate of IAA loss in the tissue was the same when the tissue was placed in 3 or 15 ml of [^{14}C]IAA. Similarly, the decline of internal IAA in pea roots was found to be independent of the size of the external pool (27). The conjugation system in the subapex was more active than the apex, whereas the decarboxylation system in these two tissues was the same. Kang *et al.* (69) showed that decarboxylation continued longer in the apex because the external source of labeled IAA for the apical tissue was greater than the subapical tissue. In general, their results indicated that the free IAA in the

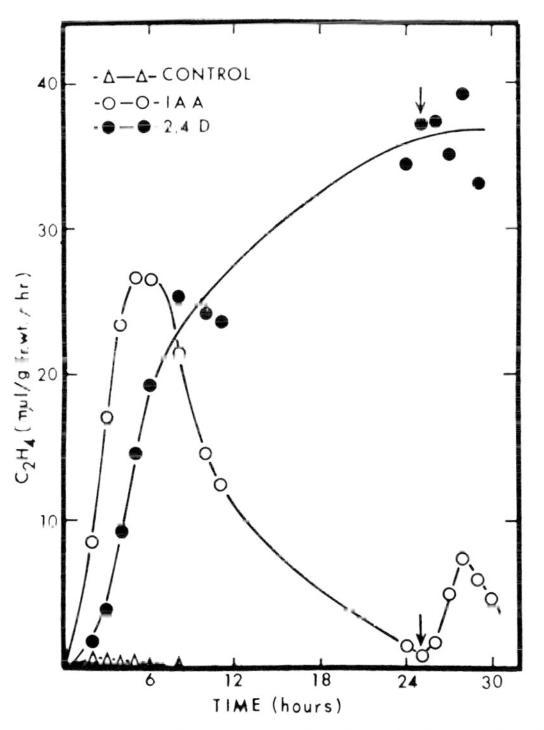

Fig. 4-1. Ethylene production in the subapex after 10 μM IAA or 10 μM 2,4-D has been applied initially. The arrows indicate when sections were transferred to fresh auxin solutions. [Courtesy of Kang *et al.* (69).]

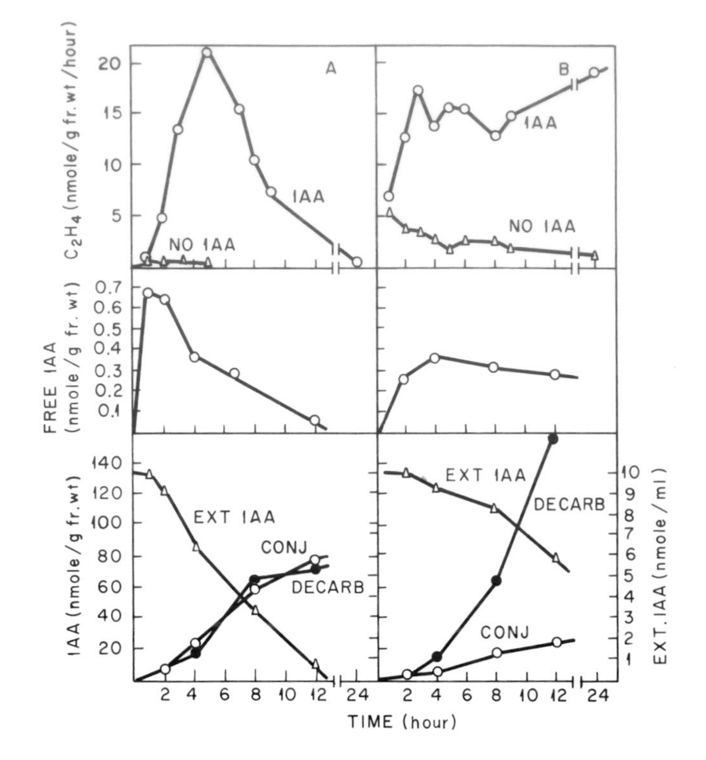

Fig. 4-2. Comparison of ethylene production (upper), the free IAA level (middle), and conjugation, decarboxylation and exogenous IAA (lower) in the subapex (A) and apex (B) of tissue treated with 10 μM [^{14}C]IAA. Apex includes all the tissue above the hook. [Courtesy of Kang *et al.* (69).]

tissue depended on the activity of the conjugating system and the initial concentration of applied IAA. It did not depend on the rate of IAA decarboxylation nor on the amount of exogenous IAA.

The increase in ethylene production induced by auxin is due to an increase in enzymes associated with the methionine to ethylene pathway. Studies with inhibitors have shown that the increase in ethylene production induced by auxin can be blocked by the addition of inhibitors of RNA and protein synthesis (1a, 121, 124). Shingo and Imaseki (124) have shown that continued synthesis of protein was required for high rates of ethylene production. Addition of cycloheximide to auxin-treated sections caused an immediate reduction in the rate of ethylene production. If cycloheximide was added after auxin was removed, the decrease in the rate of ethylene production was enhanced further. Burg and Clagett (22) demonstrated that the methionine pathway operated only when pea sections were treated with auxin. Untreated sections did not convert methionine into ethylene while auxin-treated ones did. A somewhat different result was obtained with bean leaves. In this case, auxin increased ethylene production and the rate of methionine conversion into ethylene increased as anticipated. However, unlike results reported for peas, endogenous ethylene was also produced from methionine (1). The system outlined here is not specific for auxin. We have found that nonhormonal stress-inducing chemicals such as $CuSO_4$ and Endothal also speeded up the conversion of methionine into ethylene. In addition, cycloheximide inhibited the increased rate of ethylene production induced by these chemicals.

C. Role of Auxin-Induced Ethylene Production

The suggestion by the Boyce Thompson scientists that auxin-induced ethylene may account for the ability of auxin to induce epinasty, root initiation, inhibition of growth, swelling, and anesthesia of *Mimosa* pulvini has been confirmed and extended by others. We now know that auxin-induced ethylene production also plays a role in (*1*) abscission (6), (*2*) inhibition of flowering in *Xanthium* (3), (*3*) inhibition of stem elongation (17, 20), (*4*) promotion of flowering in bromeliads (18, 25), (*5*) inhibition of bud growth (19), (*6*) fading of orchid flowers (23, 43), (*7*) inhibition of root elongation (26, 27), (*8*) isocoumarin formation in carrots (28), (*9*) latex flow in rubber trees (38), (*10*) hook opening (70, 71), (*11*) swelling of onion leaf bases (82), (*12*) induction of phenylalanine ammonium lyase in parsnip root (116), and (*13*) changing the sex of flowers in cucurbits (123). On the other hand, ethylene does not play an intermediary role in (*1*) auxin-induced cell elongation, (*2*) inhibition of abscission and ripening, (*3*) promotion of growth in tissue cultures (80), and (*4*) inhibition of filament elongation in ferns (99).

Proof for an auxin-ethylene relationship depends on five criteria:

1. Demonstration of auxin-induced ethylene production. The kinetics are critical in verifying a role for ethylene auxin action. An increase in the rate of ethylene production must occur before the phenomenon in question takes place.
2. Demonstrate that ethylene mimics the effect of auxin. It should be possible to observe the same phenomenon with both ethylene and auxin.
3. It should be possible to demonstrate that any additional effect of auxin is lost, reduced, or masked when a saturating concentration of ethylene is applied to the tissue. The additional effects of auxin observed under these conditions are ones not associated with enhanced ethylene production.
4. Removal of ethylene by absorption, flushing with air, or use of hypobaric conditions should decrease the effectiveness of auxin. It is impossible to remove all of the ethylene using these techniques since

the tissue is constantly making more, but generally, some effect of reduced ethylene levels should be apparent. Using rhizobitoxin (109) or compounds like it that interfere with the formation of ethylene from methionine may be another way to reduce ethylene levels in an experimental system. Stewart and Freebairn (129) used heat (40°C for 4 hours) to destroy the ethylene-producing system and demonstrated that while auxin would no longer cause epinasty, ethylene would.

5. Carbon dioxide acts as a competitive inhibitor of ethylene action in many systems. It should be possible to demonstrate a reversal of auxin action except in those cases where CO_2 has secondary toxic or noncompetitive effects that would complicate and compound the system.

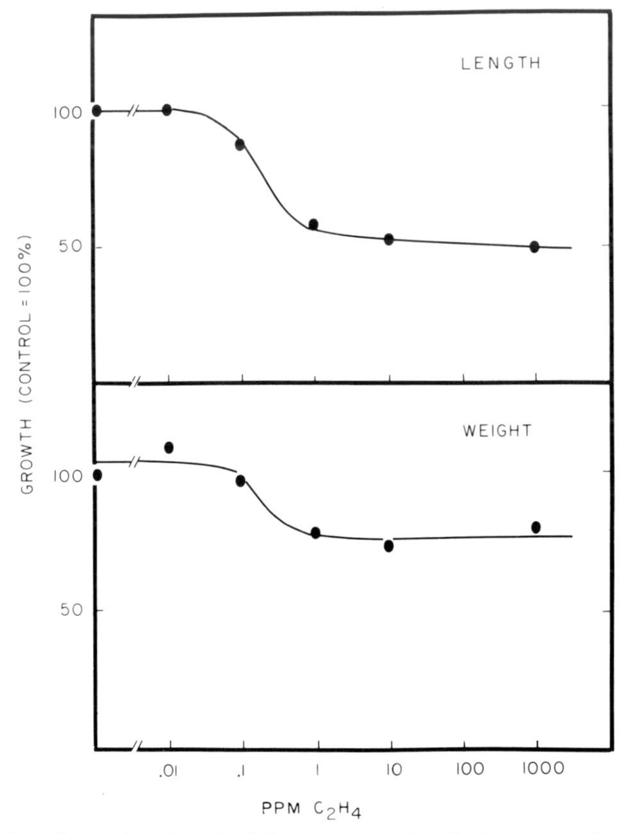

Fig. 4-3. Growth as a function of ethylene concentration during an experiment with intact pea roots. [Courtesy Chadwick and Burg (27).]

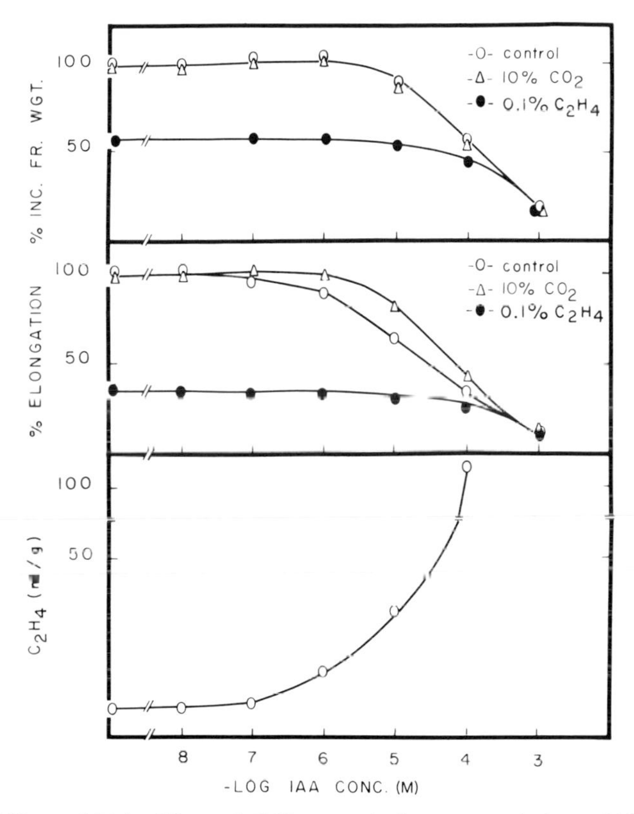

Fig. 4-4. Effects of IAA, CO_2, and C_2H_4 on excised pea roots during a 24-hour growth period in a 2-ml solution volume. Growth expressed as percentage of net growth in the absence of added IAA, CO_2, or C_2H_4. [Courtesy Chadwick and Burg (27).]

The fact that ethylene duplicates auxin action in a number of the systems studied does not mean that ethylene plays a role in the normal operation of these biological processes. For the most part, these systems are usually not exposed to auxin, or if they are, the concentrations involved are generally lower. However, Chadwick and Burg (26, 27) have presented evidence that auxin-regulated ethylene production plays a role in the response of roots to gravity.

Figure 4-3 shows that ethylene caused a reduction in elongation and weight of pea roots. While the dose-response curves were somewhat similar, it was apparent that growth in terms of elongation was more sensitive to ethylene than increase in weight. Figure 4-4 summarizes data on the effect of IAA on ethylene production, elongation, and increase in fresh weight of excised pea roots. The elongation portion of growth was found to be more sensitive to auxin (10^{-7} *M*) than the increase in weight (10^{-5} *M*). In addition, CO_2 reversal appeared to be limited to the effect on elongation and not to the increase in weight. When the system was exposed to a saturating dose of ethylene (1000 ppm) the effect of auxin was the same for both aspects of growth. In other words, under conditions in which any effect due to ethylene was masked by supraoptimal concentrations of the gas, the effect of auxin was shifted to higher concentrations (10^{-5} *M* and above) in terms of elongation but not in increases in weight. Chadwick and Burg postulated that this secondary effect of auxin was not due to the regulation of ethylene production but rather a nonspecific acid toxicity, since IAA in the presence of ethylene, and benzoic acid alone, caused equal reductions in weight and length at equal concentrations of the acids.

The kinetics of auxin-induced ethylene production and auxin-induced inhibition of elongation were similar. At low concentrations of IAA, for example 10^{-7} *M*, the increase in ethylene production and decrease in elongation both lasted for 4 hours. At the next higher concentration of auxin, 10^{-6} *M*, growth and ethylene production were altered for about 8 hours, and at 10^{-5} *M*, 12 hours. Measurements on the decrease of internal levels of free IAA following application of these concentrations of auxin showed that about as much time was required to reduce levels of free IAA to 10^{-8} *M* as was required for growth and ethylene production to return to normal. The IAA appeared to be primarily deactivated by conjugation with aspartic acid and glucose.

In order for ethylene to act as an intermediate in this system, the inhibition of elongation must be rapid and reversible. Figure 6-5 (p. 113) shows time-course data showing that inhibition and recovery to normal rates of elongation take place soon after the addition and removal of ethylene. Chadwick and Burg postulated that ethylene mediated root geotropism, assuming that the Cholodny-Went theory of geotropism was correct. As mentioned before, this theory states that auxin levels increase in the lower side of horizontal stems and roots. In the case of stems, growth promotion occurs, and in the case of roots, inhibition of growth occurs. If ethylene is the cause of this inhibition, then it should be possible to demonstrate this effect by exposing horizontal roots to CO_2 or ethylene. Moderately high concentrations of CO_2 (5–10%), which have no effect on pea root growth or very slightly stimulate it, markedly retard the geotropic curvature of roots (see Fig. 4-5). Since CO_2 is a competitive inhibitor of ethylene action, these data suggest that ethylene mediates root geotropism. Applied ethylene which would mask any gradient of ethylene production also immediately and completely prevented the development of the geotropic response in pea roots (Fig. 4-5). The mechanism was reversible since

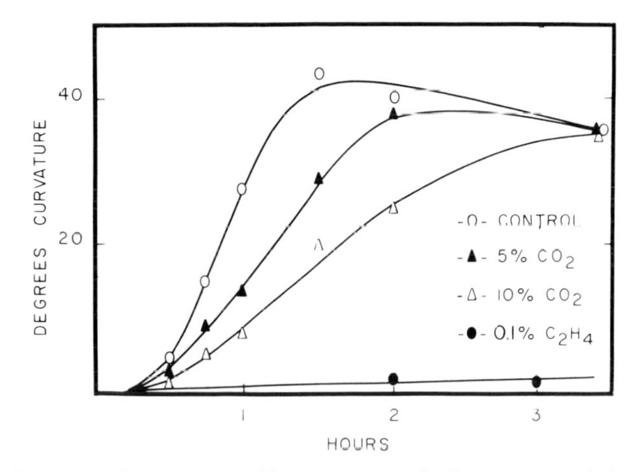

Fig. 4-5. Time course for curvature of intact pea roots in the presence and absence of ethylene and of various concentrations of CO_2. [Courtesy Chadwick and Burg (27).]

the roots regained their geotropic sensitivity once ethylene was removed. Andreae *et al.* (9) have also examined the role of auxin root elongation. They found that ethylene inhibited root growth after a 3-hour lag. On the other hand, Chadwick and Burg (26, 27) reported no lag in ethylene inhibition of growth. Andreae *et al.* (9) also showed that inhibition of root growth by auxin was completely reversible while it did not return to normal after the removal of ethylene. There was no obvious reason for the differences reported. However, at the present time, most evidence indicates that ethylene does play a role in auxin-mediated root growth inhibition and geotropism.

II. Gibberellins

Gibberellins have variable effects on ethylene production. In some cases workers have failed to observe any effect. Fuchs and Lieberman (52) reported that gibberellins had no effect on ethylene production by pea seedlings, and Takayanagi and Harrington (130) similarly reported no effect on ethylene production by rape seed. However, in a majority of cases, workers have observed a relatively small but promotive effect on ethylene production. Examples include bean (6) and citrus explants (83, 111), bean seedlings (6), orange fruit and leaves (30), blueberry flowers (50), peanut seeds (74), and potato tubers (113). Gibberellins have also been found to cause a 50% reduction of ethylene production induced by 2,4-D. This effect has been observed in soybean seedlings (65) and citrus fruit (83). Gibberellins can also reduce ethylene production induced by excising banana slices (134).

III. Cytokinins

Cytokinins usually cause a two- to fourfold increase in ethylene production. This is low when compared with the effect of auxins, which increase ethylene production by tenfold or greater. An increase in ethylene production has been found in most experiments in which the effect of cytokinins has been measured. Plants showing a positive response include bean (5), blueberries (50), radish (115), sorghum (109), banana (136), and pea (19, 52). The concentration required for a maximum effect was 10^{-4} *M*, and threshold effects were seen at 10^{-8} *M* (52). Compared to auxin, which has an induction time of less than an hour, Fuchs and Lieberman (52) reported that a noticeable increase in ethylene production occurred after a 9-hour delay. The ability to respond to cytokinin varied from one part of the seedling to another. Kinetins caused a fourfold increase in ethylene production from pea plumules but only a twofold increase from radicles.

Similar to results obtained with auxin and stress-inducing chemicals, the increase in ethylene production caused by cytokinins could be blocked by an inhibitor of protein synthesis, cycloheximide (52). Additional support for the idea that kinetin-induced ethylene production occurred by the same pathway used normally under conditions of auxin induction came from the observation that methionine enhanced the activity of cytokinin. The methionine itself had no effect (52).

A number of investigators have measured the effect of cytokinins on auxin-induced ethylene production (19, 52). They found that cytokinins enhanced the ability of auxin to increase ethylene production. The effect was synergistic since the amount of ethylene produced was greater than the quantities produced by each treatment singly. In addition to enhancing the effect of auxin, cytokinins caused the auxin effect to persist for a longer period of time, which suggests that the cytokinins might interfere with the auxin degradation mechanism. The effect of adding gibberellin and cytokinin together was also measured, but little or no interaction was observed.

IV. Abscisic Acid

Abscisic acid has been shown to promote ethylene production by leaves (2, 32, 111) and fruit (30). In these cases, the abscisic acid was tested for its ability to regulate abscission, and the authors noted that the ability to increase abscission correlated with an increase in ethylene production. However, ethylene production increased only by a factor of two, small compared to the effects of auxin and stress-inducing agents. Because of this, the role of ethylene in abscisic acid action was tested further. Craker and Abeles (32) found that abscisic acid promoted abscission in the presence of 10 ppm ethylene, a concentration that should have masked any effect of ethylene production due to the presence of abscisic acid. Palmer *et al.* (111) obtained similar results in experiments measuring the effect of abscisic acid on citrus explants. These results suggested that abscisic acid had a regulatory role separate from any effect on ethylene production. Abscission involves aging and the synthesis of cellulase, and it was thought that abscisic acid may also influence the rate of these processes. It was found that abscisic acid increased the rate of cellulase synthesis but had no effect on tissue aging.

In other tissues, workers have reported that abscisic acid inhibited ethylene production. Abscisic acid inhibited ethylene production by cell suspension cultures (54), peanut seeds (75), and soybean seedlings (10). Again the effects were small, and a complete inhibition of ethylene production was not observed.

V. Growth Retardants

A wide variety of chemicals, many of them quaternary ammonium compounds, are known to reduce growth, resulting in plants that look normal except for a dwarfed or compact appearance. These compounds are collectively called "growth retardants" and, in addition to reducing growth, have been shown to partially inhibit ethylene production. Abeles and Rubinstein (6) reported that 2-chloroethyltrimethyl ammonium chloride (CCC), *N*-dimethylaminesuccinamic acid (B9, Alar), and 2,4-dichlorobenzyltributyl phosphonium chloride (phosphon) reduced ethylene production and growth of bean seedlings by 50% when applied as a soil drench. However, Forsyth and Hall (50) reported that they failed to observe an effect of Alar on ethylene production by blueberry flowers and found a small promotive effect of CCC. Looney (84–86) made the observation that Alar retarded both ethylene production and ripening of apples. He showed that the inhibition of ripening was primarily due to a reduction in ethylene production, since fruit treated with Alar and ethylene could ripen normally (see Fig. 4-6). Abeles and Rubinstein (6) and Looney (84, 85) have summarized research by others showing that plants treated with growth retardants usually contain less auxin. Since endogenous auxin levels can control ethylene production, they proposed that the reduction in ethylene production following growth retardant application was an indirect result of reduced auxin levels.

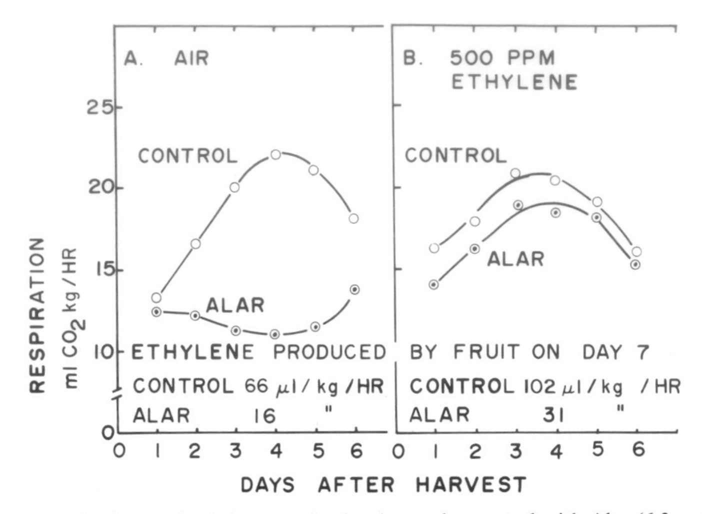

Fig. 4-6. Respiration and ethylene production by apples treated with Alar (6.3×10^{-2} *M*). Alar was sprayed on Tydeman's Early apples 14 days after full bloom. Fruit was harvested 13 weeks later. Fruits in A were exposed to air and those in B to 500 ppm ethylene. Ethylene production by the apples was measured 30 hours after the termination of the air or ethylene treatment. [Data modified from Looney (85).]

VI. Regulation of Ethylene Production by 2,4-D and Other Herbicides

A. Background

The phenoxyacetic acids 2,4-dichlorophenoxyacetic acid (2,4-D) and 2,4,5-trichlorophenoxyacetic acid (2,4,5-T) were developed during World War II as vegetation-control agents by the U. S. Army. However, the war ended before they were used and the surplus was released for agricultural use. They were readily adapted as weed killers because of their ability to selectively kill dicotyledonous weeds without damaging monocotyledonous plants such as corn. Early workers suspected that 2,4-D regulated ethylene production because plants treated with the herbicide resembled those exposed to ethylene. The first report on the effect of 2,4-D on ethylene production was published by Hansen in 1946 (62). He found that 2,4-D caused Bartlett pears to ripen and produce ethylene. Originally Morgan and Hall (102) proposed that the differential sensitivity of monocots and dicots to 2,4-D was caused by differences in the rate of ethylene production following the application of the herbicide. They found that 2,4-D increased ethylene production in cotton but only slightly in sorghum. However, we now know that while monocots are less sensitive, 2,4-D promotes ethylene production from both types of plants (4). Formulation was important and addition of a surfactant improved penetration and activity of the herbicide. However, there are some cases in which 2,4-D does not promote ethylene production. Workers have failed to observe an increase in ethylene production following treatment of peanut seeds (75) and climacteric pear fruit (62) with 2,4-D. As discussed earlier, similar results were obtained with auxin.

B. Relative Effects on Ethylene Production

The ability of 2,4-D to increase ethylene production is probably due to the fact that phenoxyacetic acid herbicides physiologically resemble auxin. Another auxin-like herbicide, picloram (4-amino-3,5,6-trichloropicolinic acid), also promotes ethylene production (11, 101). Support for the idea that 2,4-D action is due to its auxin-like activity stems from the following observations. Compounds that prevented or inhibited the ability of IAA to increase ethylene production had a similar effect on 2,4-D (1a). The growth inhibitor Alar, which reduces endogenous ethylene production, also reduced 2,4-D-induced ethylene production (63). Rubinstein and Abeles (110) found that the ability of 2,4-D analogs to promote ethylene production was similar to the ability of these compounds to promote growth and inhibit abscission.

Auxin levels are regulated by detoxification mechanisms involving conjugation with aspartic acid and glucose. The phenoxyacetic acids are either not deactivated by these mechanisms or if so at a slower rate because physiologically active concentrations persist in tissues for longer periods of time. This phenomenon can be demonstrated by the duration of a high rate of ethylene production and the persistence of physiological processes dependent on ethylene production. Chalutz *et al.* (28) reported that 2,4,5-T was more effective than auxin in increasing levels of isocoumarin in carrot tissues. Both 2,4-D and IAA delayed ripening in banana slices but the effects of 2,4-D were more pronounced (133). Curtis (37) reported that relative rates of ethylene production after treatment with IAA and 2,4-D were the same but that the effect of 2,4-D lasted longer. Warner (137) treated pea tissue with IAA, NAA, and 2,4-D. Ethylene production continued for longer periods of time with NAA and 2,4-D than with the IAA. All of these results suggested that 2,4-D was not metabolized by tissues in the same manner as IAA. Kang *et al.* (69) demonstrated that IAA-induced ethylene production in pea shoots parallels the free IAA level in the tissue, which in turn depends upon the rate of IAA conjugation and decarboxylation. The increase in ethylene production due to 2,4-D remained higher for longer periods of time, indicating that the conjugation mechanism was less effective with this substance (see Fig. 4-1).

C. Role of Ethylene in Herbicide Action

Ethylene has been shown to act as the intermediate in a number of effects attributed to 2,4-D. At sublethal concentrations, 2,4-D inhibits growth and causes abnormal swelling and enlargement of tissues and prevents normal leaf expansion. In addition, the herbicide causes changes in the RNA, DNA, and protein content of the seedlings. Holm and co-workers (4, 64, 65) have shown that increased rates of ethylene production accounted for most of the sublethal effect of 2,4-D observed. However, lethal effects and growth promotive effects on isolated sections of tissue were not due to enhanced ethylene production. Ethylene can double latex flow from rubber trees. D'Auzac and Ribaillier (38) have shown that 2,4-D could also be used. Formation of tylosis, promotion of cambial activity, and gum exudation are typical symptoms of woody tissue treated with excessive amounts of ethylene. Bradley *et al.* (14) have found that 2,4,5-T had similar effects when applied to apricot branches. Auxins, including 2,4,5-T were shown to promote isocoumarin formation in carrot root tissue. Chalutz *et al.* (28) demonstrated that ethylene had a similar effect and probably acted as an intermediate.

Leaf epinasty, senescence, and abscission are typical ethylene effects. Promotion of fig maturation and ripening also occurs following ethylene treatment. Maxie and Crane (93, 94) have shown that fig leaves and fruit sprayed with 2,4,5-T exhibited these symptoms and that the action of the herbicide was via increased ethylene production. Similarly, ethylene was found to be a factor in the promotion of leaf senescence and abscission of *Euonymus* leaves by 2,4-D (61).

Wochok and Wetherell (143) reported that Ethrel and 2,4-D prevented organization of cultured wild carrot tissue and prevented the development of organized embryos from suspension cell cultures. However, it is not clear if ethylene is an intermediate in this system since they did not demonstrate that ethylene had a similar effect or that 2,4-D promoted ethylene production from the carrot cells. The effect of 2,4-D on ethylene production by cell cultures has been shown to vary. Gamborg and LaRue (53, 54) found that 2,4-D did not increase ethylene production from cultures of horseradish, sweet clover, or rose cells. However, it did increase ethylene production by *Ruta* cells.

In common with auxin, phenoxyacetic acids act as juvenility agents and block or retard ethylene action. For example, phenoxyacetic acids have been found to prevent abscission of apricot (14) and citrus fruit (83), delay ripening of bananas (91, 133, 135), and prevent leaf senescence (108). This has been graphically demonstrated in the case of *Euonymus* leaves. Hallaway and Osborne (61) reported that droplets of 2,4-D caused yellowing of leaf tissue surrounding the treated areas while the 2,4-D-treated areas themselves remained green. The promotion of fruit ripening also demonstrates both promotion and inhibitive effects of 2,4-D (41). Delayed color changes occur on the surface of the fruit while the fleshy portion softens. The explanation is probably the same as that employed for *Euonymus* leaves. The herbicide does not translocate and has a rejuvenating effect on the site of application while the ethylene diffuses into the internal fruit tissue and promotes softening.

Other herbicides promote ethylene production but their action is probably due to their ability to chemically stress the tissue. Active compounds include Endothal (60, 67), sodium cyanamide ($NaHCN_2$), potassium cyanate (KOCN), sodium chloroacetate ($ClCH_2COONa$), sodium chlorate ($NaClO_3$), and ammonium thiocyanate (NH_4SCN) (67). Even though the mode of action of these compounds is different from that of the phenoxyacetic acids, the methionine pathway for ethylene production appears to be employed for both classes of herbicides (1).

VII. Ethrel and Other Ethylene-Releasing Chemicals

A. Introduction

The application of ethylene to plants could be used to regulate a number of physiological processes which have commercial or practical value. However, gases are difficult to use under field conditions compared with other plant regulators which are usually applied as aqueous sprays. Nevertheless, attempts to use ethylene in the field have been made. Wilson (141) fumigated the soil in which orange trees were grown by inserting a hollow tube supplied with ethylene under pressure into the soil at the base of the trees. This treatment accelerated fruit abscission but it also caused complete defoliation of the tree and its subsequent death. Winchester (142) demonstrated that he could synchronize the flowering of pineapples by spraying them with water saturated with acetylene. Another method of applying ethylene to plants involves an ethylene-clatherite mixture (96). In this system ethylene is released slowly from some absorbant such as bentonite. However, these techniques were of limited usefulness because of the small amounts of ethylene released. The practical answer to finding a means of treating plants in the field with ethylene came with the introduction of compounds capable of releasing ethylene by means of some chemical reaction. The first compound used for this purpose was β-hydroxylethylhydrazine (BOH) (see structure I). In 1955 Gowing and Leeper (56) reported that BOH induced the flowering of pineapples. In a subsequent study (57) they evaluated the effectiveness of a series of hydrazine analogs and found that other active compounds include: ethylhydrazine (II), *sym*-diethylhydrazine (III), *unsym*-bis-(2-hydroxylethyl)hydrazine (IV), aminomorpholine (V), 2-hydroxyl-*N*-(2-hydroxylethyl)carbazinate (VI), and 2-(2-hydroxyethyl)semicarbazine (VII).

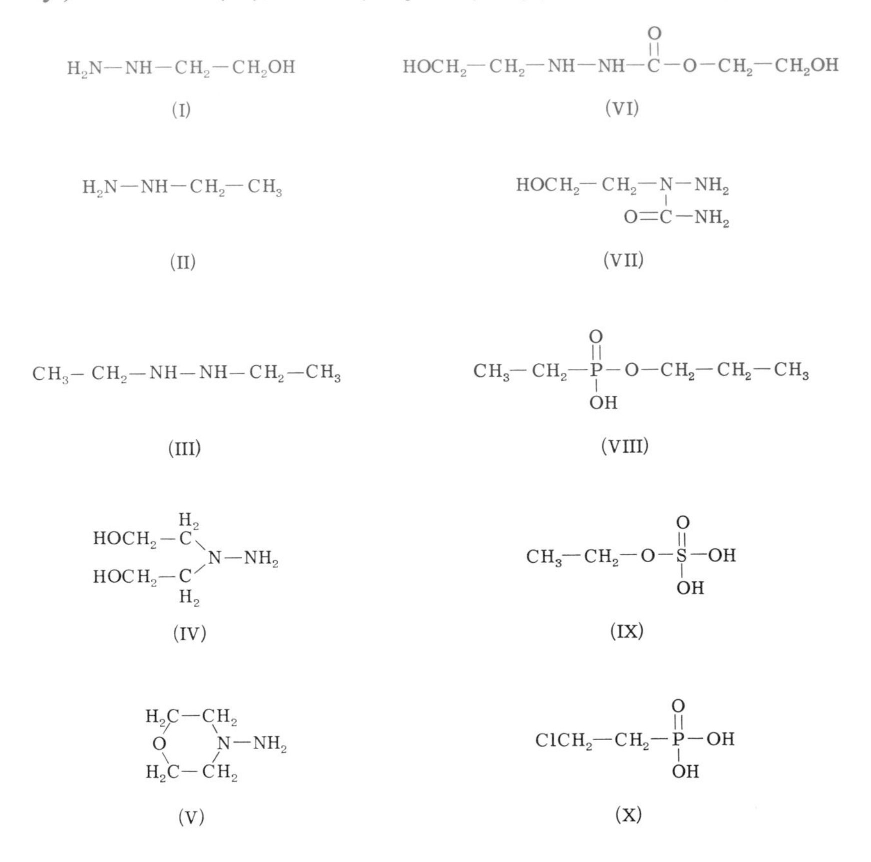

While Gowing and Leeper concluded that the physiologically active component was related to the structure $—CH_2—CH_2—NH—NH—$, it is more likely that these compounds released sufficient quantities of ethylene to promote flowering. Cathey and Downs (25) described the use of BOH to promote flowering in other members of the pineapple family. In their experiments, a 0.1–0.4% solution of Omaflora (a trade name for BOH) was poured in the well formed by the upper leaves of the plant. Block and Young (12) reported that they failed to observe the production of ethylene from BOH and suggested instead that the ethylene was released from impurities. However, Palmer *et al.* (112) reported that the percent conversion of BOH into ethylene was 1% and the rate of decomposition was influenced by pH. Other compounds reported to form ethylene are ethylpropyl phosphonate (Niagara 10637) (43a) (VIII), monoethyl sulfate (78) (IX), and Ethrel (2-chloroethylphosphonic acid, CEPA, Ethephon) (X). For reviews on the use and action of Ethrel, see references 10 and 42.

The reaction mechanism for the formation of ethylene from Ethrel was described by Maynard and Swan (95) and is probably similar for other compounds in this group. According to Yang (145) a $—CH_2—CH_2$ grouping in the center of the molecule with one end an electron-withdrawing center and the other an electron donor is capable of producing ethylene. The mechanism of ethylene production was thought to proceed by the following scheme.

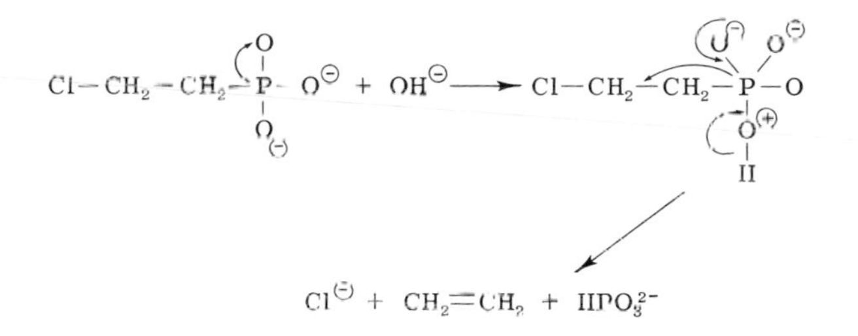

Yang postulated that a similar mechanism probably operated during the conversion of methionine to ethylene. In this case the electron acceptor is $CH_3—S—$ and the electron donor $—CHNH_2—COOH$.

Of the compounds described above, Ethrel appears to be the most effective ethylene generator and has been the one most extensively studied. The synthesis of Ethrel was first described by Kabachnik and Rossiyskaya (68) in 1946, and Warner and Leopold (138) were the first to describe its use as a plant regulator. The release of ethylene appears to be a simple base-catalyzed reaction and does not involve any enzymatic activity on the part of the plant (39, 87, 145). Ethrel is stable in the acid form but breaks down to form ethylene at a pH of 3.5 and above. The rate of ethylene formation increases as the pH is raised (10, 46). Most of the Ethrel applied to plants is eventually converted into ethylene. Yamaguchi *et al.* (144) reported that 50% of the Ethrel was converted into ethylene and the release of ethylene ceased after 6 days. Martin *et al.* (91) reported that only 1% of the Ethrel applied to walnut plants was recovered in the tissue 5 days after application. The amount of Ethrel recovered from grape plants depended on the site of application. Weaver *et al.* (139) found that when Ethrel was injected into grape berries only 10% of the original activity was recovered. When placed as a drop at the base of the leaf only 0.7% was recovered. As a drop on the surface of the berry a greater quantity was recovered, 30%. The surface

Ethrel was not fixed and could be readily washed off. These data suggest that the conditions for the breakdown of Ethrel were most favorable on the leaf surface and in the interior of fruits.

[^{14}C]Ethrel can be translocated throughout the plant. Yamaguchi *et al.* (144) found that 7 days following treatment of tomato leaves, about 15% of the ^{14}C was translocated to developing fruits and a lesser amount to other parts of the plant. After 2 days from 3–9% was translocated from the site of application to other tissues in squash seedlings. Twenty-five days after application to cucumber leaves, the fruits contained only 0.3% of the applied [^{14}C]Ethrel. Autoradiographic studies with grapes (139) revealed that Ethrel moved in the phloem in a source-to-sink relationship. It was readily translocated from mature leaves and not from young expanding leaves. In tomato tissue all the radioactivity was present as [^{14}C]Ethrel, but in the squash seedling tissue much of the radioactivity was present in a new compound (144). Metabolism of [^{14}C]Ethrel into [^{14}C]CO_2 was small. Yamaguchi *et al.* (144) reported that [^{14}C]CO_2 amounted to 0.1% of the [^{14}C]Ethrel applied. In the case of grapes, no metabolites of Ethrel were identified and all the ^{14}C was associated with Ethrel (139).

B. Uses of Ethrel

1. Promotion of Fruit Ripening

Apples treated with Ethrel late in the season ripened sooner than controls. When Ethrel was applied earlier, 10 days after bloom, the apples were reduced in size but contained more seeds, 6.5 for treated fruit versus 5.3 for controls (47). Ethrel-treated bananas were found to ripen sooner than untreated controls (120).

Application of Ethrel to field-grown cantaloupes resulted in early abscission and ripening of immature fruit (72). Practical applications of this technique were limited because even though the number of ripe fruits per acre were increased, the fruit were undersized and contained less sugar than controls. Apparently premature abscission prevented the fruits from attaining full size and sugar content (72). Ethrel application also caused premature abscission of leaves (72, 114).

Ethrel has been used to degreen citrus. Concentrations of 1000 ppm caused citrus to attain satisfactory color in 7 days while 5000 ppm caused rind damage and delayed the degreening process (51). In another study, 50–200 ppm used as a preharvest spray resulted in less postharvest degreening time for acceptable color and also resulted in less fruit decay in storage. Control fruit had a decay incidence of 25% while of those which were completely degreened by Ethrel only 4% of the total fruit had decayed (147).

As a side-effect, 100–200 ppm Ethrel caused abscission of some older leaves (147).

Application of 100, 500, and 1000 ppm Ethrel before bloom had an adverse effect on yield and size of Early Black cranberries. However, it caused the vines to rapidly synthesize anthocyanins, giving the vines a red color. Application of Ethrel 2 weeks before harvest significantly accelerated pigment development but had no effect on size and yield (40).

Fig fruits treated with Ethrel during period I (initial rapid growth) ceased growth and abscissed. Treatment during the early part of period II (slow growth) stimulated growth and maturation but quality was below that obtained with controls. Fruit treated during the later part of period II matured 3 weeks sooner than controls and were not significantly different from controls in terms of size and quality. Ethrel applied at a rate of 500 ppm to Mission figs caused leaf epinasty and abscission and growth of apical and lateral buds. No vegetative responses were noticed with the Calimyrna cultivar (33, 34). The authors also noted that Ethrel applied to leaves was able to induce fig maturation. The authors presumed that ethylene diffused from the leaves to the fruit, though it is possible that the Ethrel itself was translocated to the fruit.

Ethrel applied to Shirez grapes during the slow growth phase hastened the starting of ripening. When it was applied earlier, during the second half of the first rapid growth phase or at the start of the slow growth phase of berry ripening, it delayed ripening (59).

Application of Ethrel at rates of 250–1500 ppm during the final swell of Babygold-9, a clingstone peach, accelerated maturation of the fruit. High rates were phytotoxic and detrimental to fruit quality. Increased immature abscission and 90% defoliation occurred at high rates of Ethrel. Trees subsequently exuded gum and had a high bloom 6 weeks later. In no case was the treatment fatal to the trees. Individual terminals however failed to leaf out the following spring (117).

Ethrel at 30 and 50 ppm was applied as whole-tree sprays to Early Amber and Flordasun peaches (16, 76). The authors noted that Ethrel prevented browning of the puree and sliced peaches. This effect was due to a reduction in polyphenoloxidase content of the tissue which is responsible for darkening of flesh upon slicing. The effect, however, was not specific for Ethrel, since Alar and gibberellic acid had the same effect.

The development of red chili peppers was hastened by Ethrel (125). However, high rates caused defoliation of the leaves.

Foliage sprays on field tomatoes with high rates of Ethrel resulted in complete leaf abscission within 3 days of treatment. With lower rates, there was slight leaf yellowing and enhanced fruit ripening. Even though the Ethrel caused leaf drop at the higher concentrations it did not cause fruit drop (114).

2. *Induction of Abscission and Fruit Thinning*

A number of investigators have evaluated the use of Ethrel as a defoliating agent for nursery stock (36, 79) and snap beans (110). Two-year-old apple trees were treated in a nursery with 2000, 3700, and 5000 ppm Ethrel (36). The 2000 ppm concentration was an effective defoliant but delayed subsequent growth. The higher concentration resulted in reduced shoot growth and initiation of adventitious roots. In another study, Ethrel damage in terms of hypertrophies at the bases of buds, dieback of stems, and delayed growth was associated with sprays containing 500 to 1000 ppm Ethrel (79). Addition of Bromodine to Ethrel enhanced its activity as a defoliant (79). The action of Bromodine appeared to depend on its ability to damage the leaf. Apparently damaged leaves undergo accelerated senescence which results in an enhanced susceptibility to the ethylene released from the Ethrel. Ethrel has also been used as a preharvest treatment to facilitate mechanical harvesting of snap beans (79). At concentrations of 250–4000 ppm it caused leaf abscission, but at the higher concentrations a reduction of yield due to the yellowing of pods was noted. Damage was reduced by lowering the concentrations used and by postponing treatment close to the anticipated harvest time.

Ethrel applied to peach trees early in the development of the fruit acts as a thinning agent (15, 92, 128). The concentrations used have to be carefully regulated since high dosages cause severe gumming and reduction in the size of leaves arising from the buds. The appropriate stage for the application of Ethrel has been variously described as 80% full bloom (15), 4- to 9-mm ovule length (92), or the end of stage I of fruit growth (128). Ethrel has also been used to remove flowers and fruit from olive and Victorian box trees. The authors indicated that this treatment was a useful way to control fruit formation in landscape trees.

Mechanical harvesting of fruits has become an increasingly important agricultural technique as the supply of cheap labor has become scarce. To facilitate mechanical harvesting, workers have tested and evaluated a number of chemicals, including Ethrel, as a means of reducing the attachment force of fruits. Data on the results of the effectiveness of Ethrel on cherries (8), apples (44, 45), pears (58), oranges (111), and grape (140) have been presented. Concentrations used varied between 500 and 2000 ppm and, as reported for other systems, exudation, premature ripening, and defoliation were undesirable side effects when high doses were used. Most investigators concluded that chemical loosening facilitated mechanical harvesting, which suggests that while Ethrel may not be the final agent of choice for this technique, chemical harvest aids will become important as the degree of mechanization increases.

3. *Promotion of Latex Flow*

The auxins IAA and 2,4-D have been used to promote latex flow in rubber trees (38). With the demonstration that some physiological effects of auxin were due to an increase in ethylene production, research in the potential of Ethrel as a means of increasing rubber flow was initiated. Results of Ethrel application (7, 38) have been a two- to fourfold increase in latex flow, primarily due to a longer duration of flow. After the treatment with Ethrel, the stimulated condition falls off in 10–15 days, and after 2 months, the flow rate returns to that of untreated trees. Until this time retreatment with Ethrel is without effect. According to Leopold (81), the effect of Ethrel treatment appears to be a deferral of latex coagulation and hence slower clotting of the flow. It is not clear whether there is any similarity or relationship between Ethrel-induced gummosis in peaches and cherries and latex flow in rubber.

4. *Flowering*

Induction of flowering in bromeliads by ethylene-releasing compounds is well known (25, 29, 56, 112), and, in fact, every bromeliad tested has shown a positive response. While a promotion of flowering appears to be limited to this family, exceptions have been reported. Nitsch and Nitsch (106) demonstrated that another short-day plant, *Plumbago indica*, could also be induced to flower under noninductive long days with Ethrel treatment. An obvious practical application for Ethrel would be its use in inducing synchronous development of pineapple fruit.

5. *Sex Expression*

Ethrel has been shown to promote the development of female flowers in a wide variety of cucurbits, including *Luffa acutangula* (13), squash (31, 88, 119) (see Fig. 4-7), cucumber (55, 66, 88, 119, 123), pumpkin (126), and melon (119). However, it was not equally effective for all cultural varieties of cucumbers. Ethrel had little effect on Marketmore and Tokyo cultivars as opposed to other varieties tested (55). While the effect of Ethrel on sex determination is known primarily for cucurbits, an induction of female flowers in male plants of *Cannabis sativa* has also been reported (100). The use of Ethrel in cucurbits has been suggested as a tool for hybridizers (31, 88, 119) and as a means of increasing fruit production. Ethrel would be useful in the production of hybrids because it would eliminate the expense of hand pollination or removal of male flowers. Ethrel applied at 250 ppm at nine weekly intervals increased fruit numbers and decreased fruit weight of New Hampshire Butternut squash. However, the total fruit yield did not differ greatly among treatments (31). In another study (126) Ethrel was

Fig. 4-7. Effect of Ethrel on flowering of squash. Upper, control. Lower, 500 ppm Ethrel. [Photograph courtesy of Rudich *et al.* (119).]

applied to pumpkin plants under commercial conditions, and while the total number of fruit in terms of weight was not increased, Ethrel did increase the number of fruits.

6. Tuber and Bulb Formation

Catchpole and Hillman (24) reported that ethylene-treated potato sprouts ceased elongating and developed tubers at the end of each stolon as opposed to controls which developed tubers at only a few stolons. Even though swellings similar to tubers were initiated, the swollen tissue was free of starch as opposed to control tubers. Application of Ethrel to the stolon also promoted tuber development and, as reported above, the tubers were free of starch.

Bulbing in onions is indicated by swelling of leaf bases and an increase in cell size accompanied by translocation of assimilates to these tissues. Ethrel has been used to increase swelling of leaf bases and bulb initiation in onion and leek (82). In field experiments, seedlings of five onion cultivars were sprayed once or several times with Ethrel. Except for one very early cultivar, which showed bulbing due to long days, the treatment resulted in early bulb initiation and in greatly increased rate of bulbing. Rates of 5000 and 10,000 ppm Ethrel were more effective than lower concentrations (500 ppm) but they also caused retarded leaf growth and a decrease in the final bulb size.

7. Seed Germination

Promotion of seed germination by Ethrel has been observed by a number of investigators. The effect has been reported for witchweed (*Striga lutea*) (48), 49), peanut (74, 75), and rape (130). Witchweed is a parasitic weed of corn that infests more than 200,000 acres of land in North and South Carolina. The Department of Agriculture and United States farmers now spend more than 1 million dollars a year to eradicate this pest from infested areas. Witchweed plants are easy to kill after germination of the seeds. However, dormant seeds are very difficult to exterminate. They do not normally germinate unless they are in close proximity to the roots of host or nonhost plants that exude a stimulant that induces germination. Present control programs are largely restricted to application of herbicides to the emerged plants. The seeds that do not germinate remain viable for many years. Such seeds can grow and produce seed that can reinfest a field for as long as 15–20 years (48) Egley and Dale (49) incorporated 100 mg Ethrel per kilogram of soil and increased germination from 0% for controls to 73% for treated soil. They proposed that this could be an effective method for controlling this weed.

8. Breaking of Bud Dormancy

Ethrel was used to defoliate and subsequently increase the growth of inactive basal buds of honey mesquite, a troublesome weed on rangeland in the Southwest (104). Morgan, Meyer, and Merkle (104) proposed that the induced growth of basal and lateral buds of woody plants may allow the killing of more plants by subsequent herbicide applications.

Shanks (122) reported that Ethrel aided in developing flower buds in azalea plants that had been chemically pinched with methyl pelargonate. Ethrel has also been used to increase corm production in gladiolus. Magie (89) reported that Ethrel increased corm production by 50% and cormel production by 90%.

9. Promotion of Root Initiation

A number of reports on the effects of Ethrel on plants have included the observation that induction of adventitious roots and rhizomes was promoted. This effect has been seen on blueberry (73), mung bean (77), and tomato (107). The action of Ethrel in mung bean was enhanced if IAA was added after the addition of Ethrel (77). However, Shanks (122) reported that Ethrel was not effective in inducing rooting in carnations, poinsettia, and chrysanthemums.

10. Dwarfing

Ethylene reduces internode elongation, giving rise to a dwarfed and more compact form of growth. A number of investigators have tested Ethrel as a means of accomplishing the same effect. A reduction of growth following Ethrel application has been observed on snapdragons, marigold, zinnia (122), grass (*Poa pratensis*) (132), and sweet potato (131). The advantage of dwarfed grass could be a significant factor in reducing the cost of lawn or turf management and may represent a useful application of ethylene-releasing compounds.

11. Other Uses

Other potential uses of Ethrel include the acceleration of ripening of tobacco. Treatment of flue-cured tobacco with Ethrel caused mature leaves to lose their green color and turn yellow (127). The dollar value per hundred weight of cured leaf from treated plants was higher than that of untreated plants but the average weight per leaf from treated plants was lower. During the curing process, Ethrel-treated leaves can probably be subjected to shorter yellowing periods than untreated leaves. Steffens *et al.* (127) concluded that ethylene-releasing chemicals can be used to hasten senescence of tobacco leaves, shortening the time used to ripen and cure the crop.

Ethrel has also been tested as a means of modifying cold-hardiness of citrus (146). However, it was found to be ineffective.

In conclusion, Ethrel or other ethylene-releasing compounds have been used to increase production or efficiency in certain agricultural practices such as latex production. Progress in the areas of thinning, abscission, accelerated ripening, and control of witchweed or other weeds suggests that additional practices will be effectively controlled by ethylene-releasing agents.

References

1. Abeles, A. L., and Abeles, F. B. (1972). *Plant Physiol.* **50,** 496.
1a. Abeles, F. B. (1966). *Plant Physiol.* **41,** 585.
2. Abeles, F. B. (1967). *Physiol. Plant.* **20,** 442.
3. Abeles, F. B. (1967). *Plant Physiol.* **42,** 608.
4. Abeles, F. B. (1968). *Weed Sci.* **16, 498.**
5. Abeles, F. B., Holm, R. E., and Gahagan, H. E. (1967). *Plant Physiol.* **42,** 1351.
6. Abeles, F. B., and Rubinstein, B. (1964). *Plant Physiol.* **39,** 963.
7. Abraham, P. D., Wycherley, P. R., and Pakianathan, S. W. (1969). *J. Rubber Res. Inst. Malaya* **20,** 291.
8. Anderson, J. L. (1969). *HortScience* **4,** 92.
9. Andreae, W. A., Venis, M. A., Jursic, F., and Dumas, T. (1968). *Plant Physiol.* **43,** 1375.
10. Anonymous. (1969). "Ethrel Technical Service Data Sheet H-96." Amchem Products Inc., Ambler, Pennsylvania 19002.
11. Baur, J. R., and Morgan, P. W. (1969). *Plant Physiol.* **44,** 831.
12. Block, M. J., and Young, D. C. (1971). *Nature (London)* **231,** 288.
13. Bose, T. K., and Nitsch, J. P. (1970). *Physiol. Plant.* **23,** 1206.
14. Bradley, M. V., Marei, N., and Crane, J. C. (1969). *J. Amer. Soc. Hort. Sci.* **94,** 316.
15. Buchanan, D. W., and Biggs, R. H. (1969). *J. Amer. Soc. Hort. Sci.* **94,** 327.
16. Buchanan, D. W., Hall, C. B., Biggs, R. H., and Knapp, F. W. (1969). *HortScience* **4,** 302.
17. Burg, S. P., and Burg, E. A. (1966). *Proc. Natl. Acad. Sci. U.S.* **55,** 262.
18. Burg, S. P., and Burg, E. A. (1966). *Science* **152,** 1269.
19. Burg, S. P., and Burg, E. A. (1968). *Plant Physiol.* **43,** 1069.
20. Burg, S. P., and Burg, E. A. (1968). *In* "Biochemistry and Physiology of Plant Growth Substances" (F. Wrightman and G. Setterfield, eds.), p. 1275. Runge Press, Ottawa.
21. Burg, S. P., and Burg, E. A. (1969). *Qual. Plant Mater. Veg.* **19,** 185.
22. Burg, S. P., and Clagett, C. O. (1967). *Biochim. Biophys. Res. Commun.* **27,** 125.
23. Burg, S. P., and Dijkman, M. J. (1967). *Plant Physiol.* **42,** 1648.
24. Catchpole, A. H., and Hillman, J. (1969). *Nature (London)* **223,** 1387.
25. Cathey, H. M., and Downs, R. J. (1965). *Exch. Flower, Nursery Gard. Cent. Trade* **143,** 27.
26. Chadwick, A. V., and Burg, S. P. (1967). *Plant Physiol.* **42,** 415.
27. Chadwick, A. V., and Burg, S. P. (1970). *Plant Physiol.* **45,** 192.
28. Chalutz, E., DeVay, J. E., and Maxie, E. C. (1969). *Plant Physiol.* **44,** 235.
29. Cooke, A. R., and Randall, D. I. (1968). *Nature (London)* **218,** 974.
30. Cooper, W. C., Rasmussen, G. K., Rogers, B. J., Reece, P. C., and Henry, W. H. (1968). *Plant Physiol.* **43,** 1560.
31. Coyne, D. P. (1970). *HortScience* **5,** 227.
32. Craker, L. E., and Abeles, F. B. (1969). *Plant Physiol.* **44,** 1144.
33. Crane, J. C., Marei, N., and Nelson, M. M. (1970). *Calif. Agr.* **24,** 8.
34. Crane, J. C., Marei, N., and Nelson, M. M. (1970). *J. Amer. Soc. Hort. Sci.* **95,** 367.
35. Crocker, W., Hitchcock, A. E., and Zimmerman, P. W. (1935). *Contrib. Boyce Thompson Inst.* **7,** 231.

36. Cummins, J. N., and Fiorino, P. (1969). *HortScience* **4,** 339.
37. Curtis, R. W. (1969). *Plant Cell Physiol.* **10,** 909.
38. D'Auzac, J., and Ribaillier, D. (1969). *C. R. Acad. Sci.* **268,** 3046.
39. Dennis, F. G., Wilczynski, H., de la Guardia, M., and Robinson, R. W. (1970). *HortScience* **5,** 168.
40. Devlin, R. M., and Demoranville, I. E. (1970). *Physiol. Plant* **23,** 1139.
41. Dewey, D. H., and Uota, M. (1953). *Proc. Amer. Soc. Hort. Sci.* **61,** 246.
42. de Wilde, R. C. (1971). *HortScience* **6,** 364.
43. Dijkman, M. J., and Burg, S. P. (1970). *Amer. Orchid Soc., Bull.* p. 799.
43a. Dollwet, H. H. A., and Kumamoto, J. (1970). *Plant Physiol.* **46,** 786.
44. Edgerton, L. J. (1968). *N. Y. Food Life Sci.* **1,** 19.
45. Edgerton, L. J. (1968). *Proc. N. Y. State Hort. Soc.* **113,** 99.
46. Edgerton, L. J., and Blanpied, G. D. (1968). *Nature* (*London*) **212,** 1064.
47. Edgerton, L. J., and Greenhalgh, W. J. (1969). *J. Amer. Soc. Hort. Sci.* **94,** 11.
48. Egley, G. H., and Dale, J. E. (1970). *Proc. S. Weed Sci. Soc.* **23,** 327 (abstr.).
49. Egley, G. H., and Dale, J. E. (1970). *Weed Sci.* **18,** 586.
50. Forsyth, F. R., and Hall, I. V. (1968). *Naturaliste Can.* **95,** 1165.
51. Fuchs, Y., and Cohen, A. (1969). *J. Amer. Soc. Hort. Sci.* **94,** 617.
52. Fuchs, Y., and Lieberman, M. (1968). *Plant Physiol.* **43,** 2029.
53. Gamborg, O. L., and LaRue, T. A. G. (1968). *Nature* (*London*) **220,** 604.
54. Gamborg, O. L., and LaRue, T. A. G. (1971). *Plant Physiol.* **46,** 399.
55. George, W. L. (1971). *J. Amer. Soc. Hort. Sci.* **96,** 152.
56. Gowing, D. P., and Leeper, R. W. (1955). *Science* **122,** 1267.
57. Gowing, D. P., and Leeper, R. W. (1961). *Bot. Gaz.* **123,** 34.
58. Griggs, W. H., Iwakiri, B. T., Fridley, R. B., and Mehlschau, J. (1970). *HortScience* **5,** 264.
59. Hale, C. R., Coombe, B. G., and Hawker, J. S. (1970). *Plant Physiol.* **45,** 620.
60. Hall, W. C. (1952). *Bot. Gaz.* **113,** 310.
61. Hallaway, M., and Osborne, D. J. (1969). *Science* **163,** 1067.
62. Hansen, E. (1946). *Plant Physiol.* **21,** 588.
63. Hicks, J. R., and Brown, D. S. (1968). *Proc. Amer. Soc. Hort. Sci.* **92,** 755.
64. Holm, R. E., and Abeles, F. B. (1967). *Planta* **78,** 293.
65. Holm, R. E., and Key, J. L. (1969). *Plant Physiol.* **44,** 1259.
66. Iwahori, S., Lyons, J. M., and Sims, W. L. (1969). *Nature* (*London*) **222,** 271.
67. Jackson, J. M. (1952). *Arkansas Acad. Sci. Proc.* **5,** 73.
68. Kabachnik, M. I., and Rossiyskaya, P. A. (1946). *Izv. Akad. Nauk. SSSR., Ser. Kh.* No. 4, p. 403.
69. Kang, B. G., Newcomb, W., and Burg, S. P. (1971). *Plant Physiol.* **47,** 504.
70. Kang, B. G., and Ray, P. M. (1969). *Planta* **87,** 206.
71. Kang, B. G., Yocum, C. S., Burg, S. P., and Ray, P. M. (1967). *Science* **156,** 958.
72. Kasmire, R. F., Rappaport, L., and May, D. (1970). *J. Amer. Soc. Hort. Sci.* **95,** 134.
73. Kender, W. J., Hall, I. V., Aalders, L. E., and Forsyth, F. R. (1969). *Can. J. Plant Sci.* **49,** 95.
74. Ketring, D. L., and Morgan, P. W. (1970). *Plant Physiol.* **45,** 268.
75. Ketring, D. L., and Morgan, P. W. (1971). *Plant Physiol.* **47,** 488.
76. Knapp, F. W., Hall, C. B., Buchanan, D. W., and Biggs, R. H. (1970). *Phytochemistry* **9,** 1453.
77. Krishnamoorthy, H. N. (1970). *Plant Cell Physiol.* **11,** 979.
78. Kumamoto, J., Dollwet, H. H. A., and Lyons, J. M. (1969). *J. Amer. Chem. Soc.* **91,** 1207.
79. Larsen, F. E. (1971). *Hortscience* **6,** 135.

80. LaRue, T. A. G., and Gamborg, O. L. (1971). *Plant Physiol.* **48,** 394.
81. Leopold, A. C. (1971). *In* "What's New in Plant Physiology" (G. J. Fritz, ed.), Vol. 3, No. 3. Univ. of Florida Press, Gainesville.
82. Levy, D., and Kedar, N. (1970). *HortScience* **5,** 80.
83. Lewis, L. N., Palmer, R. L., and Hield, H. Z. (1968). *In* "Biochemistry and Physiology of Plant Growth Substances" (F. Wrightman and G. Setterfield, eds.), p. 1303. Runge Press, Ottawa.
84. Looney, N. E. (1968). *Plant Physiol.* **43,** 1133.
85. Looney, N. E. (1969). *Plant Physiol.* **44,** 1127.
86. Looney, N. E. (1971). *HortScience* **6,** 238.
87. Lougheed, E. C., and Franklin, E. W. (1970). *Can. J. Plant Sci.* **50,** 586.
88. Lower, R. L., and Miller, C. H. (1969). *Nature* (*London*) **222,** 1072.
89. Magic, R. O. *HortScience* **6,** 351.
90. Marth, P. C., and Mitchell, J. W. (1949). *Bot. Gaz.* **110,** 514.
91. Martin, G. C., Abdel-Gawad, H. A., and Weaver, R. J. (1972). *J. Amer. Soc. Hort. Sci.* **97,** 51.
92. Martin, G. C., Nelson, M. M., and Nishijima, C. (1971). *HortScience* **6,** 169.
93. Maxie, E. C., and Crane, J. C. (1967). *Science* **155,** 1548.
94. Maxie, E. C., and Crane, J. C. (1968). *Proc. Amer. Soc. Hort. Sci.* **92,** 255.
95. Maynard, J. A., and Swan, J. M. (1963). *Aust. J. Chem.* **16,** 596.
96. Mehrlich, F. P. (1941). U. S. Patent 2,245,867.
97. Michener, H. D. (1935). *Science* **82,** 551.
98. Michener, H. D. (1938). *Amer. J. Bot.* **25,** 711.
99. Miller, P. M., Sweet, H. C., and Miller, J. H. (1970). *Amer. J. Bot.* **57,** 212.
100. Mohan Ram, H. Y., and Jaiswal, V. S. (1970). *Experientia* **26,** 214.
101. Morgan, P. W., and Baur, J. R. (1970). *Plant Physiol.* **46,** 655.
102. Morgan, P. W., and Hall, W. C. (1962). *Physiol. Plant.* **15,** 420.
103. Morgan, P. W., and Hall, W. C. (1964). *Nature* (*London*) **201,** 99.
104. Morgan, P. W., Meyer, R. E., and Merkle, M. G. (1969). *Weed. Sci.* **17,** 353.
105. Morgan, P. W., and Powell, R. D. (1970). *Plant Physiol.* **45,** 553.
106. Nitsch, C., and Nitsch, J. P. (1969). *Plant Physiol.* **44,** 1747.
107. Orion, D., and Minz, G. (1969). *Nematologica* **15,** 608.
108. Osborne, D. J. (1968). *Sci.* (*Soc. Chem. Ind., London*) *Monogr.* **31,** 236.
109. Owens, L. D., Lieberman, M., and Kunishi, A. (1971). *Plant Physiol.* **48,** 1.
110. Palevitch, D. (1970). *HortScience* **5,** 224.
111. Palmer, R. L., Hield, H. Z., and Lewis, L. N. (1969). *Proc. Int. Citrus Symp., 1st,* Vol. 3, p. 1135.
112. Palmer, R. L., Lewis, L. N., Hield, H. Z., and Kumamoto, J. (1967). *Nature* (*London*) **216,** 1216.
113. Poapst, P. A., Durkee, A. B., McGugan, W. A., and Johnston, F. B. (1968). *J. Sci. Food Agr.* **19,** 325.
114. Rabinowitch, H. D., Rudich, J., and Kedar, N. (1970). *Isr. J. Agr. Res.* **20,** 47.
115. Radin, J. W., and Loomis, R. S. (1969). *Plant Physiol.* **44,** 1584.
116. Rhodes, M. J. C., and Wooltorton, L. S. C. (1971). *Phytochemistry* **10,** 1989.
117. Rom, R. C., and Scott, K. R. (1971). *HortScience* **6,** 134.
118. Rubinstein, B., and Abeles, F. B. (1965). *Bot. Gaz.* **126,** 255.
119. Rudich, J., Halevy, A. H., and Kedar, N. (1969). *Planta* **86,** 69.
120. Russo, L., Dostal, H. C., and Leopold, A. C. (1968). *BioScience* **18,** 109.
121. Sakai, S., and Imaseki, H. (1971). *Plant Cell Physiol.* **12,** 349.
122. Shanks, J. B. (1969). *HortScience* **4,** 56.

123. Shannon, S., and de le Guardia, M. D. (1969). *Nature* (*London*) **223,** 186.
124. Shingo, S., and Imaseki, H. (1971). *Plant Cell Physiol.* **12,** 349.
125. Sims, W. L., Collins, H. B., and Gledhill, B. L. (1970). *Calif. Agr.* **24,** 4.
126. Splittstoesser, W. E. (1970). *Physiol. Plant.* **23,** 762.
127. Steffens, G. L., Alphin, J. G., and Ford, Z. T. (1970). *Beit. Tabakforsch.* **5,** 262.
128. Stembridge, G. E., and Gambrell, C. E. (1971). *J. Amer. Soc. Hort. Sci.* **96,** 7.
129. Stewart, E. R., and Freebairn, H. T. (1969). *Plant Physiol.* **44,** 955.
130. Takayanagi, K., and Harrington, J. F. (1971). *Plant physiol.* **47,** 521.
131. Tompkins, D. R., and Bowers, J. L. (1970). *HortScience* **5,** 84.
132. van Andel, O. M. (1970). *Naturwissenschaften* **57,** 396.
133. Vendrell, M. (1969). *Aust. J. Biol. Sci.* **22,** 601.
134. Vendrell, M. (1970). *Aust. J. Biol. Sci.* **23,** 553.
135. Vendrell, M. (1970). *Aust. J. Biol. Sci.* **23,** 1133.
136. Wade, N. L., and Brady, C. J. (1971). *Aust. J. Biol. Sci.* **24,** 165.
137. Warner, H. L. (1970). Ph.D. Thesis, Purdue University, Lafayette, Indiana.
138. Warner, H. L., and Leopold, A. C. (1967). *BioScience* **17,** 722.
139. Weaver, R. J., Abdel-Gawad, H. A., Martin, G. C. (1972). *Physiol. Plant* **26,** 13.
140. Weaver, R. J., and Pool, R. M. (1969). *J. Amer. Soc. Hort. Sci.* **94,** 474.
141. Wilson, W. C. (1966). *Proc. Fla. State Hort. Soc.* **79,** 301.
142. Winchester, O. R. (1941). *Proc. Fla. State Hort. Soc.* **54,** 138.
143. Wochok, Z. S., and Wetherell, D. F. (1972). *Plant Cell Physiol.* **12,** 771.
144. Yamaguchi, M., Chu, C. W., and Yang, S. F. (1971). *J. Amer. Soc. Hort. Sci.* **96,** 606.
145. Yang, S. F. (1969). *Plant Physiol.* **44,** 1203.
146. Young, R. (1970). *HortScience* **5,** Sect. 2, 67th Meet. Abstr., p. 306.
147. Young, R., Jahn, O., Cooper, W. C., and Smoot, J. J. (1970). *HortScience* **5,** 268.
148. Zimmerman, P. W., and Wilcoxon, F. (1935). *Contrib. Boyce Thompson Inst.* **7,** 209.

Chapter 5

Stress Ethylene

Williamson and Dimock (77) were the first to note that ethylene production is often higher in injured or dying cells. A significant body of literature now exists that indicates that ethylene production increases rapidly following trauma caused by chemicals, insect damage, temperature extremes, drought, γ irradiation, disease, and mechanical wounding. While the function for higher rates of ethylene production is known in only a few cases, the basic phenomenon has been well studied and is commonly called wound ethylene. A better and more general term is stress-induced ethylene production or, more simply, stress ethylene.

I. Insects

The accelerated senescence and shedding of plant organs following insect infestation has been observed in a variety of plants. Infestation of cotton bolls with the cotton weevil causes abscission of the boll when the weevil reaches an appropriate developmental stage. E. E. King* has reported that the insect releases cellulolytic and proteolytic enzymes which in turn cause increased ethylene production. Galil (26) has shown that wasps stimulate ethylene production and ripening of figs when they mature

*Unpublished results (1968).

and escape from the fruit. Rose leaves infested with red spider mites produced more ethylene than uninfested ones (76).

II. Temperature

Abscission of citrus fruits and leaves often follows a damaging freeze. Vines *et al.* (73) reported that ethylene levels in grapefruit increased after fruit were exposed to freezing temperatures. However, the effect was not uniform and some fruit escaped damage and appeared normal following the cold treatment. Young and Meredith (79) examined the effects of subfreezing temperatures on citrus leaves. The difference between lethal and sublethal temperatures was extremely small. At $-6.7°C$ irreversible injury resulted, causing termination of respiration, ethylene production, and an inhibition of abscission. At $-6.1°C$ ethylene production and respiration increased, followed by leaf abscission. The increase in ethylene production started within 12 hours and increased gradually over a 3-day period. In addition, an immediate increase in electrolyte leakage followed the cold treatment. Depending on the severity of the cold treatment, it was possible to increase ethylene production without increasing abscission.

III. Water

Drought has been reported to cause abscission of cotton. According to McMichael *et al.* (52) the rate of ethylene production of cotton increased when the plants were subjected to a water deficit and returned to normal when they were watered. It seems reasonable that the function of drought-induced ethylene production is to promote abscission and reduce further water loss by transpiration.

IV. γ Irradiation

In the 1960's a number of investigators studied the feasibility of using γ radiation to preserve fruits and vegetables. However, Maxie and his co-workers reported that irradiation damaged plant material in terms of appearance and caused off-flavors. In addition, rates of ethylene production from irradiated tissue were higher than those observed from controls. Increased ethylene production has been observed in peas (40, 69), wheat (58), peaches, pears, lemons (43), nectarines (47), avocado (78), and tomato (1). The dose-response curve is variable and depends on the age of the tissue.

A 10-day lag in ethylene production occurred after irradiation of green tomato fruit, immediate production in breaker fruit, and no effect of radiation on ripe fruit (1). In oranges, mature fruit required 400 krad for induction while immature fruit had a threshold of 50 krad and a maximum response at 200 krad (28). (A rad is equal to 100 ergs of energy absorbed per gram of tissue. In man, 100 rad causes radiation sickness and 500 rad, death.) Lemons also had a threshold of 50 krad (45) and tomatoes a maximum response at 600 krad (1). Rates of ethylene production varied from one part of the fruit to another. The peel of lemons produced more ethylene than the pulp (48). An immediate response was observed in tomatoes (1), while a 30-minute lag was reported for lemons (46).

The mechanism involved in high energy radiation-induced ethylene production may not be a physiological one. Even though oxygen was required for ethylene production, Maxie *et al.* (48) noted that autoclaved lemons also produced ethylene. Irradiation of certain chemical components of fruit such as alcohols (49), Krebs cycle acids, and linolenic acid (48) resulted in ethylene production.

V. Disease

Viral and fungal diseases often give rise to symptoms such as yellowing, epinasty, and abscission. A number of investigators suspected that these were not due directly to the pathogen but resulted instead from increased rates of ethylene production from the infected plants.

In the case of viral diseases the increase in ethylene production appears to be due directly to the extent of lesion formation. This was first observed by Ross and Williamson in 1951 (66). They noted that leaves of *Physalis floridiana* infected with potato virus were epinastic and abscised 13 days after inoculation and that greater quantities of ethylene were produced by leaves with necrotic lesions. When lesions were not formed, the leaves produced less ethylene, for example, when leaves were placed under conditions of high temperature or when leaves were infected with systemic viruses. Ross and Williamson confirmed their idea that ethylene production was closely associated with the lesions themselves since plants shredded with a knife or treated with $CuSO_4$ produced more ethylene than controls.

Since that time other workers have made similar findings in a variety of diseases and plants. Balazs *et al.* (6) reported that ethylene production was high in tobacco plants infected with tobacco mosaic virus, a local lesion pathogen, while it was unchanged in plants infected with cucumber mosaic virus, a systemic virus. Similar results were reported by Nakagaki *et al.* (53). Data showing the increase in ethylene production following lesion

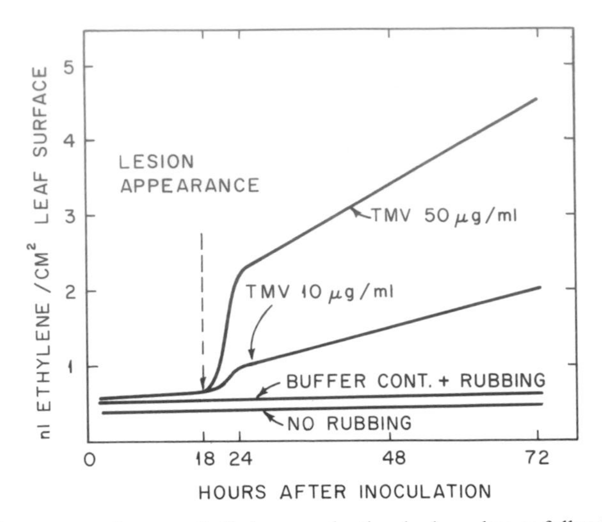

Fig. 5-1. Time course changes of ethylene production by bean leaves following inoculation with tobacco mosaic virus. Data show total accumulation of ethylene during the intervals indicated. Individual experimental points are deleted for the sake of clarity. [Courtesy of Nakagaki *et al.* (53).]

development are shown in Fig. 5-1. Recently Olson *et al.* (55) reported that when healthy tangerine plants were grafted on root stalks infected with stubborn virus, fruit abscission, but not leaf abscission, accelerated. This was in contrast to the lack of a similar effect when lemon or limes were grafted on infected seedlings. Internal levels of ethylene in healthy fruit were 0.01 ppm, compared with 1.1 ppm in virus-infected fruit. Ethylene production was the same for healthy and infected leaves.

A number of investigators have shown that plants infected with fungi produced large quantities of ethylene. Usually the host tissue produces the ethylene, although in some cases the fungus is the source of gas production. Williamson (75, 77) was the first to clearly recognize that typical disease symptoms such as abscission and yellowing were due to an increase in ethylene production by infected tissue. He reported that high levels of ethylene were associated with blackspot of rose (*Diplocarpon rosae*), cherry infected with *Coccomyces hiemalis,* chrysanthemum flowers infected with *Ascochyta chrysanthemi,* snapdragon rust, chrysanthemum rust, *Septoria* leaf spot of chrysanthemum, and *Alternaria* leaf spot of carnations. The causative fungus in the cases examined did not produce ethylene. However, because he did note that *Penicillium digitatum* produced large quantities of ethylene, he proposed that tissue damage by disease was the source of ethylene. This idea was supported by the observation that shredded rose leaves produced more ethylene than uninjured ones.

Subsequently, other investigators have made similar findings. Smith *et al.* (70) were concerned with the cause of damage to carnations transported in ships. They found that plants infected with *Botrytis* species produced sufficient levels of ethylene (0.06 ppm) in the holds to account for the damage. No ethylene was produced by the fungus itself. In brown-eyespot disease of coffee (*Cercospora coffeicola*), early symptoms were the appearance of small necrotic lesions which were followed later by yellowing and epinasty and culminated in leaf abscission. Using tomato epinasty as a bioassay, Subramania and Sridhar (72) demonstrated ethylene production from the infected leaves.

Defoliating (T-9) and nondefoliating (SS4) strains of *Verticillium albo-atrum,* a disease of cotton, are known. Wiese and DeVay (74) demonstrated that while plants infected with either strain showed some ethylene damage initially (stunting, epinasty, and chlorosis), a greater production of ethylene was subsequently associated with plants infected with the defoliating strain than with the nondefoliating one (Fig. 5-2). In addition to changes in ethylene production, higher concentrations of abscisic acid were found in leaves infected with the defoliating strain. Most of the ethylene must have been produced by the host since only trace amounts were produced by *Verticillium albo-atrum.*

Stahmann *et al.* (71) reported that ethylene was not detected from *Ceratocystis fimbriata* (black rot of sweet potato) but was readily produced by diseased sweet potato tissue. This observation was confirmed by Imaseki *et al.* (32). Nonpathogenic strains of *Ceratocystis* caused host tissue to produce more ethylene than pathogenic strains. This suggested the idea

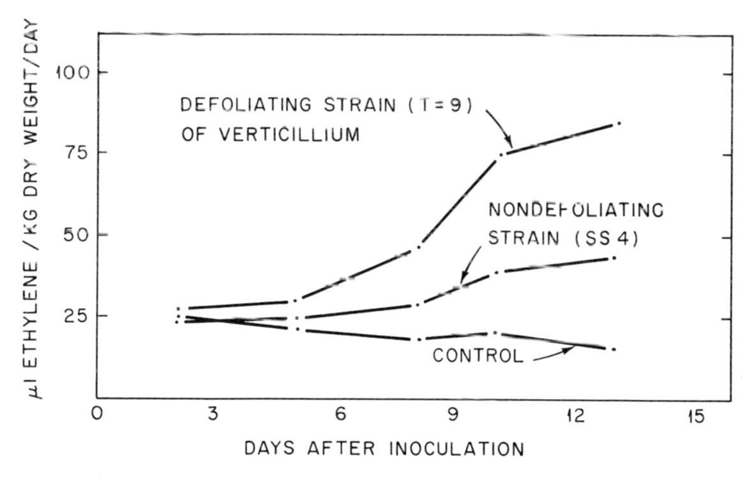

Fig. 5-2. Ethylene production by healthy and *Verticillium*-infected plants at intervals after inoculation. [Courtesy of Wiese and DeVay (74).]

that ethylene may play a role in the resistance of sweet potato tissue to *Ceratocystis*. According to Stahmann *et al.* (71), ethylene may serve as a part of the defense mechanism of plants. They suggested that it served as a stimulus for localized metabolism leading to both necrotic and hypersensitive reactions in plants following infection. Substantiating evidence was obtained when they found that ethylene increased the levels of a number of enzymes, including peroxidase, in sweet potato tubers. The exact role of peroxidase in disease resistance was not known, although there was a relationship between varietal susceptibility of sweet potato cultivars to *Ceratocystis* and levels of peroxidase. Generally speaking, resistant cultivars have greater levels of peroxidase than susceptible ones (25, 71). The role of peroxidase was thought to be involved in the production of aromatics that had antifungal (phytoalexin) activity.

However, Chalutz and DeVay (13) obtained different results. They found that all the strains of *Ceratocystis* they studied produced ethylene, although 100-fold differences in the rate of production were noted. Ethylene production was associated with the active growth of the fungus although the media also had some effect on the rate of production. Rates of ethylene production from fungi growing on sweet potatoes were different from the rates obtained with artificial media. Strains that produced low rates of ethylene on agar produced high rates on tissue slices and vice versa. Unlike the results obtained by Stahmann *et al.* (71), ethylene had no effect on disease development on sweet potato roots challenged with spores of *Ceratocystis*. Another example of an ethylene-producing pathogen is *Fusarium oxysporum*, the cause of *Fusarium* wilt of tomatoes. According to Dimond and Waggoner (23), infected tomatoes had epinastic leaves and produced a significant number of root initials. They stated that these symptoms are probably due to the ethylene produced by the fungus.

The soft rot bacterium, *Erwinia carotovora*, did not produce ethylene, but caused infected cauliflower tissue to do so (41). The effect was apparently due to the production of pectic enzymes by the microorganism since ethylene-inducing activity was heat labile and as high in the cell-free culture liquid as in the bacterial culture itself. The specific enzymes involved were shown to be pectate lyase and polygalacturonase. These enzymes were responsible for the solubilization and subsequent activation of glucose oxidase from the cauliflower cell wall. Glucose oxidase generated hydrogen peroxide during the oxidation of glucose, and according to Lund and Mapson (41), hydrogen peroxide was the limiting factor in the production of ethylene from methionine in cauliflower tissue. The reaction sequence entailed the formation of the oxo acid from methionine by a transaminase. Carbons 3 and 4 of the deaminated methionine were then converted into ethylene by a reaction utilizing peroxidase and other cofactors (*p*-hydroxybenzoate and methyl sulfinic acid).

The biochemistry of stress ethylene production during fungal invasion is not known. However, Sakai *et al.* (68) compared the formation of labeled ethylene from various substances such as [^{14}C]glucose, [^{14}C]acetate, and [^{14}C]pyruvate in freshly cut and black rot-infected sweet potato tissue. The radioactive precursors were converted primarily into CO_2 and only 0.001% was converted into ethylene. However, the data showed that there were clear differences between normal and diseased tissue in the conversion of these substrates into ethylene.

VI. Mechanical Effects

Ethylene production by plants can be controlled by a variety of mechanical stimuli such as separation of organs, incision, bruising, and pressure. Although the biochemistry behind the operation of mechanical stimuli is unknown, man has been taking practical advantage of this phenomenon for a long time.

According to Galil (26), ripening of the sycomore fig (*Ficus sycomorus*) of the Middle East can be induced by gashing or piercing immature fruit (about 16 days old) or by treating them with ethylene. Figure 5-3 shows the growth and ethylene production of normal and gashed figs. Figure 5-4 shows the appearance of immature figs before and immediately after gashing and the size of parthenocarpic fruit which ultimately result from gashing. The development of fruit of the common fig (*Ficus carica*) can also be accelerated by ethylene. The practice of gashing figs is ancient and has been traced back to early Egyptian civilizations. Examples of dried gashed figs have been found in tombs or depicted in bas-reliefs dating back to 1100 BC. According to Theophrastus (372–287 BC), "it [the sycomore fig] cannot ripen unless it is scraped, but they scrape it with iron claws, the fruit thus scraped ripens in four days." An interesting sidelight on the topic of fig-gashing is the riddle of the prophet Amos's (eighth century BC) occupation. The Bible (*Amos* 7:14) quotes him as saying, "I was no prophet, neither was I a prophet's son; but I was a herdsman and a gatherer of sycomore fruit." In the Hebrew version of the Bible, Amos's occupation is given as "Boless Shikmin." "Shikma" is sycomore, but "Boless" appears only once and its meaning is obscure. In the Septuaginta, the early Greek translation of the Bible (made in Alexandria about 200 BC), "Boless Shikmin" was translated as "Knizon Sycamina," namely a piercer, not a gatherer, of sycomore fruits. The difficulty in accepting this translation

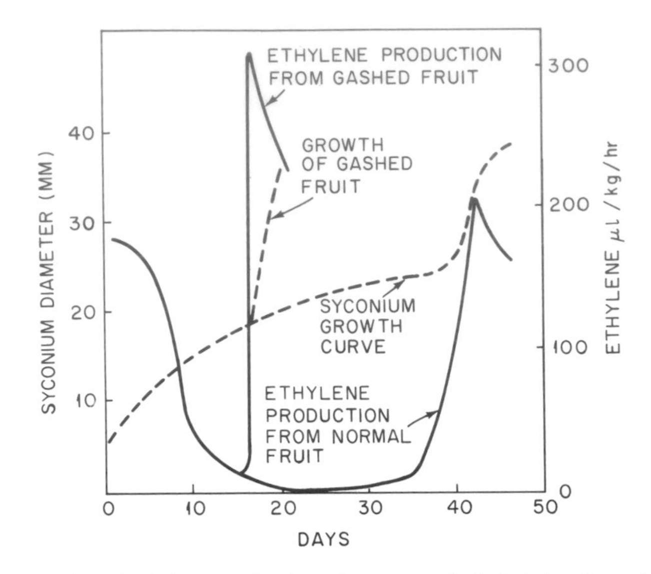

Fig. 5-3. Growth and ethylene production of sycomore fig fruit. The effect of gashing fruit on day 16 on ethylene production and fruit development is shown. [Courtesy of M. Zeroni and J. Galil, unpublished data, 1970.]

Fig. 5-4. Effect of gashing on ripening of sycomore figs. Left to right: 16-day-old fig before gashing, immediately after gashing, and ripe parthenocarpic fruit after 4 days. [Courtesy of Galil (26).]

stemmed from the fact that present-day sycomore fig fruits in Israel produced edible fruit without gashing. According to Galil the explanation for the change in horticultural practices arose from the selection of varieties that could form parthenocarpic fruit "vegetatively."

The sycomore fig tree is native to central Africa where the flowers are pollinated by a specific sycophilous wasp. Because the sycomore fig grew well in the arid Near East and produced valuable wood, shade, and fruit, it was widely cultivated in Egypt, Israel, and surrounding countries. However, the appropriate wasps for pollination were not present in these countries. Some wasps that live in these countries are able to complete their life cycle in these fruits but result in the formation of seedless fruits. Because of this, the sycomore trees are propagated solely by cuttings. In Cyprus, appropriate wasps are totally absent and the only way to obtain edible fruit is by the gashing technique which is still practiced. In Egypt, wasps enter the figs, complete their life cycle, and result in the formation of fruit filled with insects. Apparently the gashing technique was developed to produce edible fruit free of insects. In present-day Israel, the wasps are also present but only a small percentage complete an entire developmental cycle. Most figs swell quickly after the initial penetration of the wasp and a sweet rose-colored fruit develops "vegetatively" after a few days. According to Galil, gashing was important in Amos's time because either no fig wasps were present or because those that were resulted in the formation of insect-filled fruit. In the intervening centuries, varieties of figs were selected that respond to wasps by developing fruit that no longer required gashing.

A number of workers have reported that slicing fruits and vegetables promotes ethylene production. This effect has been observed in citrus, apple, tomato, banana, cucumber, sweet potato, carrot, potato, and cantaloupe. Hall (29) and others (8, 65) reported that ethylene production by oranges was increased when they were cut into segments. According to Riov *et al.* (65), stress ethylene was thought to control the synthesis of the enzyme phenylalanine ammonium lyase in citrus. Induction of stress ethylene was prevented by the addition of cycloheximide. Rhodes *et al.* (63, 64) found that excision of preclimacteric apple peel disks initiated the development of an enzyme system called the malate effect after a 6-hour lag. The lag was absent if ethylene was added to the system. Excision of the disk also set up a burst of ethylene production; the authors postulated that this was the triggering mechanism. Addition of cycloheximide to the disks blocked the increase in ethylene production, suggesting that protein synthesis was required. Other investigators have reported that ethylene production by ripe apple decreased as the tissue was subdivided (12, 39). The physiological maturity of the fruit apparently played a role in the kind of response observed.

In contrast with mature apples, mature tomato fruit showed an increase in ethylene production as the tissue was subdivided (38, 39). The increase in tomatoes was relatively large compared with other systems. Lee *et al.* (38) reported a 100-fold increase in ethylene production 5 hours after disks were cut from the fruit.

Other examples of stress ethylene production in fleshy tissue induced by incision include potato and green banana (50), cantaloupe (51), cucumber (24), and sweet potato and carrot (13, 31, 34). Both ethylene and incision were found to increase the activity of phenylalanine ammonium lyase in some of these tissues (cucumber and sweet potato) and various workers postulated that ethylene plays an intermediary role in enzyme induction following excision. Although smaller amounts were produced, leaf and stem tissue also produced stress ethylene when excised or shredded. Stress ethylene from bean (3, 5, 20, 34), pea (10), *Physalis floridiana* (66), and rose tissue (75) has been reported. In addition to ethylene production, Curtis (20) found that large quantities of ethane were produced by homogenized bean tissue.

Stress ethylene may also play a role in senescence. Kaltaler and Boodley (36) found that ethylene production by rose flowers removed from the plant was greater than those still attached. Burg and Dijkman (11) reported that removal of the pollinia from orchids set off a sharp rise in ethylene production followed by fading of the blossom. While the rise in ethylene production may be due to the mechanical separation of tissues, removal of plant parts from a supply of juvenility substances may also be important. Rough treatment is known to decrease the storage life of fruit. A number of workers have found that bruising citrus (17), tomatoes (42), bananas (44), apples (57), and grapefruit (73) increases the rate of ethylene production.

For the most part it is possible to infiltrate leaf tissue with water without causing damage or increasing ethylene production (20). However, Kawase (37) found that woody tissue produces more ethylene when infiltrated with water and that the increase was capable of inducing adventitious roots. These observations were made during a study designed to determine the cause of root initiation induced by placing cuttings of willow (*Salix fragilis*) in a centrifuge. Originally it had been thought that centrifugation either drove out inhibitors or concentrated root promoters in the base of the tissue. However, Kawase demonstrated that centrifuging stem sections forced water into the tissue which in turn increased ethylene production and root initiation.

Goeschl *et al.* (27) have suggested that the mechanical loading of soil may cause the epicotyls of germinating pea seeds to produce additional quantities of ethylene and in turn regulate the force the emerging shoot can exert on the soil. To test this idea pea epicotyls were enclosed in chambers in which their elongation was restricted by means of a foam neoprene stopper or by a medium of glass beads. These treatments increased ethylene evolution and resulted in reduced length and increased diameter of the internodes (Fig. 5-5). These responses increased with increasing degree of restriction and a time course study showed that the increase in ethylene production preceded the reduction in elongation. As the epicotyls elongated through the glass bead medium and less resistance was encountered, evolution of ethylene declined and rapid elongation was resumed. Figure 5-5

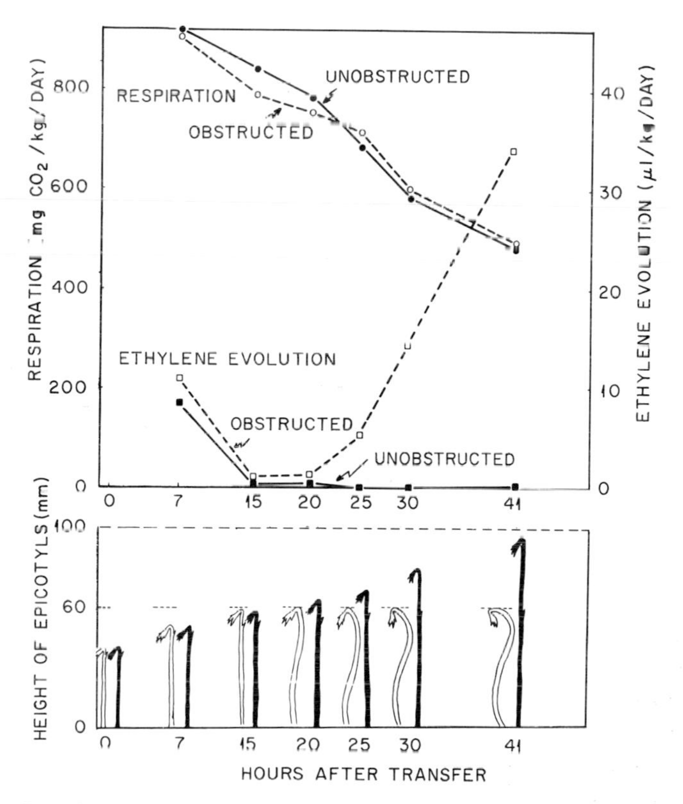

Fig. 5-5. Growth pattern and gas exchange of pea epicotyls with and without obstructions. Unshaded figures (below) represent growth habit of epicotyls that encountered a foam neoprene stopper (position indicated by short dashed lines) about 18 hours after transfer. Black figures represent the unobstructed or control epicotyls. Upper figure shows ethylene evolution and CO_2 evolution by the same epicotyls. [Courtesy of Goeschl *et al.* (27).]

shows the relationship between obstruction and the rate of ethylene production. The morphological and anatomic effects of a 120-mm column of glass beads were duplicated by an applied ethylene concentration of 0.2 ppm or less. According to Goeschl *et al.* (27), the ability of pea stem tissue to bear a load was proportional to the fourth power of the radius, while the ability to exert pressure against the medium was equal to the square of the radius. In the case of peas, the data indicate that ethylene production was increased by nonwounding physical stress and that ethylene acts as an endogenous growth regulator, decreasing elongation and increasing diameter in response to increasing increments of stress.

Neel and Harris (54) have reported that manually manipulating or shaking sweet gum or corn reduced their growth. They also cite literature in which the reduction of growth of *Bryonia dioica* and cucumber due to handling or manipulation has been reported. They noted that ethylene is known to reduce growth, and while no direct evidence was available, suggested that mechanically disturbed plants produced greater quantities of ethylene and that this hormone was responsible for the effects observed.

The coiling of pea tendrils is also thought to involve production of stress ethylene. Jaffe (35) reported that mechanical stimulation of pea tendrils resulted in an increase in ethylene production within 60 minutes. Asymmetrical applications of Ethrel to pea tendrils resulted in coiling within 30 minutes. The Ethrel was effective only when combined with dimethyl sulfoxide (DMSO) and applied to the concave or ventral side. Absence of DMSO or a symmetrical application of Ethrel resulted in little coiling. The results suggest that coiling involves an increase in ethylene production due to the mechanical stimulus followed by the gas regulating the size of the cells on the ventral side of the tendril.

VII. Chemical Effects

The kinds of chemicals capable of increasing ethylene production fall into two classes: hormonal and phytotoxic. The phytotoxic compounds include inorganic salts, organic acids, herbicides, fungal exudates, and other compounds such as cycloheximide. The hormonal chemicals, such as IAA and 2,4-D, are discussed elsewhere.

The substances Cu^{2+} (7, 14, 17, 53, 66), Fe^{3+} (7, 17, 59), Hg^{2+} (53), and KI (3, 67) are effective promoters of ethylene production. Even NaCl when supplied in excess in irrigation water will promote ethylene production (61). The mechanism of action probably involves damage and disorganization of the cell. Palmer *et al.* (56) suggested that hydrogen peroxide might play an intermediary role in the action of these metals. They indicated that ethylene was produced from ethyl acetate and ethyl butyrate when mixed with $CuSO_4$ and ascorbic acid. The reaction was thought to involve generation of H_2O_2 from Cu^{2+} and ascorbate. As described elsewhere (41), *in vitro* formation of ethylene from methionine required H_2O_2 and other cofactors.

Monoiodoacetamide, trichloroacetic acid, and sodium ethyl mercurithiosalicylate in low concentrations promoted ethylene production (33). Increasing concentrations of these and other chemicals (53) on the other hand stopped ethylene production and killed the tissue. Other chemicals examined for their ability to increase ethylene production included ascorbic acid and derivatives (15, 17, 59, 62), iodoacetic acid (17), and abscisic acid (3, 17).

The herbicide Endothal is an effective promoter of ethylene production (3, 67) and has been widely used as a defoliant in cotton. Malformins, a metabolic product of *Aspergillus niger*, are cyclopeptides consisting of cysteine, valine, leucine, and *allo*-isoleucine (19, 21, 22). According to Curtis, malformins are capable of inducing abscission, epinasty, and root coiling, and it appears that the increase in ethylene production may play a role in the first two processes. However, the coiling of roots, while mimicked by ethylene, probably does not involve ethylene as an intermediate. Since a number of fungal diseases involve increased ethylene production, it is possible that the pathogen exudates equivalent to malformin may be involved. Other compounds shown to stimulate ethylene production include rape seed oil (30), an inducer of fig ripening; 8-hydroxyquinoline sulfate (36), a cut flower preservative; and ethyl hydrogen 1-propyl phosphate (9).

Craker has shown that some of the effects of ozone are similar to those reported for ethylene, namely epinasty, abscission, premature senescence, and increased respiration in fruits. Craker (18) demonstrated that tobacco and other plants treated with ozone produced more ethylene than controls. There were varietal differences in ethylene production from tomatoes treated with ozone. A variety known to be more susceptible to ozone damage produced more ethylene than an ozone-resistant variety. Some unpublished experiments by Abeles and Heggestad have shown that ozone-induced abscission of bluebell leaves of varying maturity was closely correlated with rates of ethylene production. Leaves that abscised most readily also produced the most ethylene.

Of the chemicals studied, cycloheximide perhaps is the most interesting. We (4) observed that cycloheximide could either promote or inhibit abscission depending on the manner the antibiotic was applied. The increase in abscission was associated with an increase in the rate of ethylene production and this fact has been subsequently explored as a practical means of causing fruit drop or loosening in citrus (15, 16, 60, 79). Cycloheximide not only promotes ethylene production, but can also block ethylene production induced by auxin (15), slicing (65), $CuSO_4$, Endothal, and itself (2). This effect is thought to be due to its ability to block protein synthesis in tissues and suggests that stress ethylene production in plants is due to enzyme synthesis. This idea is supported by the observation that stress ethylene induced by $CuSO_4$, Endothal, and ozone was derived from carbons 3 and 4 of methionine in a manner similar to that observed for normal or auxin-induced ethylene production.

VIII. Conclusions

In conclusion, stress ethylene is a product of tissue subjected to trauma or damage induced by a variety of sources. However, stress ethylene is a product of living tissue since it ceases when the damage is extreme enough to kill the tissue. Data presently available suggest that stress ethylene is produced by induction of the enzymatic system which normally controls ethylene production (2). This interpretation is based on experiments utilizing metabolic inhibitors and ^{14}C-labeled intermediates, and further verification depends upon the isolation and identification of the enzymes that convert methionine into ethylene. Assuming that the same system is used, it seems remarkable that one system is used to synthesize ethylene under stress, hormonal, and normal conditions. As far as we know, the primary function of stress ethylene is to accelerate abscission of organs damaged by disease, insect, drought, and temperature extremes. It may also play a role in regulating the growth of seedling tissue through soil and may also be a part of the disease-resistance mechanism.

References

1. Abdel-Kader, A. S., Morris, L. L., and Maxie, E. C. (1968). *Proc. Amer. Soc. Hort. Sci.* **92,** 553.
2. Abeles, A. L., and Abeles, F. B. (1972). *Plant Physiol.* **50,** 496.
3. Abeles, F. B. (1967). *Physiol. Plant.* **20,** 442.
4. Abeles, F. B., and Holm, R. E. (1967). *Ann. N. Y. Acad. Sci.* **144,** 367.
5. Abeles, F. B., and Rubinstein, B. (1964). *Plant Physiol.* **39,** 963.
6. Balazs, E., Gaborjanyi, R., Toth, A., and Kiraly, Z. (1969). *Acta Phytopathol.* **4,** 355.
7. Ben-Yehoshua, S., and Biggs, R. H. (1970). *Plant Physiol.* **45,** 604.
8. Ben-Yehoshua, S., and Eaks, I. L. (1969). *J. Amer. Soc. Hort. Sci.* **94,** 292.
9. Boe, A. A. (1971). *HortScience* **6,** 399.
10. Burg, S. P., and Burg, E. A. (1966). *Proc. Nat. Acad. Sci. U.S.* **55,** 262.
11. Burg, S. P., and Dijkman, M. J. (1967). *Plant Physiol.* **42,** 1648.
12. Burg, S. P., and Thimann, K. V. (1960). *Plant Physiol.* **35,** 24.

13. Chalutz, E., and DeVay, J. E. (1969). *Phytopathology* **59,** 750.
14. Chalutz, E., and Stahmann, M. A. (1969). *Phytopathology* **59,** 1972.
15. Cooper, W. C., Henry, W. H., Rasmussen, G. K., and Hearn, C. J. (1969). *Proc. Fla. State Hort. Soc.* **82,** 99.
16. Cooper, W. C., Rasmussen, G. K., and Hutchison, D. J. (1969). *BioScience* **19,** 443.
17. Cooper, W. C., Rasmussen, G. K., Rogers, B. J., Reece, P. C., and Henry, W. H. (1968). *Plant Physiol.* **43,** 1560.
18. Craker, L. (1971). *Environ. Pollut.* **1,** 299.
19. Curtis, R. W. (1968). *Plant Physiol.* **43,** 76.
20. Curtis, R. W. (1969). *Plant Physiol.* **44,** 1368.
21. Curtis, R. W. (1969). *Plant Cell Physiol.* **10,** 909.
22. Curtis, R. W. (1971). *Plant Physiol.* **47,** 478.
23. Dimond, A. E., and Waggoner, P. E. (1953). *Phytopathology* **43,** 663.
24. Engelsma, G., and Van Bruggen, J. M. H. (1971). *Plant Physiol.* **48,** 94.
25. Gahagan, H. E., Holm, R. E., and Abeles, F. B. (1968). *Physiol. Plant.* **21,** 1270.
26. Galil, J. (1968). *Econ. Bot.* **22,** 178.
27. Goeschl, J. D., Rappaport, L., and Pratt, H. K. (1966). *Plant Physiol.* **41,** 877.
28. Guerrero, F. P., Maxie, E. C., John, C. F., Eaks, I. L., and Sommer, N. F. (1967). *Proc. Amer. Soc. Hort. Sci.* **90,** 515.
29. Hall, W. C. (1951). *Bot. Gaz.* **113,** 55.
30. Hirai, J., Hirata, N., and Horiuchi, S. (1967). *J. Jap. Soc. Hort. Sci.* **36,** 36.
31. Imaseki, H., Asahi, T., and Uritani, I. (1968). *Phytopathol. Soc. Jap.* p. 189.
32. Imaseki, H., Teranishi, T., and Uritani, I. (1968). *Plant Cell Physiol.* **9,** 769.
33. Imaseki, H., Uritani, I., and Stahmann, M. (1968). *Plant Cell Physiol.* **9,** 757.
34. Jackson, M. B., and Osborne, D. J. (1970). *Nature* (*London*) **225,** 1019.
35. Jaffe, M. J. (1970). *Plant Physiol.* **46,** 631.
36. Kaltaler, R. E. L., and Boodley, J. W. (1970). *HortScience* **5,** Sect. 2, 355.
37. Kawase, M. (1971). *Physiol. Plant.* **25,** 64.
38. Lee, T. H., McGlasson, W. B., and Edwards, R. A. (1970). *Radiat. Bot.* **10,** 521.
39. Lieberman, M., and Kunishi, A. (1971). *HortScience* **6,** 355.
40. Luchko, A. S., and Porutskii, G. V. (1964). *Sov. Plant Physiol.* **11,** 46.
41. Lund, B. M., and Mapson, L. W. (1970). *Biochem. J.* **119,** 251.
42. Lyons, J. M., and Pratt, H. K. (1964). *Proc. Amer. Soc. Hort. Sci.* **84,** 491.
43. Maxie, E. C., and Abdel-Kader, A. S. (1966). *Advan. Food Res.* **15,** 105.
44. Maxie, E. C., Amezquita, R., Hassan, B. M., and Johnson, C. F. (1968). *Proc. Amer. Soc. Hort. Sci.* **92,** 235.
45. Maxie, E. C., Eaks, I. L., and Sommer, N. F. (1964). *Radiat. Bot.* **4,** 405.
46. Maxie, E. C., Eaks, I. L., Sommer, N. F., Rae, H. L., and El Batal, S. (1965). *Plant Physiol.* **40,** 407.
47. Maxie, E. C., Johnson, C. F., Boyd, C., Rae, H. L., and Sommer, N. F. (1966). *Proc. Amer. Soc. Hort. Sci.* **89,** 91.
48. Maxie, E. C., Rae, H. L., Eaks, I. L., and Sommer, N. F. (1966). *Radiat. Bot.* **6,** 445.
49. Maxie, E. C., Sommer, N. F., Muller, C. J., and Rae, H. L. (1966). *Plant Physiol.* **41,** 437.
50. McGlasson, W. B. (1969). *Aust. J. Biol. Sci.* **22,** 489.
51. McGlasson, W. B., and Pratt, H. K. (1964). *Plant Physiol.* **39,** 128.
52. McMichael, B. L., Jordan, W. R., and Powell, R. D. (1972). *Plant Physiol.* **49,** 658.
53. Nakagaki, Y., Hirai, T., and Stahmann, M. A. (1970). *Virology* **40,** 1.
54. Neel, P. L., and Harris, R. W. (1972). *Science* **175,** 918.
55. Olson, E. O., Rogers, B. J., and Rasmussen, G. K. (1970). *Phytopathology* **60,** 155.
56. Palmer, R. L., Hield, H. Z., and Lewis, L. N. (1969). *Proc. Int. Citrus Symp., 1st.* Vol. 3 p. 1135.

57. Phan, C. T. (1965). *C. R. Acad. Sci.* **260,** 5089.
58. Porutskii, G. V., Luchko, A. S., and Matkovskii, K. I. (1962). *Sov. Plant Physiol.* **9,** 382.
59. Rasmussen, G. K., and Cooper, W. C., (1968). *Proc. Amer. Soc. Hort. Sci.* **93,** 191.
60. Rasmussen, G. K., and Cooper, W. C. (1969). *Proc. Fla. State Hort. Soc.* **82,** 81.
61. Rasmussen, G. K., Furr, J. R., and Cooper, W. C. (1969). *J. Amer. Soc. Hort. Sci.* **94,** 640.
62. Rasmussen, G. K., and Jones, J. W. (1969). *HortScience* **4,** 60.
63. Rhodes, M. J. C., Galliard, T., Wooltorton, L. S. C., and Hulme, A. C. (1968). *Phytochemistry* **7,** 405.
64. Rhodes, M. J. C., Wooltorton, L. S. C., Galliard, T., and Hulme, A. C. (1968). *Phytochemistry* **7,** 1439.
65. Riov, J., Monselise, S. P., and Kahan, R. S. (1969). *Plant Physiol.* **44,** 631.
66. Ross, A. F., and Williamson, C. E. (1951). *Phytopathology* **41,** 431.
67. Rubinstein, B., and Abeles, F. B. (1965). *Bot. Gaz.* **126,** 255.
68. Sakai, S., Imaseki, H., and Uritani, I. (1970). *Plant Cell Physiol.* **11,** 737.
69. Shah, J., and Maxie, E. C. (1965). *Physiol. Plant.* **18,** 1115.
70. Smith, W. H., Meigh, D. F., and Parker, J. C. (1964). *Nature* (*London*) **204,** 92.
71. Stahmann, M. A., Clare, B. G., and Woodbury, W. (1966). *Plant Physiol.* **41,** 1505.
72. Subramania, S., and Sridhar, T. S. (1966). *Riv. Patol. Veg.* **2,** 127.
73. Vines, H. M., Grierson, W., and Edwards, G. J. (1968). *Proc. Amer. Soc. Hort. Sci.* **92,** 227.
74. Wiese, M. V., and DeVay, J. E. (1970). *Plant Physiol.* **45,** 304.
75. Williamson, C. E. (1949). *N. Y. State Flower Growers, Bull.* **49,** 3.
76. Williamson, C. E. (1950). *Phytopathology* **40,** 205.
77. Williamson, C. E., and Dimock, A. W. (1953). *Yearb. Agr. (U. S. Dep. Agr.)* p. 881.
78. Young, R. E. (1965). *Nature* (*London*) **205,** 1113.
79. Young, R. E., and Meredith, F. (1971). *Plant Physiol.* **48,** 724.

Chapter 6

Growth and Developmental Effects of Ethylene

I. Dormancy

A. Seeds

The first report of an effect of ethylene on seed germination was that of Nord and Weicherz in 1929 (155), when they demonstrated that ethylene and acetylene increased the rate of germination and growth of barley. Since then a number of other investigators have observed ethylene-promoted germination of a variety of cereals (12, 76, 77, 110, 174). In addition to a stimulation of seed germination, Hale *et al.* (77) observed that ethylene would prolong the longevity of high-moisture wheat in terms of subsequent germination. They also noted that the baking performance of ethylene-treated wheat was superior to that of untreated wheat.

Toole *et al.* (194) studied the effect of ethylene on peanut germination. They observed that promotion of germination was not specific for ethylene, since CO_2 had a similar effect, and that a combination of CO_2 with ethylene resulted in the greatest rate of germination. Carbon dioxide had an additional effect on pigmentation of the cotyledons. In the absence of CO_2 the cotyledons remained white, while in the presence of CO_2 they turned green. Other investigators have also observed stimulation of seed germination by CO_2 (4, 60). The relationship between ethylene and peanut seed germination was studied further by Ketring and Morgan (104–106). They found that natural dormancy of the seed was probably due to the rate of ethylene evolution. Spanish-type seed that were nondormant produced ethylene during germination, while Virginia-type seed of a dormant variety produced little ethylene following imbibition. The dormancy of peanut seeds depends on their position in the pod. The basal seed is normally slower to germinate than the apical one. Ketring and Morgan (105) have explained this difference as due to the rate of ethylene production, the more dormant basal seed producing less ethylene than the apical one (see Fig. 6-1). A 15-day heat

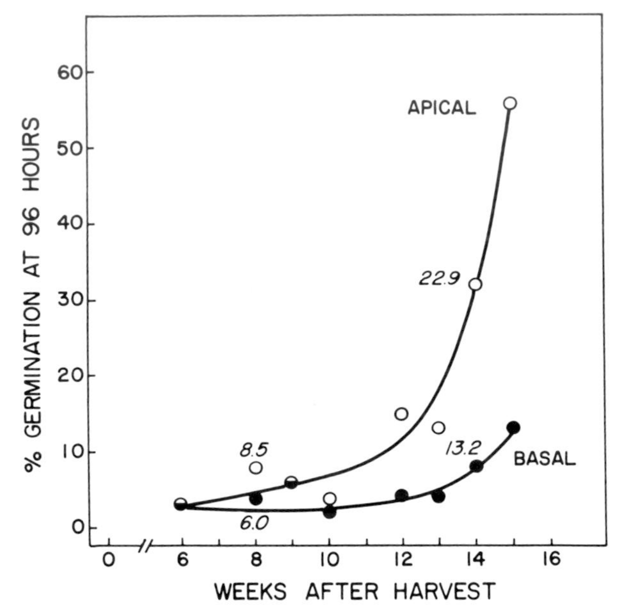

Fig. 6-1. The time course of germination of NC-13 peanut seeds as the inherent dormancy declines during storage in sealed containers at 3°C. The numbers adjacent to the points at 8 and 14 weeks after harvest indicate the ethylene produced (nl/50 seeds/hour) by the apical and basal seeds at the times indicated. [Courtesy Ketring and Morgan (105).]

shock (40°–45°C) treatment that broke dormancy of the dormant Virginia-type seed increased the rate of ethylene evolution (expressed as nanoliters per gram of dry weight per hour) from 1 to 5. Apparently most of the ethylene production was associated with the embryo. In dormant seeds, the rate of ethylene evolution of the cotyledon was 1, while the rate from the embryo was 50. Increased ethylene production probably accounted for the ability of a number of chemical treatments to promote germination. Ketring and Morgan observed that compounds such as gibberellic acid, cytokinins, and, to a lesser extent, coumarin increased both ethylene production and seed germination (105, 106).

The relationship between dormancy and rate of ethylene evolution was also studied by Esashi and Leopold (60). They found that dormant varieties of clover seeds produced less ethylene than nondormant ones, and that an increase in ethylene production preceded germination. The question of whether endogenous levels of ethylene production were capable of promoting germination was answered by placing seeds of the dormant variety in beakers of various sizes. The germination rate in 1000-ml beakers was 20%, while it increased to 80% in 50-ml beakers. Esashi and Leopold (60) proposed that ethylene may play an ecological role in seed germination. Sensitivity to ethylene may encourage germination only when seeds are suitably buried in the soil. A buildup of ethylene would occur when the seeds are sufficiently enclosed in soil but not when exposed on the soil surface, even though a supply of moisture may be temporarily available.

It is not clear whether ethylene promotes germination because it breaks dormancy or accelerates germination of those seeds no longer dormant but requiring additional factors for growth. Abeles and Lonski (4) found that ethylene increased germination of lettuce seeds. However, the gas appeared to increase the rate of germination and not break the dormancy of the seeds. When ethylene was added to freshly imbibed seeds it approximately doubled the endogenous rate of germination. When ethylene was added after a delay of 3 days, it had no effect on the remaining dormant seeds. However, the dormancy of these seeds could be readily overcome by a 15°C cold shock treatment. Similar observations were made by Takayanagi and Harrington (192). They found that ethylene accelerated germination of aged rape seeds but did not significantly improve the percentage of germination.

The ability of ethylene to promote seed germination may serve as a basis of chemical control for witchweed (*Striga lutea* Lour.), a parasitic weed of corn. Egley and Dale (59) have found that as little as 0.01 ppm ethylene stimulated germination of witchweed, and the effect saturated at 0.1 ppm. It is conceivable that the natural germination promoter may be ethylene evolved from roots.

B. Buds, Tubers, Corms, and Bulbs

The early literature on the regulation of dormancy is associated with reports on the effects of illuminating gas, smoke, and other gases on bud break of woody plants. Stone (188) reported that illuminating gas increased the development of shoots from willow cuttings, and acetylene was used by Weber (207a) in 1916 as a means of overcoming the dormancy of *Syringa vulgaris, Aesculus hippocastanum,* and *Tilia* sp. The ability of carbon monoxide to increase sprouting was tested on 108 species by Zimmerman *et al.* (214); they found that many plants gave a positive response. Galang and Agati (64) found that wood smoke promoted the flowering of mangoes.

The sprouting of potato tubers is a problem with practical implications that probably accounts for the great interest in this particular organ. Earlier workers were apparently confused between short treatments that were used to induce sprouting and prolonged effect of ethylene that inhibited subsequent growth. Rosa (169) found that a 4-week treatment after harvesting increased the rate of sprouting and the number of sprouts per seed piece. Vacha and Harvey (198) obtained variable results and found that different varieties gave different results. Burbank Russet and Bliss Triumph gave the largest response; Green Mountain and Irish Cobbler were intermediate; while Rural New Yorker and Early Ohio were almost insensitive. In general, ethylene treatment increased sprouting from 7 to 15 days and once out of the ground, the ethylene-treated seed pieces grew faster than controls. Denny (50) and Burton (32) reported that ethylene decreased sprouting of potato tubers. It is possible that they interpreted the arrested growth of the sprouts in the continual presence of ethylene as reduced sprouting. Barker (13) tried using ethylene as a means of preventing tuber sprouting in storage but found that this technique was not effective.

Promotion of tuber sprouting is not specific for ethylene. Thornton (193) and Burton (32) found that CO_2 also worked. Apparently, high concentrations of CO_2 were required, as promotive effects were noticed only at 40–60% and none at 13–30% (193).

Vacha and Harvey (198) found that ethylene promoted sprouting of gladiolus corms with a resulting advance in growth of 25–30 days. Denny (51) reported that ethylene was not effective immediately after harvest but would increase germination if the corms were aged for a month.

However, induction of corm sprouting was not specific for ethylene. Other gases such as ethyl ether and chloroform had a similar effect (198), and heat shock (3 weeks at 30°C) (51) also caused sprouting.

Haber (76) treated narcissus bulbs with ethylene for 48 hours and observed that treated bulbs bloomed 7–9 days earlier than controls. The accelerated blossoming of bulbs is not an effect on floral initiation but rather a promotion of bud break since the floral primordia are preformed in the bulbs.

A normal regulatory role for ethylene in breaking or controlling the dormancy of buds is not known. We do know that increased bud development following ethylene treatment has been observed in oaks (55), beech, birch (18), rhubarb (16), cotton (79, 86), honey mesquite, huisache (145), figs (44), and other plants (85). Table 6-1 shows that the ability of ethylene to break the dormancy of birch and beech buds increases as spring approaches and the requirement for rest is met. On the other hand, high levels of ethylene prevent subsequent growth by elongation of buds, and it is possible to consider dormancy as a case of growth repression due to high

Table 6-1

EFFECT OF ETHYLENE (APPLE GAS) ON DORMANCY OF BUDS FROM BEECH AND BIRCH[a]

Species	Date cut and treated with apple gas	Days after cutting that buds opened	
		Control	Treated
Birch	Dec. 13	—	32
	Jan. 2	—	25
	Feb.	37	19
	March 2	28	13
Beech	Dec. 6	—	—
	Dec. 19	—	—
	Jan. 2	—	—
	Feb. 1	—	28
	March	38	20

[a] Data modified from Borgström (18).

levels of endogenous ethylene. However, Burg and Burg (30) have presented evidence that ethylene does not regulate dormancy by growth supression.

C. SPORES AND POLLEN

There are only a few reports on the effects of ethylene on growth and germination of single-cell structures such as spores and pollen. Brooks (22) reported that 250 ppm ethylene stimulated the germination of *Diplodia natalensis* and *Phomopsis citri* spores. The ethylene treatment increased the poor germination of these fungal spores in water to half the rate normally observed in prune juice media controls. However, the ethylene effect was not specific, for other substances such as ethyl acetate, ethylene chlorohydrin, and methanol also promoted germination.

The release of sperm from antheridia of *Isoetes* was increased by a number of gases such as CO, acetylene, isoamylene, ethyl ether, and methyl ether (62). The effect, while not specific, was rapid, since a 5- to 10-second imbibition was sufficient to cause an effect.

Molisch (143) observed that apple gas promoted the germination of *Galanthus nivalis, Narcissus tazetta,* and *N. poeticus* pollen. In the presence of apple gas, germination started 1 hour after they had been placed in sucrose solution, while 3 hours were required for controls. However, it is not clear if ethylene was the active component of apple gas. When a container of NaOH was placed in the chamber containing the pollen, the effect of the apple gas was reduced or eliminated. Similar observations were made by Sfakiotakis *et al.* (180). The percentage germination and growth of pollen tubes were increased by adding 1–5% CO_2. No effect of ethylene on the germination of pollen was observed.

However, Buchanan and Biggs (23) reported that germination of peach pollen was increased by ethylene levels from 0.01 ppm to 1000 ppm. Pollen tube growth was also increased by ethylene, but here concentrations of 10 ppm and higher inhibited growth. In a later report, Sauls and Biggs (177) found that the promotive effect of ethylene was no longer observable when boron, another promoter of pollen germination, was added simultaneously.

II. Growth

A. Elongation

1. Stem Dicots

Inhibition of cell elongation by ethylene was first reported by Neljubov in 1901 (149). The same phenomenon was also observed by a number of other investigators studying plant growth in the presence of air contaminated by illuminating gas (141, 162–164, 184, 208). Figure 6-2 shows the effect of increasing concentrations of ethylene on the growth of etiolated peas. As Knight *et al.* (109) reported earlier, inhibition of growth elongation occurs first, followed by swelling and horizontal growth as the concentrations are increased. These three phenomena were called the triple response and led to use of the pea seedling as a diagnostic test for ethylene. However, additional phenomena also occur at the same time. These are a closure of the plumular hook and prevention of leaf expansion. Prevention of hook opening is not clearly represented in Fig. 6-2 because the controls were kept dark enough to prevent endogenous hook opening. However, inhibition of plumular leaf expansion is recognizable. Figure 6-2 also shows the characteristic dose response for ethylene: saturation at 10 ppm, half-maximal at 0.1 ppm, and the first observable effect at 0.01 ppm. Similar data have been published earlier by others (2, 24, 45). Other typical facets of ethylene action such as analog action (26, 29, 84, 108, 109, 161) and reversal by CO_2 (26, 29, 68) have also been presented. However, other compounds such as ethanol also retard elongation. Nevertheless, much larger quantities (about 10,000 ppm) are required for an effect (61). The ability of ethylene to retard elongation growth of many species has been observed and a number of workers (85, 112, 214) have surveyed the effect of ethylene on growth of a wide variety of plants.

The ability of ethylene to inhibit the rate of elongation is rapid. Van der Laan (114) measured changes in the length of various portions of the pea epicotyl with a photographic apparatus. Elongation was inhibited in the

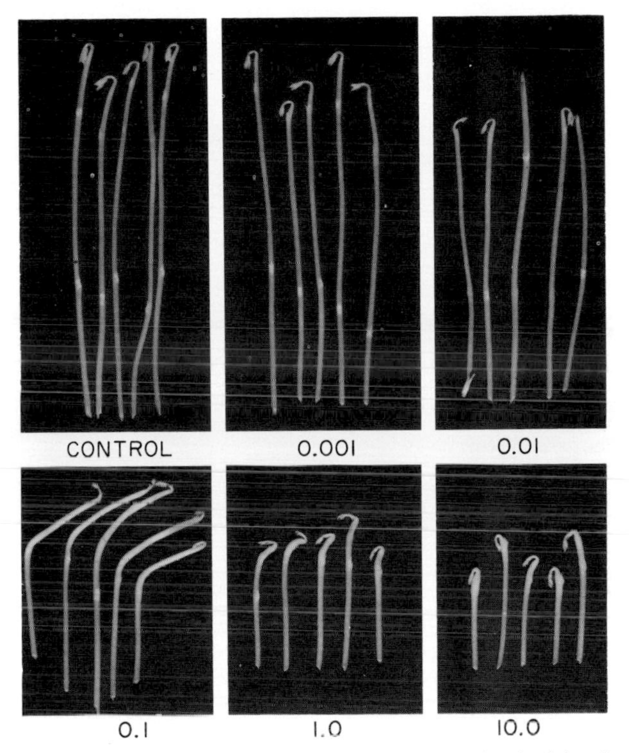

Fig. 6-2. Effect of increasing levels of ethylene on the growth of etiolated pea seedlings. Values under each group of seedlings represent the level of ethylene in parts per million applied to the tissue for 48 hours.

region 7–13 mm from the tip within 30 minutes after ethylene was introduced into the test chamber. Tissue further down the epicotyl (15 + mm) did not show decreased growth until 3 hours after the gas was added. Van der Laan (114) also pointed out that while ethylene caused a reduction in growth it did not completely inhibit elongation. The response time for ethylene action was reexamined more carefully by Warner (207) using a position-sensing transducer. Figure 6-3 shows that ethylene reduced elongation 6 minutes after it was added to the gas phase and that the growth rate returned to normal 20 minutes after ethylene was removed. Table 6-2 compares the latent time for action for a number of important growth regulators. The only compound that was faster acting under these conditions was abscisic acid, which had a latent time of 5 minutes.

Curvature of split pea stems has been used to measure the levels of auxin in test solutions. The growth of tissue in this bioassay is sensitive to ethylene, and a number of workers have shown that ethylene inhibits the curvature

Table 6-2

LATENT TIME OF VARIOUS GROWTH REGULATORS AND INHIBITORS WHEN APPLIED TO DECAPITATED ETIOLATED PEA SEEDLINGS[a]

Treatment	Concentration	Latent time (minutes)
Indoleacetic acid	10^{-5} *M*	9.3
Benzyladenine	10^{-5} *M*	11.7
Abscisic acid	10^{-4} *M*	5.1
Gibberellic acid	10^{-4} *M*	23.7
Ethylene	10 ppm	6.4
Ethylene removal	—	20.9
Actinomycin D	20 μg/ml	110.0
Dinitrophenol	10^{-4} *M*	11.2

[a] Data modified from Warner (207).

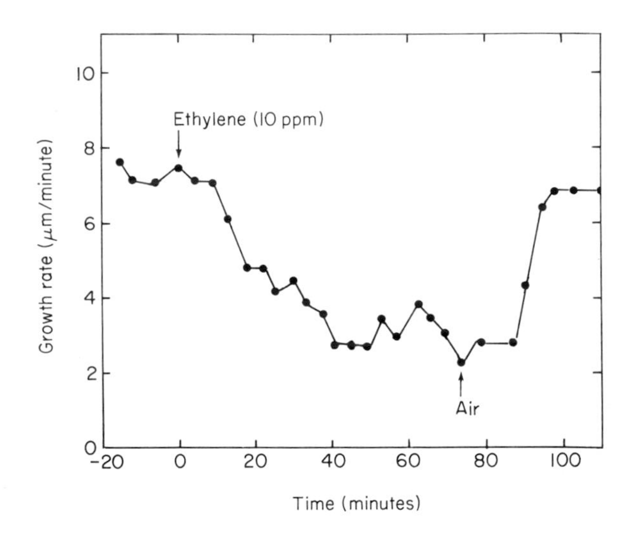

Fig. 6-3. Growth rate of intact etiolated pea seedling in μm/min versus time in minutes. The first arrow designates the application of 10 ppm ethylene and the second arrow designates the time of ethylene removal. [Courtesy Warner (207).]

of split pea stems (17, 133). Ethylene also blocks the curvature of pea epicotyl sections but this has been interpreted as being due to a loss of geotropic sensitivity (24).

The ability of auxin to increase growth of stem sections has been associated with an increase in the extensibility of the tissue. Morré and Eisinger

(147) have shown that ethylene had little effect on the extensibility of stem tissue, suggesting that growth inhibitions are not due to effects on the stretchability of the tissue. Burg (24) has shown that lack of elongation was not due to an inhibition of water uptake. In peas, the loss of elongation is associated with an increase in diameter so that what growth occurs takes place in a lateral as opposed to a longitudinal direction. Burg (24) also measured the effect of [^{14}C]glucose incorporation into the cell wall. He found that glucose incorporation, respiration, and the rate of exosmosis were unaffected by ethylene. Using polarized light microphotographs of cells from ethylene-treated plants, Burg showed a change in the orientation of cell wall microfibrils from longitudinal to radial. These observations suggest that the action of ethylene is associated in redirecting the orientation of new wall material and not a cessation of wall metabolism.

The ability of ethylene to control cell elongation is less effective in light-grown tissue than in dark-grown tissue (31, 69). Burg and Burg (31) have shown that light reduced the sensitivity of pea, sunflower, and oat seedling tissue to ethylene in terms of tissue elongation. In addition to a change in the effect of ethylene on elongation, light-grown pea tissue fails to demonstrate the horizontal growth characteristic of etiolated tissue. However, inhibition of plumular expansion is the same for both dark- and light-grown tissue (69).

2. Stem Monocots

Unlike dicotyledonous stems, ethylene can promote as well as inhibit growth of monocots. In most cases, ethylene was found to reduce elongation of monocotyledonous tissue (2, 31, 84, 85, 103, 124, 148, 185, 199). In general, higher concentrations of ethylene are required to show an effect on monocots and the degree of growth inhibition is less than that observed in dicots.

There are a number of reports showing that ethylene can act as a growth promoter. Ethylene was reported to increase the growth of barley (155) and oat (116) seedlings if it was applied to the plants during initial growth stages. Promotion of growth is best known in rice. This phenomenon was first described by Asmaev (11) and Kraynev (110) in 1937. Concentrations used were high (1000–40,000 ppm) and the effect was lost after 15 days of growth. However, some investigators have failed to observe a promotive effect of ethylene on rice (85). Dose response for the promotion of rice growth is shown in Fig. 6-4. Ku *et al.* (113) reported that in their studies the optimal concentration for growth promotion was between 100 and 200 ppm. However, Suge *et al.* (191) found that 0.1 ppm was a saturating concentration. One unusual aspect of ethylene action on the growth of rice seedlings is the observation that CO_2 will also promote growth and that the simultaneous

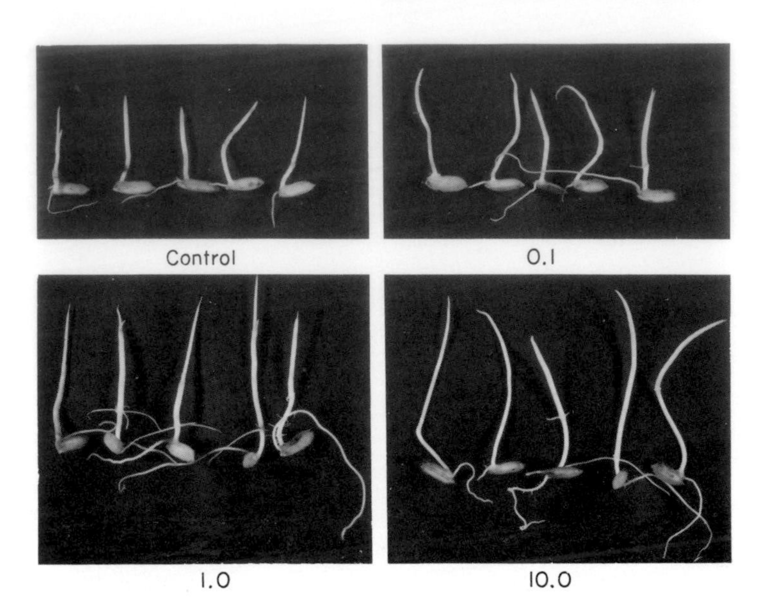

Fig. 6-4. Promotion of the growth of rice seedlings by ethylene. Seedlings were germinated and grown for 7 days in the concentration (ppm) indicated. Note that root growth was also promoted.

addition of ethylene and CO_2 results in a further increase in the rate of elongation. Ku *et al.* (113) also observed that the optimal concentration of oxygen for growth was near 4%. Addition of 10 ppm ethylene to low oxygen concentrations caused an even further increase in elongation. They also reported that ethylene increased both the rate of growth as well as the period of time the seedlings were actively elongating. Other hydrocarbons tested and found to be ineffective were methane, ethane, propane, propylene, and butane. Suge *et al.* (191) found that red light reduced the growth of rice shoots and that the same action spectra were obtained whether or not ethylene was present.

Imaseki *et al.* (93, 94) have found that ethylene had little or no effect on the growth of isolated coleoptile segments. However, ethylene enhanced the growth induced by auxin. An increase in the size of the coleoptile segments occurred as the concentration of auxin increased from 10^{-7} to 10^{-3} M, and in the presence of ethylene, there was a further enhancement. The greatest enhancement occurred at an auxin concentration of 10^{-5} M. A similar phenomenon had been observed earlier by van der Laan (114). He found that ethylene enhanced the response of oat coleoptiles to auxin when it was applied 2 hours after the start of the experiment. However, ethylene decreased the growth-promoting effect of auxin after a 6-hour incubation. Similar work has been done by Marinos (127). He found that a 24-hour pretreatment with ethylene enhanced the effectiveness of a subsequent treatment of IAA. The pretreatment with ethylene was effective with both intact oat seedling or coleoptile sections. Enhancement of auxin action by ethylene has also been reported in a number of other papers, notably the induction of adventitious roots (91, 111, 132).

3. Roots

Inhibition of root elongation has the same characteristics as other ethylene-mediated phenomena. The response is induced by a half-maximal concentration of 0.1 ppm ethylene (36, 37, 48, 160, 186) and can be reversed by CO_2 (36, 37, 160). The response to ethylene is rapid and apparently reversible. Chadwick and Burg (36, 37) have shown that the inhibition of elongation by ethylene extrapolates to zero time and removal within an 8-hour period resulted in a rapid return to the endogenous rate of elongation (see Fig. 6-5). However, Andreae *et al.* (9) studying the same phenomenon observed that inhibition of elongation occurred after a 3-hour lag and that inhibition of growth was irreversible.

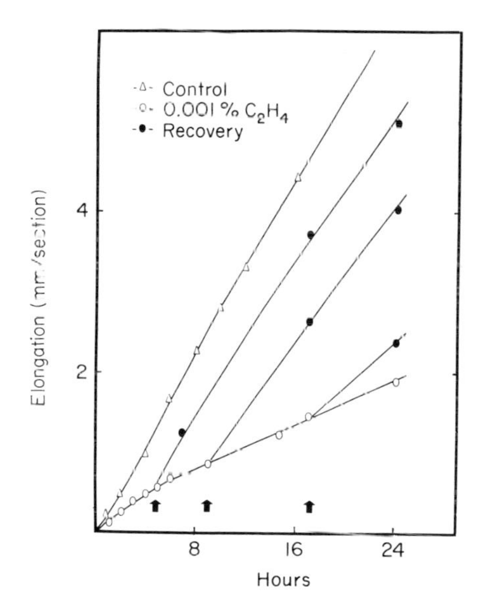

Fig. 6-5. Time course showing inhibition and recovery of elongation growth in isolated pea roots following treatment with maximally inhibitory concentrations of ethylene. Arrows denote time of transfer to control flasks. Subsequent growth of these roots is indicated by the filled circles. Fresh weight increases were similar. [Courtesy Chadwick and Burg (37).]

In addition to blocking elongation, ethylene causes a characteristic coiling of some roots (48, 82, 167). Similar coiling of roots occurs after treatment of seedlings with malformin (a metabolic product of *Aspergillus niger*) but it is not known whether ethylene mediates the malformin response. Coiling of leaf tissue of monocots (90) and stem tissue of dicots (176) has also been observed. Whether or not tissue coiling is an example of asymmetrical growth inhibition remains to be shown.

4. Other Growth Phenomena

Inhibition of growth is not limited to higher plants. Miller *et al.* (136) reported that ethylene retarded the elongation of rhizoids from fern (*Onoclea sensibilis*) gametophytes. Cell division was also inhibited but elongation of the filament was increased. Optimal concentrations were between 0.01 and 0.1 ppm. Sealing fern gametophyte cultures in chambers of limited volume produced all the effects of ethylene treatment. Gas chromatographic data later verified the fact that the gametophytes were responding to endogenously produced ethylene.

Ethylene also promotes growth of fig fruits. While ethylene inhibits growth of figs during the cell division stage (period I), it promotes growth during the growth and maturation stage (periods II and III) (130). However, increase in size and weight of fig fruits probably represents a specialized case of ethylene-induced fruit ripening since increase in growth is also associated with increased flavor and color development.

B. Swelling

Ethylene-induced swelling of stems results from an inhibition of elongation versus little or no change in water uptake (see Figs. 6-2 and 6-16). These swellings can occur in stem apices of seedlings, in roots, leaf bases in the case of bulbs, nodal tissue, and the ends of stolons in the form of tubers. The anatomic aspects of swelling involve an increase in size as opposed to increased cell division. Isaac (95) reported that swelling of the bean hypocotyl was due to an expansion of cortical cells and not a change in the cell number. He also observed that cambial activity and lignification of the pericycle and xylem were inhibited. The cell enlargement in the upper portion of leaf petioles during epinasty is another form of swelling (18). While swelling and hypertrophies both cause an increase in size, hypertrophies involve enhanced cambial activity. An increase in size is involved in both cases but the major distinctions are rapid cell enlargement in the case of swelling and a relatively longer period of cell division as well as enlargement in hypertrophies.

Marinos (127) noted that the inhibition of oat coleoptile elongation was concurrent with lateral expansion and suggested that under a given set of conditions, coleoptile cells were capable of attaining a finite volume and that the preferential lateral expansion induced by ethylene was accomplished at the expense of longitudinal extension. The swelling ratio has been used to describe this relative change in length versus volume. The swelling ratio is obtained by dividing the percentage increase in weight of a section of tissue by the percentage increase in length. For example, Chadwick and Burg (37) reported that the swelling ratio for pea roots treated with auxin was 1.54. At a saturating dose of ethylene, an essentially similar ratio of 1.58 was obtained. The auxin-induced swelling was shown to be due to increased ethylene production because CO_2 reduced the swelling ratio and no further swelling by auxin was observed when a saturating dose of ethylene was present. Holm and Abeles (92) presented similar evidence that the swelling induced by 2,4-D in soybean tissue was due to an increase in ethylene production. Figure 6-6 shows the effect of 2,4-D on ethylene production from excised soybean sections. The data in Figs. 6-7 and 6-8 show that the elongating section of soybean hypocotyls responded to 2,4-D (labeled control in the figures) in a typical fashion: growth promotion at low concentrations and growth inhibition at higher concentrations. The stimulation of ethylene production (Fig. 6-6) has the same kind of curve with the increase in ethylene production occurring at 10^{-6} M and higher.

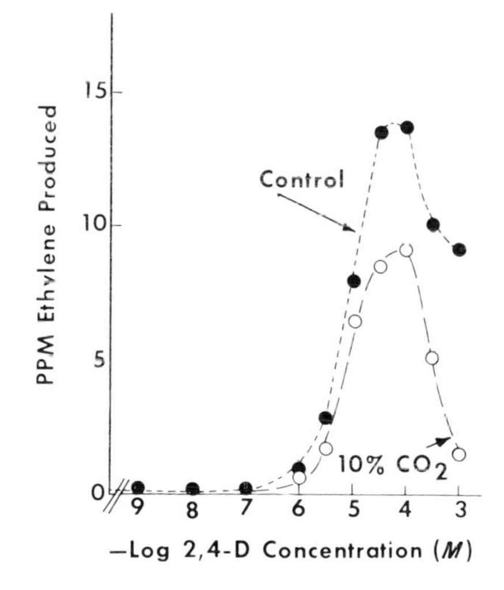

Fig. 6-6. Effect of 2,4-D on ethylene production by excised soybean sections. Elongating sections, 0.5–1.5 cm below the cotyledons from 3-day-old seedlings, were incubated with various 2,4-D concentrations for 18 hours. [Holm and Abeles (92).]

Ethylene (10 ppm) caused an inhibition of elongation and fresh weight at low 2,4-D concentrations (Figs. 6-7 and 6-8). Carbon dioxide had little effect on the weight increase induced by 2,4-D except at concentrations above 5×10^{-4} *M* where there was an inhibition. In contrast, CO_2 increased elongation at intermediate concentrations of 2,4-D. When the data in Figs. 6-7 and 6-8 were plotted as the swelling ratio (Fig. 6-9) they showed that at low 2,4-D concentrations the ratio was near unity, and 1.5 when a saturating dose of ethylene was applied. As the concentration of 2,4-D increased, the swelling ratio increased and became the same as that observed for ethylene alone. This increase in swelling ratio closely followed the data on an increased ethylene production. Carbon dioxide reduced the swelling ratio principally by overcoming 2,4-D-induced inhibition of elongation. The reduction of the swelling ratio at high 2,4-D levels was thought to be due to toxic, nonauxin effects of the herbicide.

Swelling can also be regulated by factors supplied by other parts of the plant. Oat coleoptiles failed to swell when the plants were deseeded (127). Similarly, the swelling effect of ethylene on peas was lost when the plants were derooted (133) or decapitated (166).

Pea stem tissue in the subapical region normally ceases to elongate and increase in size within 24–48 hours. In the presence of ethylene, the tissue continues to expand radially for at least 96 hours. In epidermal cells the outer wall bulges and forms hairlike structures that continue to grow for

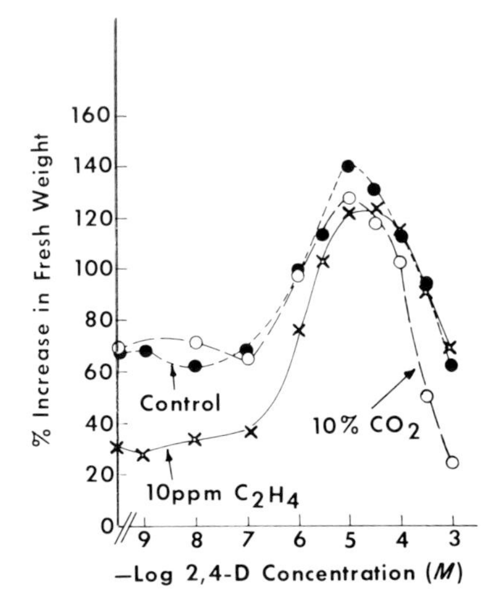

Fig. 6-7. Effect of 2,4-D, 10 ppm ethylene, and 10% CO_2 on the increase in fresh weight of excised soybean sections. [Holm and Abeles (92).]

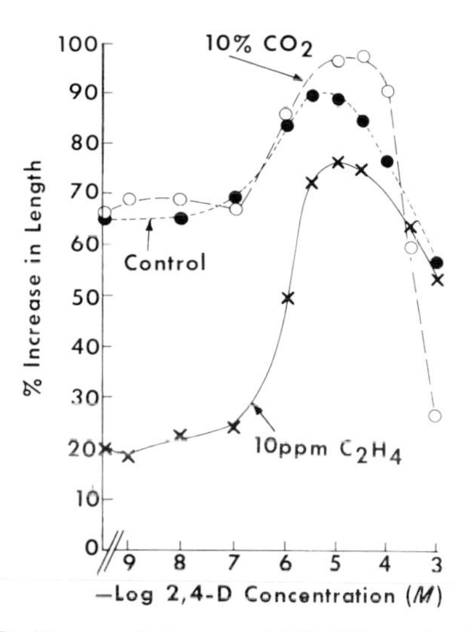

Fig. 6-8. Effect of 2,4-D, 10 ppm ethylene, and 10% CO_2 on the increase in elongation of excised soybean sections. [Holm and Abeles (92).]

48 hours. These hairs break through the waxy cuticle and increase the permeability of the tissue to water and water-soluble substances. The net effect on enhanced permeability of these epidermal hairs was equivalent to removing the cuticle with ether. This process is similar to the induction of root hairs (Fig. 6-16).

Electron micrographs of epidermal cells revealed a change in the microfibrillar pattern of epidermal cell walls. Apelbaum and Burg (10) found that swelling was due to a change from radial to longitudinal orientation of the microfibrils. The change in microfibrillar deposition probably accounts for altered appearance of cell walls as viewed under polarized light (27, 31).

Morré and Eisinger (147) reported that ethylene had no effect on the extensibility of etiolated pea stem sections. This indicates that ethylene had no effect on longitudinal stretchability of the cells, which is unexpected if ethylene causes a change in microfibrillar deposition. However, isolated stem sections were used in the extensibility study and while these tissues do not elongate in the presence of ethylene they failed to increase in diameter. If microfibrillar structure contributes to the physical characteristics of cell walls, then swollen cells should have altered extensibility characteristics.

Harvey (81) postulated that swelling might be due to increased osmotic pressure of ethylene-treated cells. He found that ethylene-treated pea seedlings had an osmotic pressure 2 atmospheres greater than that for controls and also had higher levels of soluble sugars. Using wheat as a test system, Roberts (167) reported that ethylene had no effect on water

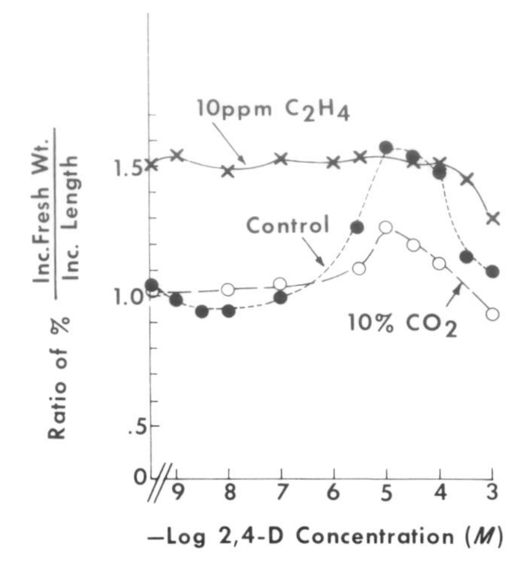

Fig. 6-9. The swelling ratio of excised soybean sections in varying concentrations of 2,4-D in the presence of 10 ppm ethylene and 10% CO_2. Data compiled from Figs. 6-7 and 6-8. [Holm and Abeles (92).]

uptake, respiration, or protein metabolism and ruled out an inhibition of cellular metabolism as the explanation for inhibition of cell elongation.

This was reexamined more carefully by Burg (24). He found that ethylene had no effect on respiration, exosmosis, or sugar content of pea tissue. In addition, he reported (10, 25) that ethylene had no effect on [^{14}C]glucose incorporation into cell wall material within 7 hours even though swelling had occurred by 3–4 hours. Also unchanged were total RNA content, incorporation of ATP into RNA, uptake of IAA, tryptophan, glucose, proline, leucine, ATP, arabinose, thymidine, and permeability of tissue to tritiated IAA. Under similar circumstances, IAA changes respiration, sugar pool sizes, exosmosis, uptake of sugars and amino acids, RNA synthesis, and incorporation of glucose, proline, and leucine into the cell walls.

However, ethylene reduced [^{14}C]proline incorporation into extensin. Apelbaum and Burg (10) postulated that this may prevent cross-linking and rigidification of the cell wall, which would explain why wall expansion and incorporation of [^{14}C]glucose into the wall continue. Control tissue stopped growing within 1–2 days and at the same time the walls became rich in hydroxyproline, still incorporated [^{14}C]proline, but only small amounts of [^{14}C]glucose. When cells were removed from ethylene, the process was reversed, [^{14}C]proline incorporation decreased, and growth quickly ceased.

Ridge and Osborne (166) compared auxin- and ethylene-induced swelling of pea stem tissue in terms of changes in cellulase content and activity.

They verified the observation that there was an increase in cellulase in auxin-treated tissue. However, no change in cellulase was observed when identical swellings were induced with ethylene. These observations suggest that cellulase does not play a role in cell wall modification during the swelling process.

Swelling may involve a change in the function or activity of microtubules. Nooden (154) found that ethylene and colchicine caused similar swellings in the elongation zone of roots. The activity of colchicine was not due to enhanced ethylene production since no change in ethylene production was noted nor did CO_2 prevent depolarization of corn root cell enlargement. Nooden pointed out that even though side effects of colchicine may limit its usefulness as a diagnostic tool, the data suggest that cell elongation involves properly functioning microtubules and that ethylene may inhibit or block their normal activity.

The function of ethylene-induced swelling varies according to its location or the kind of plant on which it occurs. In the case of seedling apices such as pea, it probably provides the support required for the penetration of the soil by seedling tissue. Goeschl *et al.* (71) pointed out that the ability of stem tissue to bear a load was proportional to the fourth power of the radius while the ability of the tissue to exert pressure was equal to the square of the radius. For example, a 20% increase in diameter increases the ability to bear a load by 100% and the ability to exert pressure by 44%. They also demonstrated that ethylene production increased as stem tissue was subjected to a physical stress.

Bulbing of onions is induced by long days under normal temperature conditions. Initiation of bulbing is indicated by the swelling of leaf bases and an increase in cell size, accompanied by translocation of assimilates to these tissues. Levy and Kedar (119) demonstrated that Ethrel promoted bulbing in onions and leeks under noninductive photoperiods and suggested that ethylene may play a role in bulb formation.

Ethylene may also play a role in tuber development at the ends of stolons. Ethylene-treated potato sprouts developed tubers at the ends of stolons as opposed to controls which developed tubers at only a few stolons (34). However, even though the swollen tissue resembled tubers, it was free of starch as opposed to control tubers.

C. Leaf Expansion

Inhibition of leaf expansion by ethylene is a specialized example of growth inhibition (31, 63, 69, 90, 134, 214). Figure 6-17 (p. 131) shows the inhibition by ethylene of *Forsythia* bud development. Analogous to other ethylene effects, a dose-response curve for bud growth inhibition shows a half-maximal effect at 0.1 ppm and saturation at 10 ppm (31). Middleton *et al.* (134) found that 0.05 ppm ethylene caused abnormal growth of marigold and 0.1 ppm caused leaf abnormalities in tomato. Figure 6-10 shows the effects of low concentrations of ethylene on the development of cucumber leaves. Hitchcock *et al.* (90) found that illuminating gas retarded or altered the development of leaves of lily, tulip, and hyacinth. In addition to preventing leaf expansion, ethylene caused curling, looping, double bending, irregular inward rolling, and inflation. With the exception of inflation, these distortions were permanent, though new leaves continued to grow normally after removal from the gas. Funke *et al.* (63) studied expansion of sunflower (*Helianthus annuus*) cotyledons. They reported that in addition to preventing normal expansion of leaf cells, ethylene changed the ratio of epidermal cells to stomates. On the upper surface, the ratio increased while on the bottom it decreased. According to Burg *et al.* (25), an inhibition of cell division accounts for the prevention of leaf expansion.

Fig. 6-10. Leaf expansion of cucumber leaves grown in the presence of various concentrations of ethylene. Leaves are from the third node of 4-week-old plants. Larger size of leaves grown in air filtered through $KMnO_4$ was due to the removal of trace amounts of ethylene normally present in urban air. Ethylene levels varied from 0 to 60 ppb in Beltsville, Md., and were due to auto emissions. (Unpublished photograph of Abeles and Heggested.)

III. Epinasty

Epinasty was originally observed as an effect of gas mixtures containing ethylene such as smoke (142, 202), illuminating gas (80, 90, 112, 178, 202), and apple emanations (18, 20, 63). However, it wasn't until Harvey's experiments in 1913 (80) that the active component was shown to be ethylene. Epinasty also resulted as an action of compounds that promote ethylene production by the plant. Included in this list of compounds are auxin (45, 126, 216), phenoxyacetic acids (126, 129), picloram (14, 144), and malformin (48). Analogs of ethylene such as CO (46, 178, 213, 214), acetylene (46, 178), and propylene and butylene (46) also induce epinasty. In addition, there have been some reports that other compounds such as rubidium (15), ethyl bromide, ethyl iodide, propyl chloride, acetonitrile (53), benzene, petroleum, ether, acetone, H_2S, ammonia (178), ethyl ether, and formaldehyde (202) cause epinasty. The significance of these observations is in doubt because Crocker *et al.* (46) failed to observe an epinastic effect of benzene, H_2S, aldehydes, acetone, and ammonia, and the purity of the material studied by earlier workers was not known. It is also difficult to eliminate the possibility that these materials induced stress or wound ethylene, which in turn would explain their action. Epinasty was also observed after rotation of plants on a clinostat and during the course of a number of diseases. It has been shown that in the case of horizontal plants (52) and infected plants (170) ethylene evolution accounts for the observed epinasty.

The minimum amount of ethylene required to cause epinasty varied between 0.01 ppm for African marigolds to 0.1 ppm for tomatoes. The fact that the response is specific for ethylene or its analogs, is sensitive to low concentrations, occurs rapidly, and requires little in the way of experimental setup to observe has made epinasty a useful bioassay for ethylene production by earlier workers (52, 54, 56, 87, 135).

Not all plants give an epinastic response after ethylene treatment (45, 46, 112, 214). Crocker *et al.* (46) tested 202 species and varieties of plants and found that 72 gave a marked response, 17 showed a slight effect, and 113 gave no response. The ability of ethylene to cause leaf curvature not only varied from species to species but also from one variety to another. For example, Crocker *et al.* (46) found that only 13 of 31 varieties of sweet pepper (*Capsicum frutescens*) demonstrated ethylene-induced epinasty.

The ability of leaf petioles to respond to ethylene has been shown to depend upon their physiological age. In general, only young leaves will bend, while older ones appear to be either less sensitive to the gas or, alternatively, less able to respond to the ethylene applied (46, 80, 212).

The response time to ethylene has been measured by a number of investigators and is in the order of 1–3 hours (46, 63, 144, 182, 212). Figure 6-11

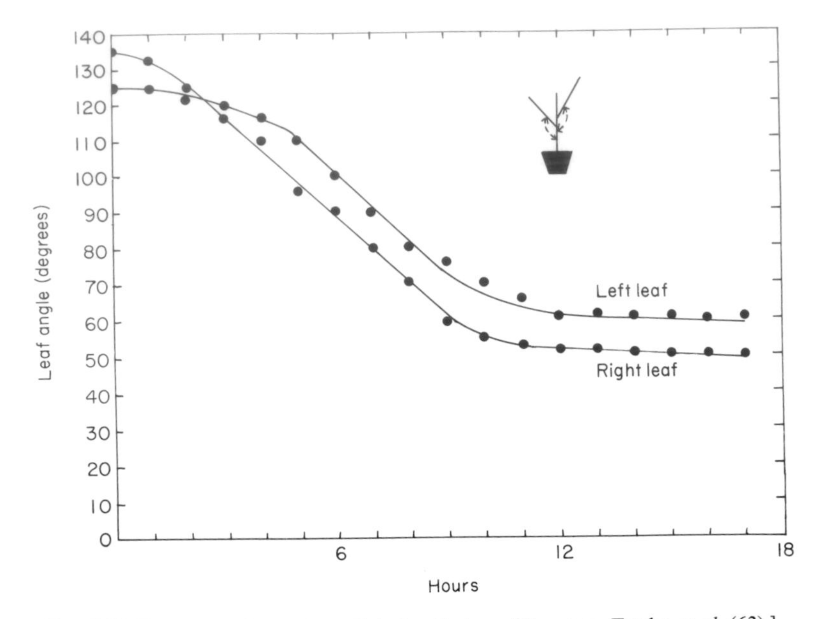

Fig. 6-11. Response of tomato petiole to ethylene. [Courtesy Funke *et al.* (63).]

shows that curvature of tomato petioles is noticeable 1 hour after the addition of ethylene and continues to bend for a 10-hour period. With younger material, curvature of the petiole is reversible. Figure 6-12 shows that tomato petioles will recover to some extent after they have been treated with ethylene.

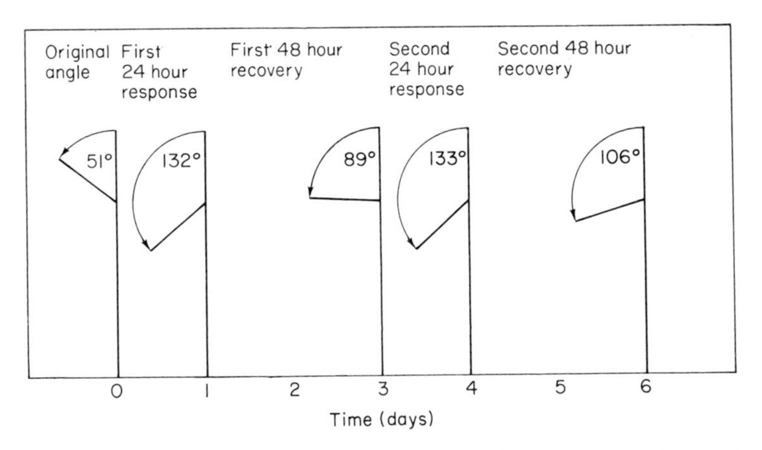

Fig. 6-12. Curvature and recovery of tomato petioles after ethylene treatment. [Courtesy Crocker *et al.* (46).]

Curvature of tomato petioles was shown to be due to swelling of the upper cells of the basal end of the petiole (17, 46, 196). This is easily demonstrated by comparing the spacing of lines made on the base of a petiole by a series of equally spaced parallel lines before and after ethylene fumigation (Fig. 6-13).

One observation, as yet unexplained, was made by Crocker *et al.* (46). They found that the ability of ethylene to cause petiole curvature is influenced by the relationship of the petiole to a gravitational field. They, and subsequently Borgström (18), found that tomato petioles failed to curve when the plants were held upside down and gave a half-maximal response when they were laid on their sides.

Fig. 6-13. Expansion of basal petiole cells demonstrated by increase in spacing of lines after ethylene fumigation. [Courtesy Crocker *et al.* (46).]

The ability of ethylene to cause curvature also depends upon a supply of auxin to the leaf. Leather *et al.* (117) have shown that petiole explants consisting of the basal portion of the tomato petiole with subtending stem tissue curved to a greater extent when auxin was applied to the cut surfaces of the petiole tissue. The requirement for auxin is anticipated when one considers the expansion of cells as a growth phenomenon. Lyon (123) has shown that ethylene alters auxin transport patterns in the petiole. He proposed that epinasty was the result of an inhibition of lateral auxin transport to the lower side of the petiole. He reported that the upper portion of petioles contained more auxin after plants had been treated with ethylene.

Little is known concerning the biochemistry involved in cell expansion during epinasty. Russian workers (126, 197) have reported an increase in ATPase activity in petioles following treatment with ethylene.

IV. Hook Opening

The terminal portion of the shoot axis of dark-grown seedlings is often hook-shaped. These hooks occur in both monocots and dicots and are presumed to facilitate the penetration of tender seedling tissue through the soil.

The bean hook has been intensively studied and it is now known that its shape is regulated by phytochrome, IAA, cytokinins, gibberellins, coumarin, and ethylene. The shape of the hook is derived from the fact that the growth on the outer convex side is faster than that on the inner concave side. Measurements of growth rates at various zones of the hypocotyl hook of intact bean plants (98, 172) indicate that the hook continuously tends to open by growth at the inner basal portion of the elbow. This is true even when the plants remain in the dark. The opening is counteracted, however, by a continuous reformation of the hook due to a greater growth at the outer apical portion of the elbow. The net result is that in dark-grown seedlings, the hook is always present but appears to move up the hypocotyl.

Light causes the hook to open. The effect is mediated by phytochrome; red light is most effective for inducing hook opening and its action can be reversed by far-red light. The hook opens because growth at the inside of the hook is greater than that at the outside (98, 171).

Auxin and ethylene cause the hook to close. In this case, growth of the inner half is inhibited while growth of the outer half is unaffected (98). Continued growth of the outer part of the hook can result in recurved tissue, a common observation in ethylene-treated seedlings. The significant processes in hook opening are summarized in Table 6-3.

Ethylene plays an intermediary role in both light- and auxin-controlled hook opening. Goeschl *et al.* (70) and Kang *et al.* (101) reported that red light decreased ethylene production by hook tissue while far-red light had the opposite effect. In addition, light increased CO_2 production (101), which suggests that CO_2 was also involved as a growth regulator through its antagonism of ethylene action. The effect of light was limited to the hook region since no effect of light on the straight portion of the tissue below the hook was observed. The effect of ethylene was rapid. Kang and Ray (99) reported that hook closure was initiated 2 hours after the addition of ethylene and that the effect was fully reversible within 2 hours after removal

Table 6-3

RELATIVE GROWTH PROCESS DURING THE OPENING AND CLOSURE OF BEAN HYPOCOTYL HOOKS

Treatment	Relative growth: Outside of hook	Relative growth: Inside of hook	Resultant hook shape
Dark	Fast	Slow	Closed
Light	Small promotion	Large promotion	Opening
Ethylene	No change	Inhibition	Recurved hooks
Light + 0.3 ppm ethylene	Small promotion	Inhibition	Enhanced closure, greater than ethylene alone

from ethylene-containing air. The effect of ethylene was reversed by CO_2 and also by Co^{2+} and Ni^{2+} ions. The mode of action of these ions was unknown but the authors pointed out that beneficial effects of Co^{2+} and Ni^{2+} on other growth systems have been described and may be due to an antiethylene effect. The reason for an inhibition of ethylene production by red light is not known. However, Kang and Ray (98) observed a small decrease in diffusible auxin in the hook region following red light treatment. Since endogenous auxin levels control ethylene production, the drop in auxin content may be responsible for the decrease in ethylene production.

Auxin increases ethylene production and simultaneously promotes hook closure. The effect of auxin has been shown to be mediated by ethylene. The increase in ethylene production is rapid following auxin application as is the inhibition of hook opening. The hook reopens about 12 hours after application of auxin and the rate of ethylene production also returns to normal at about the same time. Carbon dioxide was able to reverse the effect of auxin, as were Co^{2+} ions (99). Hook opening can also be inhibited by coumarin. Morgan and Powell (146) have shown that ethylene also mediates the action of coumarin in hook opening.

Figure 6-14 shows the dose-response curve of hook opening in the dark and in red light. The data show an anomalous effect of red light at 0.3 ppm ethylene. The hook angle was smaller, instead of greater, after red light treatment. Had the red light effect been limited solely to an inhibition of ethylene production, then light-treated tissue should be more open than ethylene-treated hooks. In other words, the ultimate shape of the hooks is regulated by the sum of external and internal ethylene levels. In the case of light-treated plants, there would be less internal ethylene so that the hooks should always be a little more open than dark ones. The accelerated closure of light-treated hooks in the presence of ethylene may be due to a light-induced promotion of growth on the outer side of the hook which is unresponsive to ethylene and an ethylene-induced growth inhibition on the inside of the hook. These data suggest that the effect of light on hook opening is not due solely to an effect on ethylene production.

Hook opening physiology is complex and in addition to control by ethylene, it is also regulated by factors from other parts of the seedling. Powell and Morgan (159) found that the presence of cotyledons was required for hook opening in cotton, while they inhibited opening in beans. With cotton, lower hypocotyl and root tissues stimulated hook opening, but with bean, the tissues below the hook section had little effect. Rubinstein (172) found that factors in the upper part of the hook were required to maintain closure in the dark while factors in the shank were required to open the hook in the light.

Protein synthesis may be required for hook closure. Kang and Ray (100) reported that cycloheximide prevented hook opening by light and hook closure by ethylene.

Hook closure and epinasty are examples of ethylene-induced asymmetrical growth. Diageotropic growth is also an asymmetrical growth phenomenon and always seems to occur on the side opposite to the hook in pea seedlings (see Fig. 6-2). However, the similarity is only superficial. Epinasty is due to a growth promotion by ethylene in terms of swelling on the upper portion of the petiole and requires auxin for its full expression. Hook opening, on the other hand, is due to an inhibition of cell elongation on the inside of the hook and auxin is not required for full expression of ethylene action.

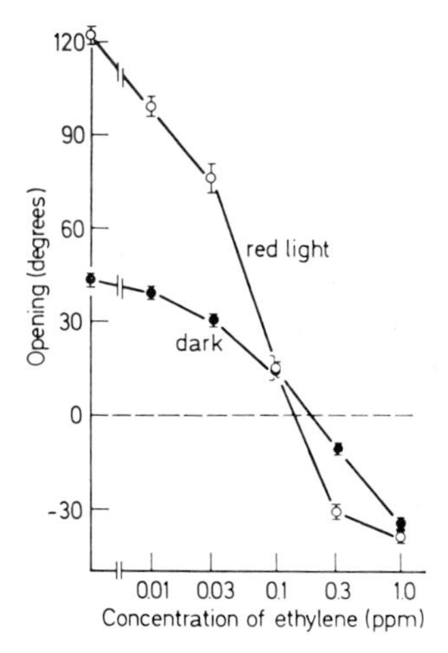

Fig. 6-14. Effect of ethylene concentrations on hook opening. [Courtesy Kang and Ray (99).]

V. Induction of Roots and Root Hairs

Ethylene can cause the initiation of roots from leaves, stems, flower stems, and preexisting roots of a variety of plants (215). The phenomenon has been observed by a number of investigators but the most thorough survey was completed by the Boyce Thompson investigators Zimmerman, Crocker, Hitchcock, and Wilcoxon (213–216). They found that roots were initiated by ethylene and its analogs CO (213, 214), acetylene, and propylene (215). As yet, reversal of root initiation by ethylene by CO_2 has not been reported. The list of plants in which root initiation has been observed includes monocots and dicots and herbaceous and woody plants. Ten parts per million of ethylene are usually required for root induction, relatively high compared with other ethylene-induced phenomena. Reports on the effect of lower concentrations are not available. Since high ethylene levels inhibit root growth and have a number of adverse side effects, such as defoliation and senescence, the most effective treatments to induce roots are 1- to 3-day applications followed by transferring the plants to air. In some cases additional treatments of ethylene are required.

Roots are initiated on various parts of the plant. In tobacco and hydrangea, roots are formed on the region of elongation. Roots are formed generally over the whole stem in coleus and tomato plants, while *Cosmos sulphureus* produced roots only around the nodes. When popcorn was treated with ethylene, prop roots were formed on the first five nodes above the ground while controls produced roots only from the basal nodes.

Rooting from the leaves of tomato, cosmos, marigold, and heliotrope is also induced by ethylene. With a few ethylene exposures, the roots arise from the midrib and with increasing exposures from the secondary veins. The oldest and youngest leaves never formed roots.

Many plants form root primordia without any treatment and usually form roots when cuttings are made of them. Application of ethylene can initiate growth from these primordia without cutting. Ethylene can also induce secondary roots from adventitious roots by additional intermittent gas treatments. The marigold plants shown in Fig. 6-15A and B demonstrate the induction of roots by ethylene and in addition show that secondary treatments of ethylene caused the formation of secondary roots on the initial set of roots. The figures also show that the secondary ethylene treatments caused the formation of root hairs on the tips of the adventitious roots.

A number of investigators have observed that ethylene increased the formation of root hairs on a variety of plants including *Pisum sativum* (18, 36), *Vicia faba, Lupinus albus, Sinapsis alba, Cheiranthus cheiri, Raphanus sativa* (18), and barley (186). However, not all of the plants treated with ethylene formed root hairs (215). Figure 6-16 is two microphotographs of

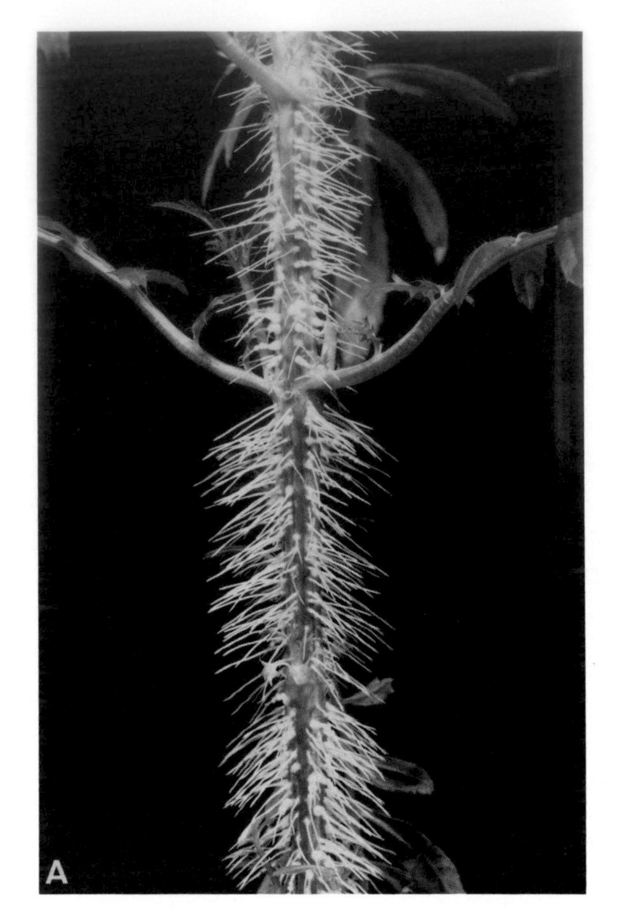

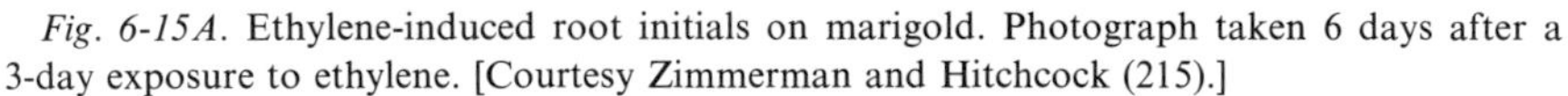
Fig. 6-15A. Ethylene-induced root initials on marigold. Photograph taken 6 days after a 3-day exposure to ethylene. [Courtesy Zimmerman and Hitchcock (215).]

pea roots showing the characteristic shortening and swelling of root cells as well as induction of hairs from the epidermal cells. It appears from these figures that a part of the superabundance of root hairs may be due to the fact that there are more epidermal cells per unit length of root after ethylene treatment, resulting in an increase in the number of root hair initials per given length of root.

Auxin is known to induce both ethylene production as well as rooting (216). However, the induction of roots appears to involve more than the simple increase in ethylene production. A number of workers (91, 111, 132) have found that simultaneous application of auxin with ethylene increased the number of roots formed. However, other workers (107) have found that in the case of sugarcane, auxin antagonized ethylene action. Characteristic of other ethylene-controlled phenomena, gibberellic acid partially reversed root induction by ethylene (111).

The use of Ethrel to increase rooting of apple seedlings (47) and blueberry (102) represents an attempt to encourage propagation of important horticultural plants by ethylene-induced root formation. Ethrel along with other chemicals may become an increasingly important tool in plant propagation.

Fig. 6-15B. Branched roots on adventitious roots induced by intermittent exposures to 0.2% ethylene. Note induction of secondary roots, root hairs, and change in orientation of new roots to gravity. [Courtesy Zimmerman and Hitchcock (215).]

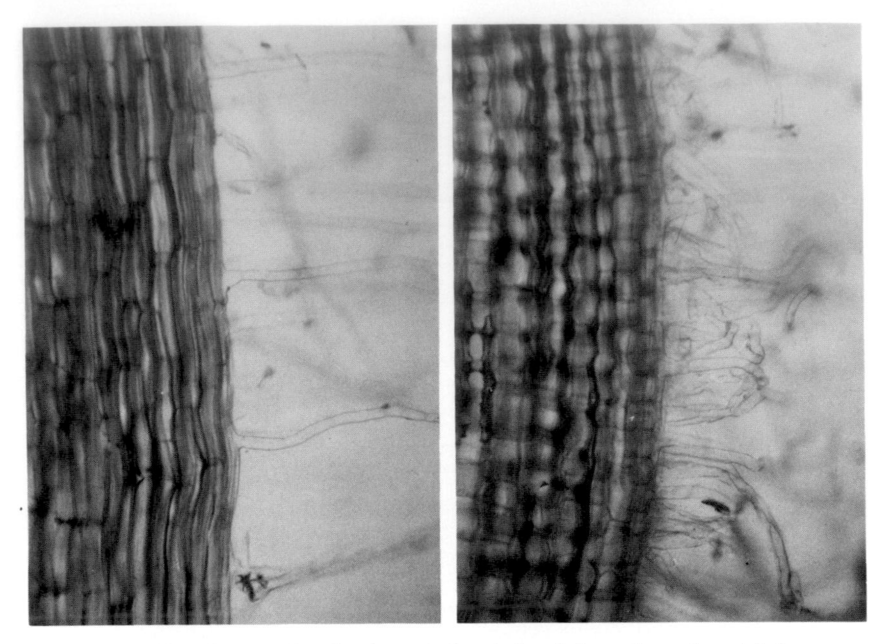

Fig. 6-16. Microphotographs showing ethylene-induced formation of root hairs on pea roots. Left, control; right, 1 ppm ethylene for 24 hours.

VI. Hypertrophy

Ethylene can cause hypertrophy of lenticels, cortex, and other tissues. Hypertrophies of lenticels are referred to as intumescences and were first described as an effect of tobacco smoke on *Sambucus nigra* stems by Molisch (142). Although a number of investigators have reported ethylene induction of intumescences (55, 75, 80, 142, 189, 210, 213, 214), only Wallace studied the phenomenon in detail (204–206). He found that about half of the species he tested formed intumescences. However, not all of the species that showed hypertrophy on other parts of the stem would form intumescences. The opposite was also observed: some species would form intumescences but would not have swollen nodal tissue, etc. Figures 6-17 and 6-18 show an example of ethylene-induced intumescences.

Only a short time was required to elicit a response; either 200 ppm for 24 hours or 2000 ppm for 2 hours caused definite intumescences. Dose-response studies showed no reaction in pure ethylene, due probably to anaerobiosis; a plateau response from 100 ppm to 75% ethylene; and a decrease in effect from 100 to 1 ppm. Minimum concentration required to elicit some response was 0.01 ppm (205). Carbon dioxide antagonism was

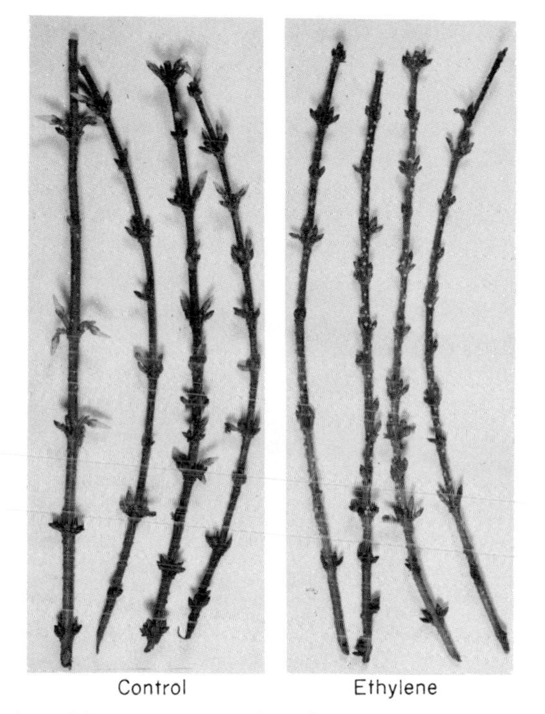

Fig. 6-17. Induction of intumescences on *Forsythia suspensa* stems by ethylene. Note also that ethylene prevented growth of buds compared with controls

also observed. Wallace found that concentrations of CO_2 equalling 25% or greater blocked the development of intumescences. There is, however, some question as to whether the effect represents competitive inhibition since the levels of ethylene during CO_2 experiments were 2%. Maximum development of intumescences took place at 15°C with progressively less development at higher temperatures. Almost no intumescence formation took place at 30°C. Induction of intumescences has also been observed after fumigation with CO (213, 214).

The most characteristic change in tissue during intumescence formation is the digestion or dissolution of cell walls (see Fig. 6-19). This results in more or less complete separation of cells from tissue continuity and rounding up of individual protoplasts. Wallace (206) and Hall (78) noted a great similarity between the formation of intumescences and the proliferation of cells at the surface of abscission separation layers following prolonged incubation in ethylene under moist and undisturbed conditions. According to Wallace, the middle lamella seems to have been dissolved just before or at the same time as the final dissolution of the secondary wall thickenings. The middle lamella stained red-brown with ruthenium red in normal walls

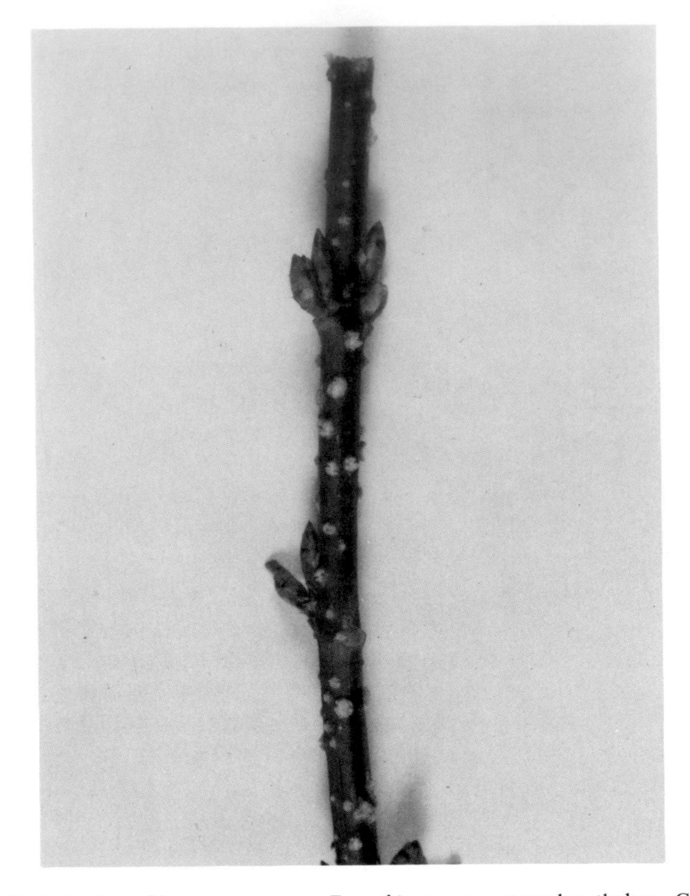

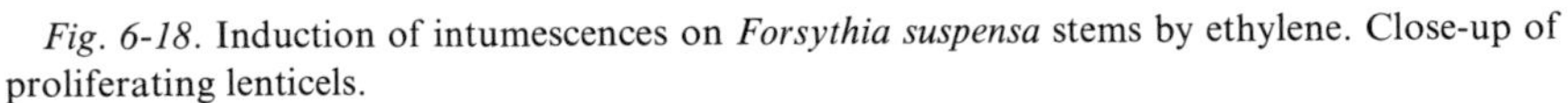

Fig. 6-18. Induction of intumescences on *Forsythia suspensa* stems by ethylene. Close-up of proliferating lenticels.

but was difficult to demonstrate in modified ones. There is an enormous increase in the size of most of the cells that make up the outgrowths. Approximately 1% of the cells become truly gigantic in size. The dimensions of a normal cortical cell are 29 μm in diameter and 26 μm in width. The gigantic cells achieve sizes of 280 μm in diameter and 54 μm in width.

Hypertrophy of cortex tissue located at nodes, internodes, and cut surfaces has also been described (204–206). These hypertrophies can occur on roots as well as on stem tissue (18, 55). In cortex tissue, promotion of cell division appears to be due to the formation of cells outside the cork cambium (phellogen) (82, 210). Ethylene-induced cork cells were unsuberized compared with those formed normally. Apparently one reason for the ability of ethylene to cause death of cortical tissue is the fact that the cells formed

Fig. 6-19. Microphotograph of an intumescence showing the change in cell shape and the sloughing off of external cells.

in ethylene-induced proliferations appear to be thin-walled and susceptible to desiccation and subsequent death.

The proliferation of cells (callus formation) of cut surfaces is similar to the phenomenon associated with callus cultures. Because ethylene causes proliferation of cortex cells and separation layers it is not surprising that some investigators have shown a promotion of callus formation in the presence of ethylene. Chalutz and DeVay (38) reported that ethylene increased development of callus formation at the cut surfaces of sweet potato roots. Stoutemyer and Britt (189) reported that Ethrel produced substantial increases of growth of cultures of Algerian ivy and Virginia Gold tobacco pith.

Nematodes cause proliferation of root tissue due to an increase in cell division. Doubt (58) reported that ethylene caused tubercles in tomato similar to those produced by nematodes. Orion and Minz (156) found that Ethrel had no effect on the number of nematodes inside of galls or the rate of nematode development. However, it did increase the size of the galls.

A number of workers have observed ethylene production from plant cell cultures (65, 66, 115, 125). However, it appears that ethylene has no effect on growth of the cultures nor did removal by absorption with mercuric perchlorate retard the growth of cultures of sycamore maple (*Acer pseudoplatanus*) (125). LaRue and Gamborg (115) observed that ethylene did not replace the effect of 2,4-D or NAA in cultures of rue (*Ruta graveolens*) or rose and, in fact, inhibited growth by 30% and 20%, respectively.

VII. Exudation

Exudation of various liquids as a result of ethylene treatment has been observed by a number of workers. These exudations fall into three types: guttation, gummosis, and latex flow. Guttation represents accumulation of xylem sap at hydathodes located at the edges of leaves. Gummosis is the secretion of gummy or resinous fluids from lenticels on the bark of trees, while latex represents a fluid consisting of lutoids, vesicles, and rubber particles released from laticifers from a variety of plants including lilies, sunflowers, and, of course, rubber trees.

Harvey (80) observed that ethylene promoted guttation by castor bean plants, and this observation has been repeated by others. Ethylene-induced guttation has been observed in a variety of plants including *Lamium purpureum, Stelluria media, Impatiens holstii, Torenia fournieri* (79), *Avena sativa, Brassica campestris, Cheiranthus cheiri, Helianthus annuus, Impatiens sultani, Lepidium sativum, Rhaphanus sativa, Sinapsis alba, Solanum lycopersicum,* and *Zea mays* (18). It is possible that other ethylene analogs will have a similar effect since CO has also been reported to induce guttation (89, 178).

Exudation of sap or gummosis was originally observed when plants were treated with illuminating gas. Shonnard (183) observed exudation of sap from lemon trees; Doubt (58) noted exudation from *Grevillia robusta* after fumigation from illuminating gas. Ethylene-induced gummosis has been commonly observed by investigators studying the effects of fruit-loosening agents such as Ethrel. Ethrel- or ethylene-induced gummosis has been reported from peach (23, 128), apricot (21), and cherry trees (8). Bradley *et al.* (21) indicated that gum pocket initials originated through differentiation of cells of newly formed xylem. Through enlargement and fusion of gum pockets the ducts were formed and fluid released.

Leopold has recently reviewed the regulation of rubber production by ethylene (118). The coagulation of rubber in latex is due to a reaction between rubber particles and vacuole-like structures called lutoids. Lutoids contain a number of negatively charged proteins and are easily disrupted when they emerge from laticifers. The release of these negatively charged proteins results in flocculation of the rubber particles and reduction of rubber flow as latex plugs form at the ends of the laticifers. Apparently, ethylene acts to defer latex coagulation by preventing clotting and permitting larger flow rates for prolonged periods of time. Originally 2,4-D was

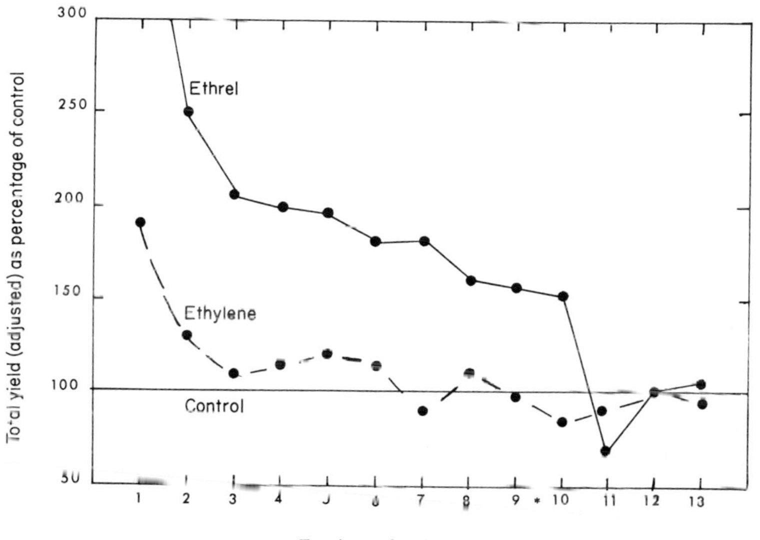

Fig. 6-20. Increase in rubber yields due to application of ethylene and ethylene-generating substances. Tappings were made once a month. [Courtesy Abraham *et al.* (6).]

used to stimulate latex flow. The increased awareness that some of 2,4-D's effects were due to an increase in ethylene production by the plant led to the use of Ethrel as a latex flow promoter (6, 49).

Not only is Ethrel an effective rubber stimulant, but it appears to be free of undesirable side effects associated with the use of 2,4-D. Ethrel is applied on a band of bark below the tapping cut, and within 3 days rates of latex production are double compared with those for controls. The increase in production appears to be due to a longer duration of latex flow as a result of the depression of the coagulation reaction. After 10 days, the greater rate of flow tapers off and, after a number of months, flow rates are similar to those for untreated controls (see Fig. 6-20). Retreatment with Ethrel is without effect until this amount of time has passed. The chemical stimulation of rubber yields with Ethrel promises to have a major impact on the rubber industry and may represent the largest agricultural application of ethylene-releasing substances.

VIII. Flowering

The effects of ethylene on flowering fall into three categories: promotion, inhibition, and sex reversal. Promotion of flowering is associated primarily with bromeliads, and sex reversal with cucurbits, while inhibition of floral initiation has been described in a number of families.

Floral initiation in pineapples with smoke was discovered in 1893 when a carpenter working in a greenhouse in the Azores accidentally set fire to a pile of shavings. To the surprise of the grower, the plants burst into flower instead of being damaged (7). Rodriguez (168) reported that ethylene as well as smoke would accelerate flowering in pineapples and subsequent workers have shown that other compounds will also promote flowering. These materials either are ethylene analogs such as acetylene (39, 42), produce ethylene chemically [Ethrel (40) and β-hydroxyethylhydrazine (35, 73, 74)], or cause the plants to evolve additional quantities of ethylene [NAA (28, 35, 41) and 2,4-D (35)].

Foster (61a) has found that other members of the Bromeliaceae would also produce flowers after ethylene fumigation and included *Vriesias, Aechmea, Orthophytum, Billbergia,* and *Quesnelia.* Cathey and Downs (35) have observed flowering in other bromeliads after treatment with β-hydroxyethylhydrazine. Figure 6-21 shows the appearance of *Billbergia* plants 6 weeks after a 24-hour 10 ppm ethylene fumigation.

Induction of flowering in pineapple requires a 6-hour treatment. Cooper and Reece (42) found that a 4-hour exposure of 1000 ppm failed to induce flowering of pineapple, while a 6-hour treatment was fully effective. The quantity of ethylene required was a function of the duration of incubation.

Fig. 6-21. Promotion of flowering of *Billbergia pyramidalis* with a 24-hour exposure of 10 ppm ethylene. Untreated control on right.

A 6-hour incubation with 100 ppm or 10 ppm resulted in 33% and 0% flowering, respectively, while both concentrations caused 100% flowering after a 12-hour incubation.

Cooper and co-workers (42, 195) found that pineapple would flower after ethylene treatment when only one leaf was left on the plant. Plants with no leaves failed to flower. If the one remaining leaf was removed 24 hours after ethylene fumigation, the plant would still flower. These results suggest that the leaf may be the primary site for ethylene action and the flowering stimulus travels from the leaf to the apex. However, I have failed to observe flowering in *Billbergia* after fumigation of a single leaf by placing it in a plastic bag and adding ethylene.

Changes in the morphology of the apex of pineapple plants occur 3 days following a 10-ppm 24-hour ethylene fumigation. By 20 days, flowers in the first row of the inflorescence are formed and by 6 weeks flowers are visible in the crown.

Synchronization of fruiting is important for commercial harvesting of pineapple plants. Growers have used smoke, acetylene, ethylene, and more recently NAA to achieve uniform fruiting to facilitate once-over harvesting (39, 42, 168, 195, 209). Ethrel has also been used for this purpose.

Ethylene can promote flowering in plants other than bromeliads. Mango trees in the Philippines tend to flower once every 2 or 3 years. However, smudging the trees with smoke for 2 weeks was reported to promote flowering (72). While ethylene is often the active ingredient of smoke, it remains to be shown that this is also true for this case. The rate of floral initiation of iris (190) and gladiolus (83) has been shown to be increased by ethylene. However, unlike the case of bromeliads where there is a shift from vegetative to floral apical meristem, the effect in iris and gladiolus may be due to breaking the dormancy of preformed floral tissue.

The work of Nitsch and Nitsch (152, 153) provides an additional example of conversion from vegetative to floral growth. Plants of *Plumbago indica* L. var. Angkor remain vegetative under 16-hour long days. However, when the plants are sprayed with Ethrel, they form flowers in the same fashion as those exposed to short days (see Fig. 6-22).

Other short-day plants such as cocklebur and *Perilla* behave in the opposite fashion. Zhdanova (211) reported that 300 ppm ethylene blocked flowering of *Perilla* under short-day conditions. Similar observations were made on cocklebur plants. The initiation of blossoms during a long night was completely inhibited with 10 ppm ethylene, while 1 ppm had a partial effect (1) (see Fig. 6-23). Earlier Campbell and Leopold (33) reported that application of CO_2 during the inductive dark period promoted flowering of cockleburs. This is the anticipated result assuming floral inhibition is a typical ethylene effect. Dostál (57) found that ethylene blocked formation of *Soja*

Fig. 6-22. Flower induction in *Plumbago indica* var. Angkor. Both plants are of the same age and grown under noninductive long days of 16 hours. Left, control; right, plant sprayed twice with Ethrel, 240 mg/liter, 3 days apart. (Courtesy C. Nitsch, Physiologie Pluricellulaire, Gif-sur-Yvette.)

hispida during an inductive long night. He also found that flowering of *Circaea intermedia, Scrophularia nodosa,* and *Lycoris europaeus* was blocked by ethylene and the plants formed runners and bulbs instead.

Ethylene can change the sex of developing flowers in members of the Cucurbitaceae, Euphorbiaceae, and Cannabinaceae. As in the case of the promotion of flowering in bromeliads, greenhouse operators in the 1860's noticed the promotion by smoke of female flowers in cucurbits (131). According to Minina (137), growers smoked cucumber plants in the three- to four-leaf stage with two 12-hour treatments which resulted in the production of female flowers and larger fruit set. These Russian workers measured the effect of a number of gases on the sex ratio of cucurbits and found that CO, acetylene, and ethylene produced the same effect as did smoke (131, 137–139).

Fig. 6-23. Inhibition of flowering in cocklebur by ethylene. Top, 10 ppm ethylene during an inductive 16-hour night; bottom, control has flowering apex.

Most of the subsequent work on promotion of female flowers has been done with Ethrel because any practical horticultural application would require some material that could be applied to plants in the field. While this is a good way to approach a practical problem it makes interpretation of experimental results difficult because it is hard to determine the duration of effective ethylene concentration. Two objectives for the use of Ethrel are to increase yield by encouraging the formation of female flowers (173, 187) and as a hybridizing tool (43, 122, 173). Splittstoesser (187) found that while Ethrel did not increase the yield of pumpkin fruits measured in terms of weight, it did increase the total number of fruits produced.

A wide variety of cucurbits including cucumber (67), squash (43) (see Fig. 6-24), pumpkins (187), melons, and gourds (122) are induced to form

Fig. 6-24A. Effect of ethylene on the sex of cucumber flowers. Control plants grown in ethylene-free air. Note presence of predominantly male flowers.

Fig. 6-24B. Effect of ethylene on the sex of cucumber flowers. Plants grown continuously in air containing 0.1 ppm ethylene. Only female flowers are present. Ethylene also caused a reduction in perianth development and premature senescence of flowers on the lower nodes.

female flowers by Ethrel. Varietal differences exist. George (67) found that Ethrel failed to increase female flowers in two of six varieties of cucumbers he examined.

The time of Ethrel application influences the formation of female flowers. Applications of Ethrel to seeds during soaking (19) or in the cotyledon stage (96) were ineffective. The effect of applying Ethrel at various times after the germination of cucumber seeds is shown in Fig. 6-25. As observed earlier by Russian growers with smoke, the most effective time to apply Ethrel is in the third leaf stage. The later the time of application, the higher the node at which the first female flower appeared. However, the total number of female flowers was the same regardless of application time. Abortion of flowers at the lower nodes is probably due to the ability of ethylene to hasten floral senescence. Plants at the first, second, and third leaf stages have already differentiated flower primordia up to the ninth, twelfth, and fifteenth nodes, respectively. At the second and third leaf stages, sex of flowers has already been determined up to the fourth and seventh nodes, respectively. These observations show that ethylene action is probably directed toward determining the sex in preformed floral primordia which have not yet advanced to the point of sexual differentiation. The gas apparently has no effect on either sexually differentiated primordia or primordia that have not been initiated.

Induction of female flowers by ethylene also occurs in other species. The Heslop-Harrisons (88) found that 1% CO for two 48-hour periods

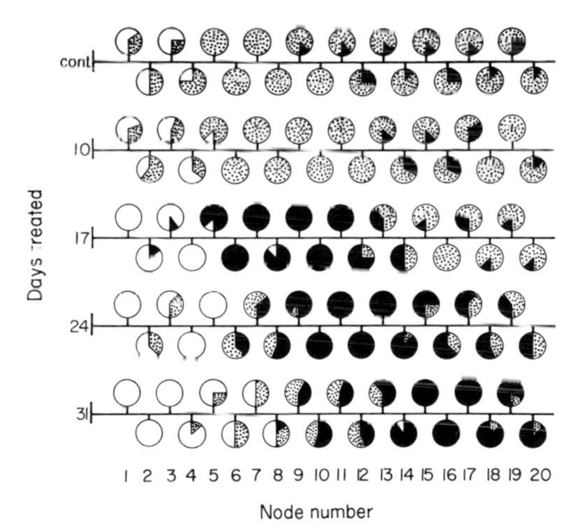

Fig. 6-25. Effects of time of application of Ethrel on sex expression of cucumber plants. Ethrel (50 mg/liter) was sprayed on certain days after seeding as indicated at the left side of the figure. The circle represents the ratio of sex of each node. Clear areas represent percentage of aborted buds; dotted and black areas represent male and female flower-bearing nodes, respectively. [Courtesy of Iwahori *et al.* (96).]

reduced the formation of male flowers and increased the number of female flowers in *Mercurialis ambigua*. Mohan Ram and Jaiswal (140) reported that Ethrel induced the formation of female flowers on male plants of *Cannabis sativa*. Flowers showing both male and female characteristics where formed as a result of sprays that contained lower concentrations of Ethrel (see Fig. 6-26). The same phenomenon had been observed earlier by the Heslop-Harrisons (89) during a study on the effect of CO on flowering of *Cannabis*.

Fig. 6-26. A, Terminal part of control *Cannabis sativa* male plant showing male flowers; × 0.225. B, Male plant treated with 960 ppm Ethrel has developed female flowers; × 0.225. C, D, and E, Flowers from Ethrel-treated male plants showing stages of transformation into female flowers; C, × 12.5; D, × 7; E, × 11.25. [Courtesy of Mohan Ram and Jaiswal (140).]

IX. Gravitational Responses

Gravity has an effect on ethylene production and the ability of plants to respond to ethylene. In addition, ethylene can also alter the response of plants to gravity.

A. Effect of Gravity on Ethylene Production

Denny (52) was the first investigator to note a similarity between some of the effects of ethylene and that observed when plants were placed on their sides. He observed epinastic leaves on horizontal tomato plants and postulated that this might be due to an increase in ethylene production following the change in the plants' position. Using a tomato petiole bioassay, he found that horizontal tomato plants produced three times as much ethylene as vertical plants. Since that time, other investigators have shown that other effects of a change in direction of a gravitational field were also indirectly due to an increase in ethylene production. Horizontal *Coleus* plants lost debladed petioles faster than those in a vertical position. It was subsequently shown that the horizontal plants produced more ethylene than the vertical plants (3). Ethylene is known to promote flowering in pineapple. Van Overbeek and Cruzado (200) reported that pineapple plants placed in a horizontal position curved upward and produced flowers. A minimum of 3 days of geotropic stimulation was required for flower induction with the Cabezona variety. Other varieties, Red Spanish and Smooth Cayenne, failed to produce flowers under similar conditions.

Ethylene pretreatment reduced auxin transport in sunflower shoots and a similar inhibition of auxin transport was observed when plants had been laid on their sides for a while (158). Horizontal orientation of the stem segments during the transport test itself had no effect on IAA transport.

The clinostat has been used to grow plants under conditions approximating zero gravity. The idea behind this instrument is that the normal gravitational pull parallel to the stem has been removed and the rotational movement would balance out any lateral effect of gravity. Plants grown on a clinostat show a number of changes from normal growth including epinasty and an inhibition of nutation. Borgström (17) noted that ethylene and growth on a clinostat both inhibited nutation of pea seedlings. Epinasty is another characteristic ethylene effect, and Leather *et al.* (117) demonstrated that increased ethylene production during clinostat experiments may cause leaf epinasty. Ethylene production of tomato plants placed on a clinostat increased during the first 2 hours of rotation, and CO_2 reversed the ability of ethylene and clinostat rotation to cause petiole curvature.

Ethylene production does not always increase by placing plants on their sides. Abeles and Rubinstein (5) found that while horizontal bean seedlings produced more ethylene on the lower side than on the upper side, the total rate of gas production by horizontal and vertical plants was the same. The mechanism responsible for increased ethylene production by horizontal plants is unknown. It may be another example of stress-induced ethylene production.

B. Effect of Gravity on the Response of Plants to Ethylene

Crocker *et al.* (46) noted that the effectiveness of ethylene was reduced if the plants were not vertical. Ethylene induced marked epinasty in tomato and African marigold plants when they were upright, but had little or no effect when the leaves were inverted and the upper sides faced the earth. Intermediate positions of the plants resulted in intermediate degrees of curvature. The weight of the leaf was not a factor in these experiments. When leaves were supported by strings to overcome the effect of their weight, ethylene still failed to induce epinasty when the plants were upside down. Secondly, they noted that the weight of the tomato leaves was 3.7 g and the force by which the petioles bent over due to epinasty was equal to 15–30 g. The epinastic force was capable of exerting a force equal to four to eight times the total weight of the leaf. These observations were later confirmed by Borgström (18).

C. Effect of Ethylene on the Response of Plants to Gravity—Diageotropism

Neljubov (150, 151) and other earlier workers (165) reported that ethylene caused pea seedlings to grow horizontally instead of upright. Neljubov reported that this horizontal growth occurred irrespective of the position in which the plant was placed. Figure 6-27 shows a repeat of Neljubov's early work and demonstrates the effect of 0.1 ppm ethylene on plants grown upside down, sideways, and right side up. After the addition of ethylene all new growth occurred horizontally. This horizontal growth has been variously called horizontal nutation, transverse geotropism, ageotropism, and diageotropism. The latter term is probably the most appropriate.

A number of earlier workers originally thought that ethylene blocked the response of plants to gravity (58). This conclusion was based on the observation that horizontal plants placed in ethylene continued to grow in their original plane instead of upright. However, it is clear from the plants shown in Fig. 6-27 that the plants do not lose the ability to sense gravity, but rather that the new growth that occurs more closely resembles that of a horizontal branch than that of a vertical stem. In peas diageotropic growth takes place only in dark-grown tissue (69). However, Hall *et al.* (79) reported that low

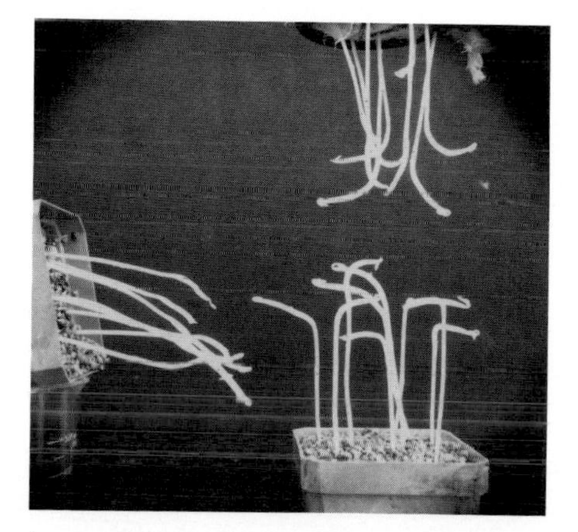

Fig. 6-27. Diageotropism in dark-grown pea seedlings. After 7-day-old seedlings were placed in various positions, growth in the presence of 0.1 ppm ethylene took place for 5 days before the photograph was taken.

concentrations of ethylene caused cotton plants to assume prostrate, vine-like growth patterns.

Burg *et al.* (25) have similarly pointed out that ethylene does not interfere with phototropic responses in plants. Ethylene blocked rapid phototropic curvature of radish, while it had no effect on phototropic curvature of monocot coleoptiles of oat and corn. Similarly, ethylene had no effect on phototropic curvature of *Phycomyces,* which apparently has the same photoreceptor as higher plants but does not use auxin transport to mediate tropistic responses.

Ethylene also alters growth patterns of roots. Zimmerman *et al.* (213, 215) noted that ethylene altered the growth patterns of a number of roots. In some species, roots grew upward instead of down. Figure 6-15B shows the effect of ethylene on the angle of root growth.

X. Effect on Motor Cells

A. Stomates

Ethylene has no effect on the opening or closing of stomates. Schwartz (178) reported that illuminating gas had no effect on stomates. Pallaghy and Raschke (157) reported that concentrations of ethylene between 1 and 100,000 ppm also had no effect on the opening and closing of corn and pea stomates. In addition, stomates did not respond to allene or nitrous oxide. On the other hand, stomates are known to be sensitive to CO_2. It would appear that while CO_2 can act as a competitive inhibitor of ethylene action the opposite is not true. Contrary to the above results. Saad (175) reported that 1% ethylene caused rapid closure of *Pelargonium zonale* stomates and that removal of ethylene caused the stomates to open again. However, concentrations below 0.1% ethylene had no effect and adequate controls were absent from the experiments utilizing the higher concentrations of ethylene.

B. PULVINI

Early workers such as Doubt (58) and Zimmerman *et al.* (214) reported that ethylene caused rigor in the sensitive plant (*Mimosa pudica*). After 18 hours of ethylene treatment, the plants were insensitive to touch or any other stimuli, but after an additional 10 hours the plants regained their sensitivity to touch. Time-lapse movies of the plants were made and, in contrast to normal movements, leaves of treated plants swept widely from side to side and moved up and down through at least 180 degrees. In one experiment, the plants were gassed for 27 hours at which time they started to lose some of their leaves. When the plants were replaced in air, the remaining leaves regained sensitivity to stimuli after 24 hours.

Baur and Morgan (14, 144) studied the effect of ethylene on leaf movement of other leguminous species. They found that a 12-hour treatment with 2 ppm ethylene was required to cause loss of leaf movement in mesquite and 6 hours in huisache. Minimum concentrations required to cause changes in leaf movement were 0.1 ppm.

C. TENDRILS

The coiling of pea tendrils may also be regulated by ethylene. Jaffe (97) has shown that mechanical stimulation of pea tendrils resulted in an increase in ethylene production within 60 minutes. Asymmetrical application of Ethrel to pea tendrils resulted in noticeable coiling within 30 minutes. The Ethrel was effective only when combined with DMSO (dimethyl sulfoxide) and applied ventrally. Absence of the DMSO and symmetrical application of Ethrel resulted in little coiling. These results suggest that tendril coiling involves an increase in ethylene production due to a mechanical stimulation followed by a regulation of cell size on the ventral side of the tendril by ethylene. The effect of ethylene and CO_2 on the coiling of *Marah fabaceus* (wild Californian cucumber) has also been studied. Ethylene had only a slight effect on coiling while CO_2 was extremely effective. However, the concentration of CO_2 used was very high, 100%.

XI. Algae and Fungi

Ethylene was reported to increase cell division in the diatom *Nitzschia putrida*. If the cells were left in the ethylene atmospheres for a long period of time they showed an increase in polyploidy which eventually led to death (203). However, the source of ethylene used in these experiments was apple emanations. It was not proven that ethylene was the active ingredient in these emanations.

Slime molds were also reported to be sensitive to ethylene. Seifriz and Urbach (179) reported that 75% ethylene increased the growth rate of *Physarum polycephalum*. Concentrations below 25% had little or no effect. Carbon dioxide proved to be an anesthetic and inhibited growth

Even though many fungi are known to produce large quantities of ethylene, only a few reports have appeared claiming to observe an effect on growth or metabolism. Similar to observations on slime molds, the concentrations involved are generally very high compared with those effective on higher plants.

Moderately high concentrations of ethylene (250 ppm) were reported to promote the germination of *Diplodia natalensis* and *Phomopsis citri* (22). In water, these spores germinated poorly while in a prune juice medium they germinated well. The effect of the ethylene was to increase the rate of germination in water so that it was about half that observed in the prune juice. The effect was not specific for ethylene. A similar effect was observed with ethyl acetate and methanol.

Lockhart *et al.* (121) reported that ethylene had no effect on the growth of *Gloeosporium album* except at very low incubation temperatures (3.3°C). Under these conditions, ethylene inhibited growth. At high concentrations of ethylene (2000 ppm) the growth of *Fusarium oxysporum* was inhibited (120). The inhibition was greater under conditions of reduced O_2 tension than in air.

Only a few reports on the metabolic effects of ethylene on fungi have appeared. Vesselov (201) reported that ethylene (1000 ppm) stimulated the respiration of *Aspergillus niger, A. fumigatus,* and *A. flavus* about 10%. Simultaneously, a slight inhibition of growth was observed. Shaw (181) reported that ethylene had no effect on the permeability of yeast cells but it did stimulate the formation of invertase.

References

1. Abeles, F. B. (1967). *Plant Physiol.* **42,** 608.
2. Abeles, F. B. (1968). *Weed Sci.* **16,** 498.
3. Abeles, F. B., and Gahagan, H. E. (1968). *Life Sci.* **7,** 653.

4. Abeles, F. B., and Lonski, J. (1961). *Plant Physiol.* **44,** 277.
5. Abeles, F. B., and Rubinstein, B. (1964). *Plant Physiol.* **39,** 963.
6. Abraham, P. D., Wycherley, P. R., and Pakianathan, S. W. (1969). *J. Rubber Res. Inst. Malaya* **20,** 291.
7. Adams, H. C. (1935). *Nat. Geogr. Mag.* **67,** 35.
8. Anderson, J. L. (1969). *HortScience* **4,** 92.
9. Andreae, W. A., Venis, M. A., Jursic, F., and Dumas, T. (1968). *Plant Physiol.* **43,** 1375.
10. Apelbaum, A., and Burg, S. P. (1971). *Plant Physiol.* **48,** 648.
11. Asmaev, P. G. (1937). *Proc. Agr. Inst. Krasnodar* **6,** 49.
12. Balls, A. K., and Hale, W. S. (1940). *Cereal Chem.* **17,** 490.
13. Barker, J. (1934). *Gt. Brit., Dep. Sci. Ind. Res., Food Invest. Bd., Rep. 1933* p. 80.
14. Baur, J. R., and Morgan, P. W. (1969). *Plant Physiol.* **44,** 831.
15. Berry, L., and Smith, O. E. (1969). *Plant Cell Physiol.* **10,** 161.
16. Bjornseth, E. H. (1946). *Proc. Amer. Soc. Hort. Sci.* **48,** 369.
17. Borgström, G. (1939). "The Transverse Reactions of Plants." Hakan Ohlson, Lund, Sweden.
18. Borgström, G. (1939). *Kgl. Fysiogr. Sellsk. Lund, Ferh.* **9,** 135.
19. Bose, T. K., and Nitsch, J. P. (1970). *Physiol. Plant.* **23,** 1206.
20. Botjes, J. O. (1933). *Tijdschr. Plantenziekten* **39,** 207.
21. Bradley, M. V., Marei, N., and Crane, J. C. (1969). *J. Amer. Soc. Hort. Sci.* **94,** 316.
22. Brooks, C. (1944). *J. Agr. Res.* **68,** 363.
23. Buchanan, D. W., and Biggs, R. H. (1969). *J. Amer. Soc. Hort. Sci.* **94,** 327.
24. Burg, S. P. (1964). *Colloq. Int. Cent. Nat. Rech. Sci.* **123,** 719.
25. Burg, S. P., Apelbaum, A., Eisinger, W. R., and Kang, B. G. (1971). *HortScience* **6,** 359.
26. Burg, S. P., and Burg, E. A. (1965). *Science* **148,** 1190.
27. Burg, S. P., and Burg, E. A. (1966). *Proc. Nat. Acad. Sci. U. S.* **55,** 262.
28. Burg, S. P., and Burg, E. A. (1966). *Science* **152,** 1269.
29. Burg, S. P., and Burg, E. A. (1967). *Plant Physiol.* **42,** 144.
30. Burg, S. P., and Burg, E. A. (1968). *Plant Physiol.* **43,** 1069.
31. Burg, S. P., and Burg, E. A. (1968). *In* "Biochemistry and Physiology of Plant Growth Substances" (F. Wightman and G. Setterfield, eds.), p. 1275. Runge Press, Ottawa.
32. Burton, W. G. (1952). *New Phytol.* **51,** 154.
33. Campbell, C. W., and Leopold, A. C. (1958). *Plant Physiol.* **33,** xix (abstr.).
34. Catchpole, A. H., and Hillman, J. (1969). *Nature* (*London*) **223,** 1387.
35. Cathey, H. M., and Downs, R. J. (1965). *Exch. Flower, Nursery Gard. Cent. Trade* **143,** 27.
36. Chadwick, A. V., and Burg, S. P. (1967). *Plant Physiol.* **42,** 415.
37. Chadwick, A. V., and Burg, S. P. (1970). *Plant Physiol.* **45,** 192.
38. Chalutz, E., and DeVay, J. E. (1969). *Phytopathology* **59,** 750.
39. Collins, J. L. (1935). *Pineapple News* **9,** 78.
40. Cooke, A. R., and Randall, D. I. (1968). *Nature* (*London*) **218,** 974.
41. Cooper, W. C. (1942). *Proc. Amer. Soc. Hort. Sci.* **41,** 93.
42. Cooper, W. C., and Reece, P. C. (1941). *Proc. Fla. State Hort. Soc.* **54,** 132.
43. Coyne, D. P. (1970). *HortScience* **5,** 227.
44. Crane, J. C., Marei, N., and Nelson, M. M. (1970). *J. Amer. Soc. Hort. Sci.* **95,** 367.
45. Crocker, W., Hitchcock, A. E., and Zimmerman, P. W. (1935). *Contrib. Boyce Thompson Inst.* **7,** 231.
46. Crocker, W., Zimmerman, P. W., and Hitchcock, A. E. (1932). *Contrib. Boyce Thompson Inst.* **4,** 177.
47. Cummins, J. N., and Fiorino, P. (1969). *HortScience* **4,** 339.
48. Curtis, R. W. (1968). *Plant Physiol.* **43,** 76.

49. D'Auzac, J., and Ribaillier, D. (1969). *C. R. Acad. Sci.* **268,** 3046.
50. Denny, F. E. (1929). *Contrib. Boyce Thompson Inst.* **1,** 59.
51. Denny, F. E. (1930). *Amer. J. Bot.* **17,** 602.
52. Denny, F. E. (1936). *Contrib. Boyce Thompson Inst.* **8,** 99.
53. Denny, F. E. (1939). *Contrib. Boyce Thompson Inst.* **10,** 191.
54. Denny, F. E., and Miller, L. P. (1935). *Contrib. Boyce Thompson Inst.* **7,** 97.
55. Deuber, C. G. (1933). *Amer. Gas. Ass. Mon.* **15,** 380.
56. Dimond, A. E., and Waggoner, P. E. (1953). *Phytopathology* **43,** 663.
57. Dostál, R. (1941). *Ber. Deut. Bot. Ges.* **59,** 437.
58. Doubt, S. L. (1917). *Bot. Gaz.* **63,** 209.
59. Egley, G. H., and Dale, J. E. (1970). *Weed Sci.* **18,** 586.
60. Esashi, Y., and Leopold, A. C. (1969). *Plant Physiol.* **44,** 1470.
61. Fergus, C. L. (1954). *Mycologia* **46,** 543.
61a. Foster, N. B. (1953). "Bromeliads." Bromeliad Soc. Inc.
62. Fujii, K. (1925). *Flora (Jena)* [N. S.] **18–19,** 115.
63. Funke, G. L., de Coeyer, F., de Decker, A., and Maton, J. (1938). *Biol. Jaarb.* **5,** 335.
64. Galang, F. G., and Agati, J. A. (1936). *Philipp. J. Agr.* **7,** 245.
65. Gamborg, O. L., and LaRue, T. A. G. (1968). *Nature (London)* **220,** 604.
66. Gamborg, O. L., and LaRue, T. A. G. (1971). *Plant Physiol.* **46,** 399.
67. George, W. L. (1971). *J. Amer. Soc. Hort. Sci.* **96,** 152.
68. Girardin, J. P. L. (1864). *Jahresber. Agrikulturchem.* **7,** 199.
69. Goeschl, J. D., and Pratt, H. K. (1968). *In* "Biochemistry and Physiology of Plant Growth Substances" (F. Wightman and G. Setterfield, eds.), p. 1229. Runge Press, Ottawa.
70. Goeschl, J. D., Pratt, H. K., and Bonner, B. A. (1967). *Plant Physiol.* **42,** 1077.
71. Goeschl, J. D., Rappaport, L., and Pratt, H. K. (1966). *Plant Physiol.* **41,** 877.
72. Gonzalez, L. G. (1924–1925). *Philipp. Agr.* **12,** 15.
73. Gowing, D. P., and Leeper, R. W. (1955). *Science* **122,** 1267.
74. Gowing, D. P., and Leeper, R. W. (1961). *Bot. Gaz.* **123,** 34.
75. Grafe, V., and Richter, O. (1912). *Sitzungsber. Kaiserl. Akad. Wiss. Wien.* **120,** 1187.
76. Haber, E. S. (1926). *Proc. Amer. Soc. Hort. Sci.* **23,** 201.
77. Hale, W. S., Schwimmer, S., and Bayfield, E. G. (1943). *Cereal Chem.* **20,** 224.
78. Hall, W. C. (1952). *Bot. Gaz. (Chicago)* **113,** 310.
79. Hall, W. C., Truchelut, G. B., Leinweber, C. L., and Herrero, F. A. (1957). *Physiol. Plant.* **10,** 306.
80. Harvey, E. M. (1913). *Bot. Gaz. (Chicago)* **56,** 439.
81. Harvey, E. M. (1915). *Bot. Gaz. (Chicago)* **60,** 193.
82. Harvey, E. M., and Rose, R. C. (1915). *Bot. Gaz. (Chicago)* **60,** 27.
83. Harvey, R. B. (1927). *Rev. Off. Bull. Amer. Gladiolus Soc.* **4,** 10.
84. Heck, W. W., and Pires, E. G. (1962). *Tex., Agr. Exp. Sta., Misc. Publ.* **MP-603.**
85. Heck, W. W., and Pires, E. G. (1962). *Tex., Agr. Exp. Sta., Misc. Publ.* **MP-613.**
86. Heck, W. W., Pires, E. G., and Hall, W. C. (1961). *J. Air Pollut. Contr. Ass.* **11,** 549.
87. Herbert, D. A., and Lynch, L. J. (1934). *Proc. Roy. Soc. Queensl.* **46,** 72.
88. Heslop-Harrison, J., and Heslop-Harrison, Y. (1957). *New Phytol.* **56,** 352.
89. Heslop-Harrison, J., and Heslop-Harrison, Y. (1957). *Proc. Roy. Soc. Edinburgh, Sect. B* **66,** 424.
90. Hitchcock, A. E., Crocker, W., and Zimmerman, P. W. (1932). *Contrib. Boyce Thompson Inst.* **4,** 155.
91. Hitchcock, A. E., and Zimmerman, P. W. (1940). *Contrib. Boyce Thompson Inst.* **11,** 143.
92. Holm, R. E., and Abeles, F. B. (1967). *Planta* **78,** 293.
93. Imaseki, H., and Pjon, C. J. (1970). *Plant Cell Physiol.* **11,** 827.

94. Imaseki, H., Pjon, C. J., and Furuya, M. (1971). *Plant Physiol.* **48,** 241.
95. Isaac, W. E. (1938). *Trans. Roy. Soc. S. Afr.* **26,** 307.
96. Iwahori, S., Lyons, J. M., and Smith, O. E. (1970). *Plant Physiol.* **46,** 412.
97. Jaffe, M. J. (1970). *Plant Physiol.* **46,** 631.
98. Kang, B. G., and Ray, P. M. (1969). *Planta* **87,** 193.
99. Kang, B. G., and Ray, P. M. (1969). *Planta* **87,** 206.
100. Kang, B. G., and Ray, P. M. (1969). *Planta* **87,** 216.
101. Kang, B. G., Yocum, C. S., Burg, S. P., and Ray, P. M. (1967). *Science* **156,** 958.
102. Kender, W. J., Hall, I. V., Aalders, L. E., and Forsyth, F. R. (1969). *Can. J. Plant Sci.* **49,** 95.
103. Kessler, H. (1934). *Landwirt. Jahrb. Schweiz* **48,** 853.
104. Ketring, D. L., and Morgan, P. W. (1969). *Plant Physiol.* **44,** 326.
105. Ketring, D. L., and Morgan, P. W. (1970). *Plant Physiol.* **45,** 268.
106. Ketring, D. L., and Morgan, P. W. (1971). *Plant Physiol.* **47,** 488.
107. Khan, M. A., and Hall, W. C. (1954). *Bot. Gaz.* (*Chicago*) **116,** 172.
108. Knight, L. I., and Crocker, W. (1913). *Bot. Gaz.* (*Chicago*) **55,** 337.
109. Knight, L. I., Rose, R. C., and Crocker, W. (1910). *Science* **31,** 635.
110. Kraynev. S. I. (1937). *Proc. Agr. Inst. Krasnodar* **6,** 101.
111. Krishnamoorthy, H. N. (1970). *Plant Cell Physiol.* **11,** 979.
112. Krone, P. R. (1937). *Mich., Agr. Exp. Sta., Spec. Bull.* **285.**
113. Ku, H. S., Suge, H., Rappaport, L., and Pratt, H. K. (1969). *Planta* **90,** 333.
114. Laan, P. A. van der. (1934). *Rec. Trav. Bot. Neer.* **31,** 691.
115. LaRue, T. A. G., and Gamborg, O. L. (1971). *Plant Physiol.* **48,** 394.
116. Lazanyi, A., and Cabulea, I. (1957). *Stud. Cercet. Agr.* **8,** 117.
117. Leather, G. R., Forrence, L. E., and Abeles, F. B. (1972). *Plant Physiol.* **49,** 183.
118. Leopold, A. C. (1971). *In* "What's New in Plant Physiology" (G. Fritz, ed.), Vol. 3, No. 3. Bot. Dept., Univ. of Florida Press, Gainesville.
119. Levy, D., and Kedar, N. (1970). *HortScience* **5,** 80.
120. Lockhart, C. L. (1970). *Can. J. Plant Sci.* **50,** 347.
121. Lockhart, C. L., Forsyth, F. R., and Eaves, C. A. (1968). *Can. J. Plant Sci.* **48,** 557.
122. Lower, R. L., and Miller, C. H. (1969). *Nature* (*London*) **222,** 1072.
123. Lyon, C. J. (1970). *Plant Physiol.* **45,** 644.
124. Mack, W. B., and Livingston, B. E. (1933). *Bot. Gaz.* (*Chicago*) **94,** 625.
125. Mackenzie, I. A., and Street, H. E. (1970). *J. Exp. Bot.* **21,** 824.
126. Madeikyte, E., and Turkova, N. S. (1965). *Liet. TSR Mokslu Akad. Darb., Ser. C* No. 2, p. 37.
127. Marinos, N. G. (1960). *J. Exp. Bot.* **11,** 227.
128. Martin, G. C., and Nelson, M. M. (1969). *HortScience* **4,** 328.
129. Maxie, E. C., and Crane, J. C. (1967). *Science* **155,** 1548.
130. Maxie, E. C., and Crane, J. C. (1968). *Proc. Amer. Soc. Hort. Sci.* **92,** 255.
131. Mekhanik, F. J. (1958). *Dokl. Vses. Akad. Sel'skokhoz. Nauk* **21,** 20.
132. Michener, H. D. (1935). *Science* **82,** 551.
133. Michener, H. D. (1938). *Amer. J. Bot.* **25,** 711.
134. Middleton, J. T., Darley, E. F., and Brewer, R. F. (1958). *J. Air Pollut. Contr. Ass.* **8,** 9.
135. Miller, E. V., Winston, J. R., and Fisher, D. F. (1940). *J. Agr. Res.* **60,** 269.
136. Miller, P. M., Sweet, H. C. and Miller, J. H. (1970). *Amer. J. Bot.* **57,** 212.
137. Minina, E. G. (1938). *Dokl. Akad. Nauk SSSR* **21,** 298.
138. Minina, E. G. (1952). "Changes in Sex of Plants Produced by Environmental Factors." Acad. Sci. USSR, Moscow.

139. Minina, E. G., and Tylkina, L. G. (1947). *Dokl. Akad. Nauk SSSR* **55,** 165.
140. Mohan Ram, H. Y., and Jaiswal, V. S. (1970). *Experientia* **26,** 214.
141. Molisch, H. (1905). *Ber. Deut. Bot. Ges.* **23,** 2.
142. Molisch, H. (1911). *Sitzungsber. Kaiserl. Akad. Wiss. Wien* **120,** 813.
143. Molisch, H. (1937). "The Influence of One Plant on Another. Allelopathy." Fischer, Jena.
144. Morgan, P. W., and Baur, J. R. (1970). *Plant Physiol.* **46,** 655.
145. Morgan, P. W., Meyer, R. E., and Merkle, M. G. (1969). *Weed Sci.* **17,** 353.
146. Morgan, P. W., and Powell, R. D. (1970). *Plant Physiol.* **45,** 553.
147. Morré, D. J., and Eisinger, W. R. (1968). *In* "Biochemistry and Physiology of Plant Growth Substances" (F. Wightman and G. Setterfield, eds.), p. 625. Runge Press, Ottawa.
148. Morton, D. J. (1970). Ph.D. Thesis, George Washington University, Washington, D. C.
149. Neljubov, D. (1901). *Beih. Bot. Zentralbl.* **10,** 128.
150. Neljubov, D. (1910). *Bull. Acad. Imp. Sci. St.-Petersbourg* **4,** 1443.
151. Neljubov, D. (1911). *Ber. Deut. Bot. Ges.* **29,** 97.
152. Nitsch, C. (1968). *In* "Biochemistry and Physiology of Plant Growth Substances" (F. Wightman and G. Setterfield, eds.), p. 1385. Runge Press, Ottawa.
153. Nitsch, C., and Nitsch, J. P. (1969). *Plant Physiol.* **44,** 1747.
154. Nooden, L. D. (1972). *Plant Cell Physiol.* **12,** 739.
155. Nord, F. F., and Weichherz, J. (1929). *Hoppe-Seyler's Z. Physiol. Chem.* **183,** 218.
156. Orion, D., and Minz, G. (1969). *Nematologica* **15,** 608.
157. Pallaghy, C. K., and Raschke, K. (1972). *Plant Physiol.* **49,** 275.
158. Palmer, O. H., and Halsall, D. M. (1969). *Physiol. Plant.* **22,** 59.
159. Powell, R. D., and Morgan, P. W. (1970). *Plant Physiol.* **45,** 548.
160. Radin, J. W., and Loomis, R. S. (1969). *Plant Physiol.* **44,** 1584.
161. Richards, H. M., and MacDougal, D. T. (1904). *Bull. Torrey Bot. Club* **3,** 57.
162. Richter, O. (1903). *Ber. Deut. Bot. Ges.* **21,** 180.
163. Richter, O. (1908). *Verh. Ges. Deut. Naturf. Arzte* **80,** 189.
164. Richter, O. (1913). *Verh. Ges. Deut. Naturf. Arzte* **85,** 649.
165. Richter, O. (1914). *Sitzungsber. Kaiserl. Akad. Wiss. Wien.* **123,** 967.
166. Ridge, I., and Osborne, D. J. (1969). *Nature (London)* **223,** 318.
167. Roberts, D. W. A. (1951). *Can. J. Bot.* **29,** 10.
168. Rodriguez, A. B. (1932). *J. Dep. Agr. P. R.* **26,** 5.
169. Rosa, J. T. (1925). *Potato News Bull.* **2,** 363.
170. Ross, A. F., and Williamson, C. E. (1951). *Phytopathology* **41,** 431.
171. Rubinstein, B. (1971). *Plant Physiol.* **48,** 183.
172. Rubinstein, B. (1972). *Plant Physiol.* **49,** 640.
173. Rudich, J., Halevy, A. H., and Kedar, N. (1969). *Planta* **86,** 69.
174. Ruge, U. (1947). *Planta* **35,** 297.
175. Saad, S. I. (1953–1954). *Bull. Inst. Egypt.* **36,** 269.
176. Sankhla, N., and Shukla, S. N. (1970). *Z. Pflanzenphysiol.* **63,** 284.
177. Sauls, J. W., and Biggs, R. H. (1970). *HortScience* **5,** Sect. 2, Abstr. 67th Meet., p. 341.
178. Schwartz, H. (1927). *Flora (Jena)* [N. S.] **22,** 76.
179. Seifriz, U., and Urbach, F. (1945). *Growth* **8,** 221.
180. Sfakiotakis, E. M., Dilley, D. R., and Simons, D. H. (1970). *HortScience* **5,** Sect. 2, Abstr. 67th Meet., p. 341.
181. Shaw, F. H. (1935). *J. Exp. Biol.* **13,** 95.
182. Shimokawa, K., Yokoyama, K., and Kasai, L. (1969). *Mem. Res. Inst. Food Sci., Kyoto Univ.* **30,** 1.

183. Shonnard, F. (1903). p. 48. Dept. Public Works, Yonkers, New York.
184. Singer, M. (1903). *Ber. Deut. Bot. Ges.* **21,** 175.
185. Smith, A. J. M., and Gane, R. (1932). *Gt. Brit., Dep. Sci. Ind. Res., Food Invest. Bd., Rep. 1931* p. 156.
186. Smith, K. A., and Russell, R. S. (1969). *Nature (London)* **222,** 769.
187. Splittstoesser. W. E. (1970). *Physiol. Plant.* **23,** 762.
188. Stone, G. E. (1913). *Mass., Agr. Exp. Sta., Bull.* **31,** 45.
189. Stoutemyer, V. T., and Britt, O. K. (1970). *BioScience* **20,** 914.
190. Stuart, N. W., Asen, S., and Gould, C. J. (1966). *HortScience* **1,** 19.
191. Suge, H., Katsura, N., and Inada, K. (1971). *Planta* **101,** 365.
192. Takayanagi, K., and Harrington, J. F. (1971). *Plant Physiol.* **47,** 521.
193. Thornton, N. C. (1933). *Contrib. Boyce Thompson Inst.* **5,** 471.
194. Toole, V. K., Bailey, W. K., and Toole, E. H. (1964). *Plant Physiol.* **39,** 822.
195. Traub, H. P., Cooper, W. C., and Reece, P. C. (1940). *Proc. Amer. Soc. Hort. Sci.* **37,** 521.
196. Turkova, N. S. (1942). *Bull. Acad. Sci. USSR* **6,** 391.
197. Turkova, N. S., Vasileva, L. N., and Cheremukhina, L. F. (1965). *Sov. Plant Physiol.* **12,** 721.
198. Vacha, G. A., and Harvey, R. B. (1927). *Plant Physiol.* **2,** 187.
199. van Andel, O. M. (1970). *Naturwissenschaften* **57,** 396.
200. Van Overbeek, J., and Cruzado, H. J. (1948). *Amer. J. Bot.* **35,** 410.
201. Vesselov, I. J. (1937). *Mikrobiologiya* **6,** 510.
202. Wächter, W. (1905). *Ber. Deut. Bot. Ges.* **23,** 379.
203. Wagner, J. (1970). *Protoplasma* **70,** 225.
204. Wallace, R. H. (1926). *Bull. Torrey. Bot. Club.* **53,** 385.
205. Wallace, R. H. (1927). *Bull. Torrey Bot. Club* **54,** 499.
206. Wallace, R. H. (1928). *Amer. J. Bot.* **15,** 509.
207. Warner, H. L. (1970). Ph.D. Thesis, Purdue University, Lafayette, Indiana.
207a. Weber, F. (1916). *Sitzungsber. Akad. Wiss. Wien, Math.-Naturwiss. Kl., Abt. 1* **125,** 189.
208. Wehmer, C. (1917). *Ber. Deut. Bot. Ges.* **35,** 135.
209. Winchester, O. R. (1941). *Proc. Fla. State Hort. Soc.* **54,** 138.
210. Woffenden, L. M., and Priestley, J. H. (1924). *Ann. Appl. Biol.* **11,** 42.
211. Zhdanova, L. P. (1950). *Dokl. Akad. Nauk. SSSR* **70,** 715.
212. Zimmerman, P. W., Crocker, W., and Hitchcock, A. E. (1930). *Proc. Amer. Soc. Hort. Sci.* **27,** 53.
213. Zimmerman, P. W., Crocker, W., and Hitchcock, A. E. (1933). *Contrib. Boyce Thompson Inst.* **5,** 1.
214. Zimmerman, P. W., Crocker, W., and Hitchcock, A. E. (1933). *Contrib. Boyce Thompson Inst.* **5,** 195.
215. Zimmerman, P. W., and Hitchcock, A. E. (1933). *Contrib. Boyce Thompson Inst.* **5,** 351.
216. Zimmerman, P. W., and Wilcoxon, F. (1935). *Contrib. Boyce Thompson Inst.* **7,** 209.

Chapter 7

Phytogerontological Effects of Ethylene

I. Flower Fading and Leaf Senescence

Leaves and flowers represent lateral terminal appendages modified for food production and reproduction. Although different in appearance and purpose, their death can be accelerated by ethylene, which suggests a common physiological ancestry. One difference associated with these organs is the length of their functional life. Leaves make and translocate food for a growing season or more, while the life of flowers varies from hours to weeks depending in part on the mechanism of pollination. These differences in longevity are possible because of differences in the regulation of juvenility factors. For example, IAA can retard or block the ability of ethylene to induce leaf senescence (11, 12, 95), but accelerate the senescence of flowers (43). The acceleration of flower and leaf sensecence by IAA is due to auxin-induced ethylene production. Hallaway and Osborne (95) reported that 2,4-D esters stimulated the senescence of *Euonymus japonica* leaf tissue in the regions surrounding the auxin-treated areas while the treated areas themselves remained green. These results are interpreted as showing that ethylene diffusing from the treated zones accelerated the senescence of surrounding tissue while the treated areas themselves remained green due to auxin-maintained juvenility. In the case of flowers, auxin inhibition of floral senescence is not known to occur and only the effect of increased ethylene production is observed. In spite of these differences in auxin activity, ethylene-induced floral and leaf senescence is probably much the same.

A. Flower Fading

The first report of ethylene-induced floral senescence—wilting or fading, as it is referred to—was made by Crocker and Knight in 1908 (56). Their studies, as many other aspects of ethylene physiology, were initiated by reports of damage by leaking illuminating gas. They reported that ethylene had a number of effects on carnations. It prevented the opening of young blossoms, caused opened flowers to close, and discolored and withered the petals. The concentrations involved were low. They found 0.5 ppm ethylene for 12 hours closed open blossoms and 1 ppm for 3 days prevented blossoms from opening. This phenomenon was called "sleepiness" and was characterized by a partial closing of the flower due to an incurving or curling of the petals. The condition progressed until the flowers were almost closed. The outer petals were markedly curled and usually showed a drying and darkening along the margins. Since that time, others have examined the dose-response relationships for floral senescence and confirmed the sensitivity of flowers to ethylene. In a time-course experiment, Nichols (168) reported that 0.2 ppm ethylene for 6 hours had no effect on carnations, after 16 hours had a noticeable effect, after 24 hours had a greater effect but the flowers recovered, and after 48 hours had an irreversible effect. Concentrations as low as 0.03–0.06 ppm for 48 hours and 0.12 ppm for 24 hours were found to have threshold effect by Uota (214).

Similar data have been obtained by investigators studying orchid senescence or "dry sepal." Dry sepal is characterized by a progressive drying and bleaching of the sepals beginning at the tips and extending toward the bases. Davidson (58) reported that dry sepal was caused by either 0.1–0.04 ppm for 8 hours or 0.02–0.002 ppm for 24 hours. No response was observed at 0.001 ppm for any length of time. Other investigators using higher concentrations have also reported accelerated orchid senescence (43, 136).

Roses were found to be more resistant to ethylene. After 7 days, 0.026–0.06 ppm had no effect; 0.2 ppm had some effect with accelerated petal drop after 5 days, and complete petal drop after treatment with 5 ppm occurred (121).

Many other species of flowers have been found to be damaged by ethylene, though differences in sensitivity exist. Cottrell (50) reported that carnations, ageratum, larkspur, and zinnias were sensitive species, while more resistant crops were dahlias, China asters, forget-me-nots, lobelia, sweet peas, viola, acacia, and calendula.

Characteristic of ethylene-regulated processes, CO_2 has been found to preserve flowers (74, 75, 168, 214). At a concentration of 10% CO_2, carnations were not damaged by 0.25 ppm ethylene, and 20% CO_2 protected blossoms against 0.5 ppm ethylene (214).

A number of workers have presented evidence that an increase in ethylene production by flowers marked the end of their functional life. Wilkins (226) and Nichols (167) found that a surge of ethylene production and an increase in respiration accompanied the wilting of carnations. Wilkins (226) reported a nitrogen flush through the gas phase delayed senescence and reduced ethylene production. Other flowers such as chrysanthemum, narcissus, and anemone were also examined, but none of these blossoms produced an equivalent surge of ethylene during wilting (167).

An increase in ethylene production also accompanies the fading of orchids (43, 62). Burg and Dijkman (43) observed an increase in ethylene production from orchid flowers when they were pollinated, treated with IAA, or had their pollinia removed. Fading usually became obvious after the rise or increase in ethylene (see Fig. 7-1). They also reported that ethylene production was autocatalytic in a fashion analogous to some fruits since orchids treated with ethylene produced larger amounts of ethylene than controls which were not gassed with ethylene. Thus the events following pollination

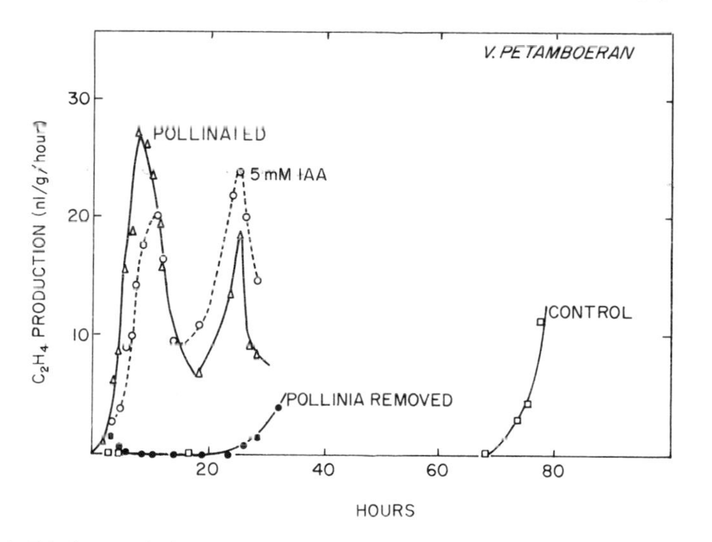

Fig. 7-1. Ethylene evolution by blossoms of *Vanda petamboeran* after pollination (with pollinia intact), application of 5 m*M* IAA in lanolin to the stigmas, or removal of the pollinia. The lower petals of blossoms which were pollinated or treated with IAA began to fade after 8–10 hours, those of emasculated flowers after about 35 hours, and controls after about 80 hours. Removal of the pollinia caused an initial transient production of ethylene, perhaps a wound response, which subsided within 6 hours. [Courtesy of Burg and Dijkman (43).]

that lead to fading of Vanda blossoms were envisioned as including the transfer of auxin from the pollen to the stigma, spread of auxin to the column and lip, induction of ethylene formation in these tissues, diffusion of ethylene from its site of production to adjacent tissues where the gas stimulates its own formation, and finally floral fading as a consequence of the endogenous ethylene production. In the case of flowers where no pollination occurs, removal of the pollinia eventually induces ethylene production and flower senescence, terminating the existence of flowers unlikely to be pollinated.

B. Leaf Senescence

The accelerated yellowing of fruits and leaves is a well-known effect of ethylene and has been reported by a large number of workers. While yellowing of citrus has been taken as an index of ripening by consumers, growers of citrus have been aware that color is not an indication of quality. Early maturing varieties of oranges are edible before they have lost their green pigment. In late varieties such as Valencia, this color change occurs long before fruits become suitable for consumption and they may regreen during the normal harvest period. Fruits of late varieties sometimes fail to lose their green color because of shade or other local conditions and may pass into senility and finally drop from the tree without having shown the characteristic orange color (158).

In an analogous fashion, leaf yellowing is often associated with abscission although a number of examples have been published showing acceleration of abscission at high ethylene levels without concomitant yellowing (233). While abscission or ripening appears to be separate from the yellowing process, factors that retard or block ethylene action such as IAA or CO_2 also prevent or delay yellowing, suggesting that the destruction of chlorophyll parallels but is separate from other processes that occur during senescence (11).

In citrus, the yellowing appears to be due to a decrease in chlorophyll content. At the same time carotenoid content may either increase (early to midseason fruit) or remain the same (late fruit) (158).

The biochemistry of chlorophyll degradation is associated with the destruction of the chloroplast (see Fig. 7-2A–E) and as such represents the

Fig. 7-2. Degradation of chloroplasts in ripening oranges. (A) Micrograph of a chloroplast and mitochondria from a deep green Valencia orange. × 29,750. (B) A developing chromoplast in a yellow Navel orange. (C) Developing chromoplast in a yellow Navel orange. Note the vesiculation of the membranes within the chromoplast. × 22,950. (D) A nearly mature chromoplast from a Navel orange. In the center of the chromoplast note the irregular aggregate of membrane remnants. × 18,700. (E) Mature chromoplast from a ripe Valencia orange. The chromoplast contains several large round osmiophilic globules (o) and a single starch (sg) grain. × 30,600. [Photographs from Thompson (211).]

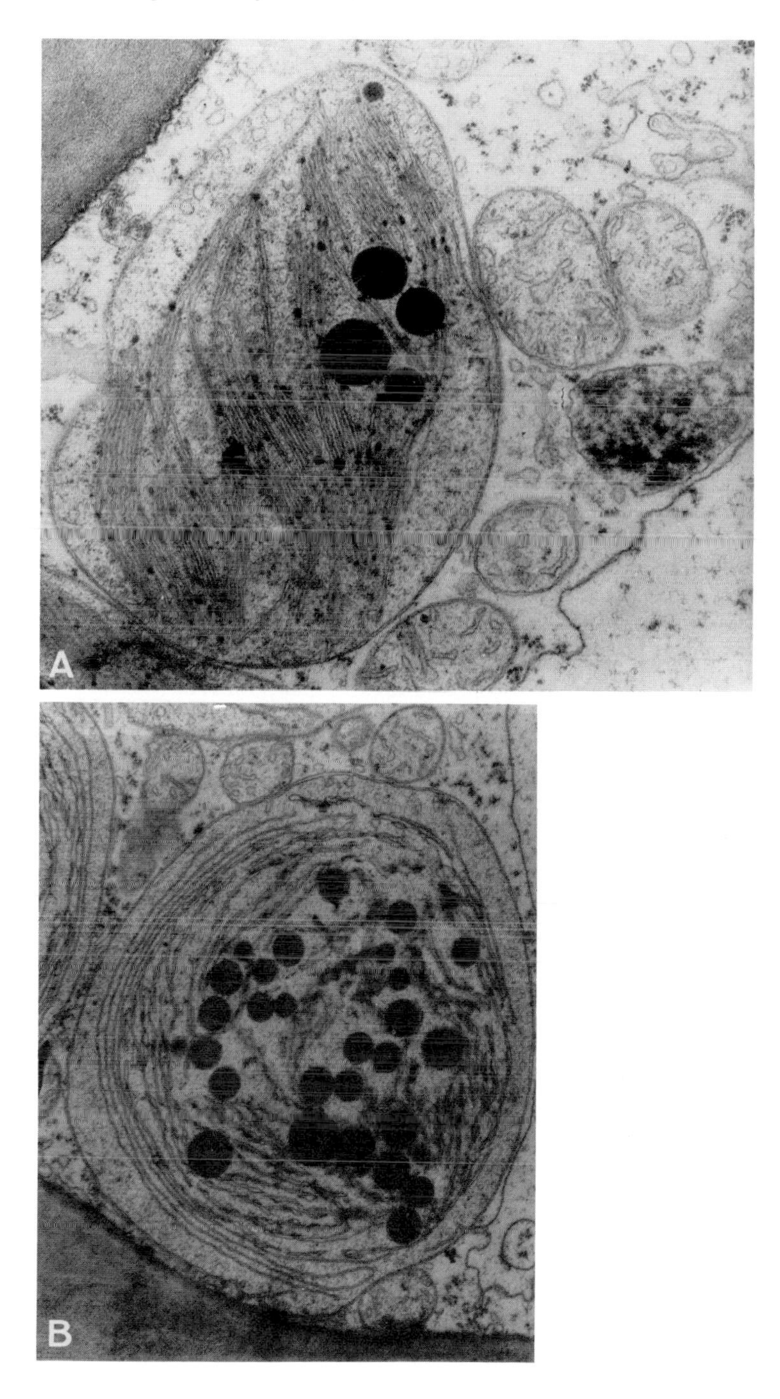
A
B

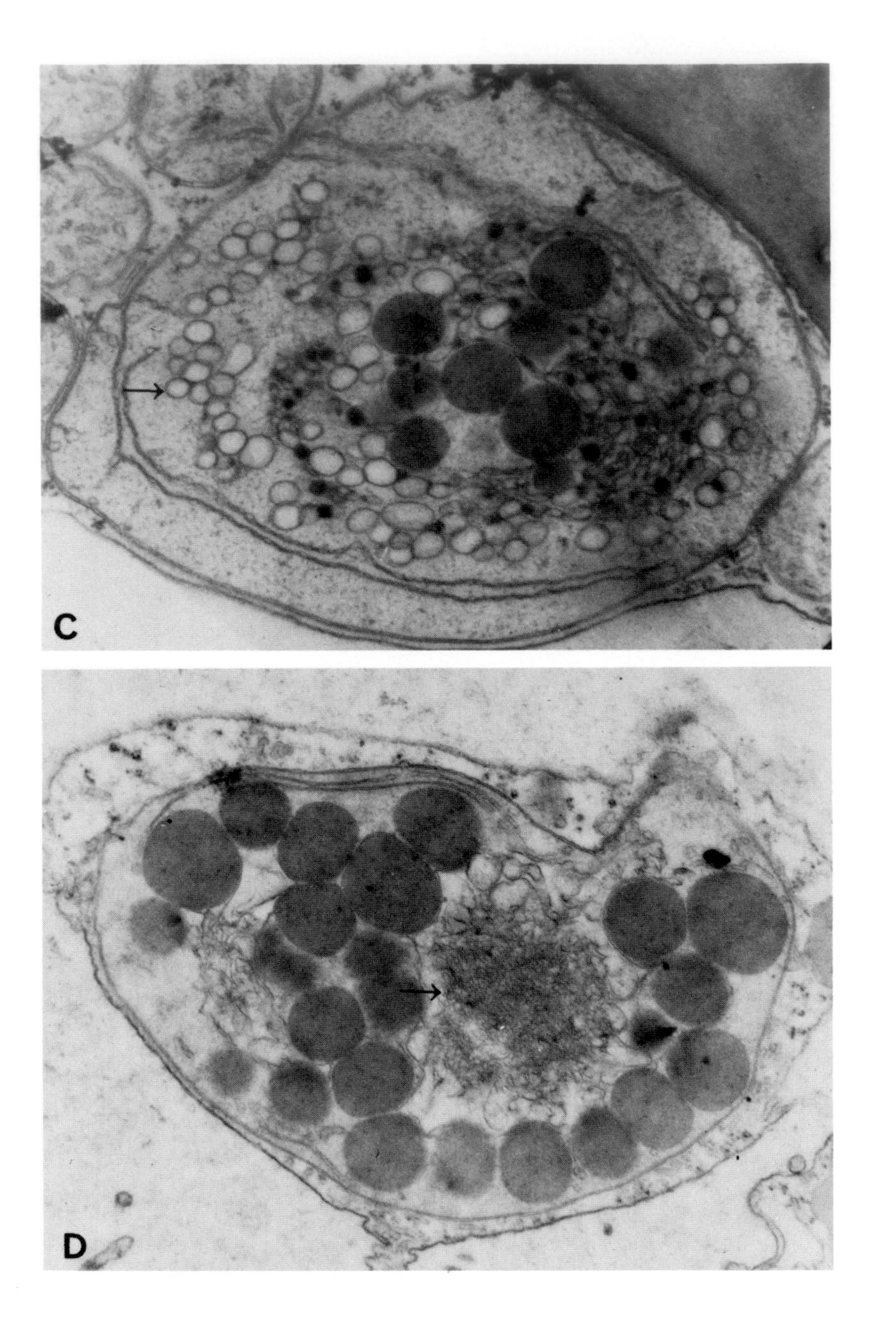

Fig. 7-2 C, D

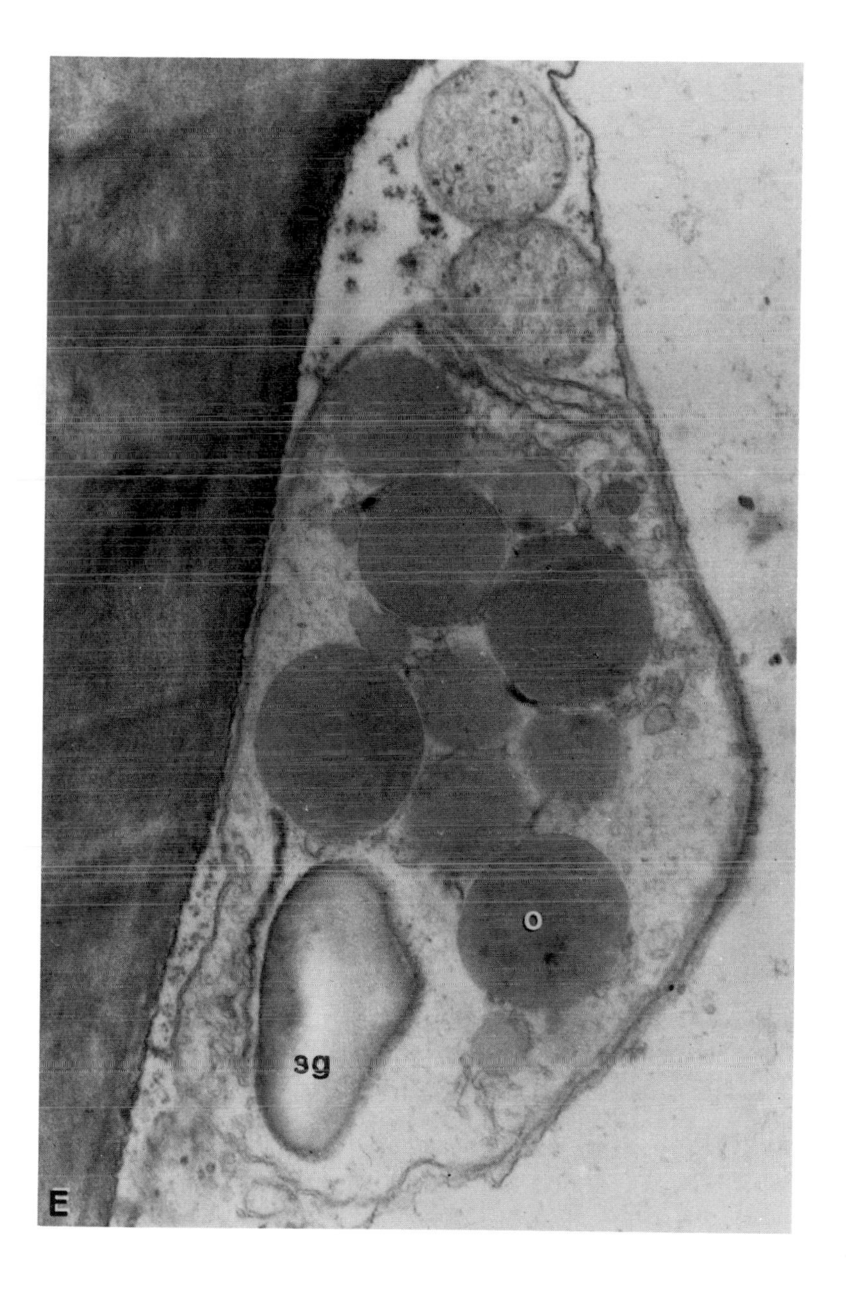

Fig. 7-2 E

action of a variety of enzymes capable of degrading various parts of the chloroplast structure. Looney and Patterson (140) reported an increase in chlorophyllase activity in ripening apples and bananas, suggesting that this enzyme plays a role in chloroplast degradation. However, many other enzymes must also take part in the breakdown of these organelles. While the exact nature of these enzymes is unknown, it appears likely that protein synthesis is involved in their appearance (12).

Senescence also involves the loss of RNA (11, 12, 193), protein (11, 193, 205), and other constituents from the leaf. The effect of ethylene on these degradative processes has been measured and the data show little effect on RNA degradation but a significant effect on the reduction of protein levels (11) in bean explants. A delay in the degradation of these processes occurred when the tissues were treated with IAA. Sacher and Salminen (193) performed similar experiments with other tissues and also observed a reduction in the rate of senescence in the presence of auxin. Ethylene decreased the rate of RNA and protein synthesis in bean pods but had no effect on *Rhoeo discolor* leaf sections. The loss of RNA from excised bean tissue was found to correlate with an increase in RNase activity. The increase in enzyme activity probably involved protein synthesis because the addition of actinomycin D and cycloheximide prevented the loss of both RNA and protein from the excised tissue. Ethylene was found to have no effect on either rate of RNA degradation or the induction of RNase (12).

The data suggest that the ability of ethylene to accelerate the degradation of leaf constituents is indirect. Ethylene appears to regulate some master reaction, for example control of auxin levels, which in turn regulates or maintains the juvenility of plant tissue. Ethylene may play an active role in determining the senescence of some leaves. Osborne and Jackson (119, 170) reported that ethylene production increased as leaves aged and may be responsible for triggering senescence.

Acceleration of leaf senescence by ethylene has received commercial attention in tobacco curing, celery blanching, and storage disorders of lettuce. A number of workers have demonstrated that ethylene increased the yellowing or curing of tobacco (21, 130, 133, 174, 205). Rossi (189) reported that ethylene reduced the time for curing by 40%. He concluded that the quality of the product was not materially changed with regard to aroma, color, and elasticity. Burn, however, was improved. Pfützer and Losch (174) reported that ethylene decreased the nicotine content and also improved the smoking quality of tobacco. Russian workers (21, 130) found that the starch content of leaves was also lowered by ethylene fumigation. Recently plants grown in the field have been treated with Ethrel to accelerate curing prior to harvest (205). Ethrel-treated leaves were found to contain lower levels of starch and protein although little difference in the alkaloid content was observed. Steffins *et al.* (205) concluded that Ethrel can be used to reduce the time required for curing the crop resulting in reduced costs in tobacco production.

Celery blanching is normally accomplished by hilling soil around the stalks or covering the plants with paper or boards. Harvey (106) discovered that ethylene would also decrease the chlorophyll content of the stalks although yellow and red pigments were not affected. Fumigation for 6 days was required to blanch light green varieties; 12 days were required for dark green ones. The action of ethylene was found to be irreversible. Only newly formed leaves could form chlorophyll. Mack and Livingston (142, 143) found that high levels of ethylene (2000 ppm) could cause secondary effects such as splitting of the inner surface of the stalks and pronounced pithiness. The ethylene treatment also caused the celery to become more susceptible to rot. Leaf maturity also played a role in the degree of blanching. Older leaves were found to be more susceptible to ethylene than younger inner leaves.

Head lettuce frequently develops a brown or russet spot disorder which detracts from its appearance and marketability. Rood (188) reported that ethylene increased the development of these spots and that the most severe damage occurred at 7°C, less at 15°C, and none at 0°C. However, apple volatiles were used as the source of ethylene, which suggests that other gases may be a critical factor. Lipton (138) reported that lettuce maintained in ethylene-free air (brominated charcoal filter) still developed the russet spots. He suggested that the russet spots might be associated with high CO_2 levels or some other factor.

II. Fruit Ripening

A. Historical

"It was shown by direct trial that the emanations from oranges stored in a chamber were found to have the effects of bringing about a premature ripening of bananas if these gases were passed through a chamber laden with this fruit. The practical lesson indicated by these experiments is that separate storage is desirable for citrus fruits and bananas when they are being transported for long distances by sea." This report of Cousins in 1910 (51) was the first observation that fruits produced an emanation that controlled their rate of maturation and is the first suggestion that ethylene might be involved in fruit ripening.

Although the active ingredient wasn't known, man has been using ethylene in the form of smoke to promote ripening of a variety of fruits. Ryerson

(192) reported that an oriental procedure to remove the astringency of persimmons was placing them in an upright bamboo cylinder of open weave in the center of a large earthen jar in which the fruits were packed. A large stick of incense was burned in the bamboo cylinder and the smoke brought about the removal of the astringency.

Von Loesecke (219) reported that dried cakes of cow dung, coke, charcoal, or cocoanut fiber were burned in earthen pots or storage sheds to ripen bananas in India. Another practice was to bury bananas in underground pits plastered with mud or covered with green leaves of certain plants.

A similar technique was used in California to facilitate the degreening of citrus. According to Chace (44), in the early days of orange shipments from California, kerosene stoves were used in packing sheds and railroad cars to avoid the danger of frost injury. After one or two fires in transit, the railroads put a stop to their use in cars. Shippers stated that at first they had no purpose other than to keep the fruit warm, but they eventually noticed that fruit heated in this way developed a better color than those packed without heating. Kerosene stoves were used solely for the reason that they were available and afforded the cheapest source of heat. Some enterprising but unfortunate individual built a heating plant using steam in the place of the cumbersome kerosene stoves and discovered to his dismay that he could get no better color in this way than if the fruit were not heated. Very soon the use of kerosene stoves became universal in citrus packing houses. Gas and the exhaust from automobiles and motorcycles were also used. In 1912, Sievers and True (196) proved that the combustion gases were responsible for the increase in fruit color. They examined the effect of various components, including CO_2. They noticed but failed to grasp the potential usefulness of the fact that fruits failed to ripen when high concentrations of CO_2 were used. Carbon monoxide was also tried but found to be a menace to the workmen. By separating the products of combustion they found that unsaturated gases were the most effective and that ethylene was more effective than acetylene. In 1923, F. E. Denny, who was working in this laboratory, took a public patent on the process.

In 1925, Harvey (107) found a promotion of ripening in bananas, pineapples, dates, persimmons, tomatoes, pears, apples, melons, mangoes, avocados, papayas, and jujubes, thus establishing the fact that the action of ethylene was widespread among a variety of fruits.

Kidd and West (125) in England first noted that the respiration of apples increased during ripening and that this increase or climacteric, as they called it, took place on and off the tree. Later (126) they found that ethylene increased fruit ripening and the climacteric and confirmed the finding of Cousins that a similar effect was obtained with emanations from other apples. These emanations also caused growth deformation in seedlings, suggesting that ethylene was the active ingredient. Two years later Gane (85) demonstrated that ethylene was among the volatile emanations of apples. By this time it was clear that ethylene was a ripening hormone, namely, a substance produced by fruits which had the specific biological action of accelerating its maturation and senescence.

Because of the economic significance of fruits, this area of ethylene physiology has been the most intensively studied and a number of excellent reviews on this subject have appeared (29, 32, 102, 144, 157, 176, 187, 204) as well as a two-volume book (113, 113a).

B. Hormonal Regulation of Ripening

1. Ethylene

There is no exception to the observation that ethylene can promote ripening as long as the tissue is in a receptive state. Removal of ethylene by ventilation, absorption, or hypobaric storage will delay ripening of most fruits, and there is no reason to believe that ripening can occur autonomously without the benefit of ethylene. On the other hand, there is no reason to believe that ripening is controlled by ethylene alone. In other words, fruits become increasingly sensitive to ethylene as they age. This is often shown by the observation that less ethylene, either in amount or length of exposure, is required to promote ripening in mature fruits than in immature fruits (39, 100, 103, 178, 221).

Ripening involves more than an increase in sensitivity to ethylene. At the time sensitivity increases, ethylene production may also increase. One of the generalized statements arising from research with fruits is that an increase in ethylene production often precedes an increase in respiration in climacteric fruits such as apples while no increase in respiration occurs in nonclimacteric fruits such as citrus. However, exceptions in both classes of fruits occur. Even though there is an increase in respiration of bean fruit during maturation, there is no increase in ethylene production (222). The opposite was observed in *Vaccinium* fruit. Here an increase in ethylene production but no increase in respiration was reported (79). In other words, the relationship between changes in ethylene production and respiration can vary from fruit to fruit and there is little advantage in trying to draw analogies between fruits of varying generic and structural origin. Each fruit or group of very similar fruits is best studied as a separate entity in order to reduce confusion and prevent the drawing of erroneous analogies.

2. Tree Factors

Tree factors are unidentified compounds produced by the parent plant which control ripening. Generally speaking, these factors delay maturation and senescence. As long as the fruit is attached to the tree, ripening is inhibited. The first report of a tree factor was by Wilkinson (227). He reported that apples picked later showed a displacement in the climacteric peak indicating that some factor from the tree was keeping the fruit juvenile. Meigh *et al.* (153) observed that the apple tree factor was an inhibitor of ethylene synthesis. They noted that ethylene production was higher in detached fruit than in those attached to the tree.

A tree factor was also identified in pears (104). As long as pears were attached to the tree the increase in respiration took place slowly. Hansen and Hartman (104) suggested that it might be useful to harvest and store fruit with sections of branches still attached.

Burg and Burg (38, 39) have also identified tree factors in bananas and mangoes. In mangoes, which do not ripen on the tree, they found that a steam girdle applied to the fruit stalk directly above the point of attachment caused fruit to ripen. However, if the steam girdle was applied to the stalk about 12 inches above the point of fruit attachment with a cluster of leaves remaining between the girdle and fruit, ripening proceeded at a reduced rate after a considerable delay.

3. *Auxin*

Auxin is known to inhibit the ability of ethylene to promote degreening and abscission of leaves. Similarly it has been shown that auxin can prevent degreening and softening of bananas (147, 216, 218, 220). Marth and Mitchell's (147) initial work demonstrated that 2,4-D both promoted and inhibited ripening. Similarly, Hansen (101) demonstrated that 2,4-D enhanced ripening of pears. It is now known that the promotive effects were due to an enhancement of ethylene production. Movement of auxins into bulky fruit tissues is slow and limited unless the auxin is vacuum-infiltrated. Vendrell (218) demonstrated that ripening was delayed when auxin was infiltrated into the tissue but not when applied as a dip. The advancement in ripening when whole fruit was dipped was attributed to the uneven distribution of auxin which caused a localized production of ethylene. While ripening was delayed in the peel, which contained high levels of auxin, ethylene diffusion from these cells triggered ripening in the surrounding pulp.

Vendrell (216) found that 2,4-D delayed degreening and softening in bananas but did not prevent a climacteric-like rise in respiration. However, fruits treated with 2,4-D recovered from an ethylene treatment and the rate of respiration and ethylene production declined and the tissue took on the characteristics of normal green fruit. He concluded that 2,4-D caused a reversion of bananas to a more juvenile state.

Hale *et al.* (91) reported that benzothiazone-2-oxoacetic acid, a compound with auxin-like activity, delayed the ripening of grapes and that ethylene partially reversed the effect of the auxin analog.

4. *Gibberellin*

Gibberellin has also been found to delay ripening and to block the ability of ethylene to act. Dostal and Leopold (63) reported that gibberellin delayed ripening of tomatoes. Ethylene was able to overcome the block against respiration, but only partially against the change in skin pigments. Gibberellin also delayed the ripening of bananas (217). Here again, as was the case with 2,4 D described above, the effect was directed primarily against changes in pigmentation, as opposed to effects on respiration.

C. Role of RNA and Protein Synthesis

Ripening is controlled by enzymes and the process probably requires the synthesis of RNA and protein. The importance of nucleic acids was first suggested by experiments on the effect of high energy radiation on fruit ripening (1, 2, 37, 149). Maxie and his colleagues and others reported that γ radiation of young preclimacteric fruit delayed and altered normal fruit ripening. It is known that the probable target of irradiation was the nucleus, causing damage to DNA structure. This dimerized and altered DNA would no longer serve as a normal template for RNA biosynthesis, which in turn caused an inhibition of normal maturation.

Confirmation for this idea arose from experiments by Holm and Abeles (110). They showed that ethylene increased the incorporation of ^{32}P into the RNA of green, immature bananas. This observation was confirmed by others studying the incorporation of [^{14}C]uridine into apple RNA (116). Marei and Romani (146) demonstrated that ethylene increased RNA synthesis in the fig and that the levels of all classes of RNA were involved. However, Sacher and Salminen (193) reported that ethylene had no effect on uridine incorporation into bean endocarp tissue. Using actinomycin D, Rhodes *et al.* (184) found that RNA synthesis was probably important in the development of the malate effect in apples.

Early workers pointed out that the protein content of apples and other fruits increased during ripening (103, 112, 128). Some of this protein was probably enzymatic. As early as 1928, Harvey (107) demonstrated that the protease content of pineapples increased following ethylene treatment. Since then, other workers demonstrated an increase in other enzymes during ripening, including catalase (22) and lipoxidase (153). However, the function of many of these enzymes is unknown and little is understood concerning the enzymology of ripening. An exception to this statement is the report that malate and pyruvate decarboxylase increase during ripening (120, 180, 184, 210). These enzymes probably participate in the increase in CO_2 production associated with the climacteric. An increase in pectinase has been reported in pears and avocado (150). This enzyme possibly contributes to softening.

Considerable evidence has accumulated that ripening and enzyme induction depend on protein synthesis since a number of workers have demonstrated that cycloheximide blocked both processes (36, 80, 84, 116, 151, 184) and that ethylene enhanced the incorporation of radioactive amino acids into protein (36, 80, 116, 146).

D. Ripening of Specific Fruits

Botanically, a fruit is a mature ripened ovary, or ovaries, together with any closely associated parts. The wall of the ovary is subdivided into three sections, the exocarp, which includes the epidermis, the mesocarp, or middle portion, and the endocarp, which is the inner cells of the ovary. Fruits are generally classified into two groups, dry or fleshy. The development of dry fruits such as legumes, capsules, and grains is not considered here. Fleshy fruits may develop by cell enlargement without cell division. This procedure usually occurs in smaller fruits such as *Rubus* and certain *Ribes* species. Cell division followed by cell enlargement occurs in *Prunus*, *Cucurbita*, and *Malus*. Continuous cell division occurs in avocado.

If the entire ground tissue is edible, the fruit is a berry, e.g., grape and tomato. The hesperidium (citrus) is closely related to a berry and the juice sacs are developed from the endocarp. If the ovary wall matures into a fruit with a stony endocarp and fleshy mesocarp, the fruit is called a drupe, for example peach and olive. The word drupe was derived from the Latin *druppa*, for overripe olive. A fleshy fruit derived from an inferior ovary such as the apple is thought to be derived from both floral tube and carpellary tissue.

The English word ripen is derived from the Old Saxon *ripi*, meaning to reap or gather. It has no precise scientific meaning, but the biochemical and physical changes that are associated with the visible process of ripening include changes in color, sweetness, astringency, and aroma. The simplest definition of ripe is the time the fruit is ready to eat.

1. Avocado

The two important Californian varieties of avocado (*Persea americana* Mill.) are Fuerte and Hass (30). Under suitable conditions avocado can be stored on the tree beyond the normal maturation season since ripening does not take place as long as the fruit is attached to the tree. Ripening of the Hass variety is associated with a skin color change from green to black, while the Fuerte retains its green color even when soft. In Florida important varieties are Lula, Booth 8, Waldin, Booth 7, and Pollock. These and the Californian varieties are products of three major races of avocado, Mexican, Guatemalan, and West Indian.

Botanically the avocado is a berry consisting of a single ovary and seed. Its development is different from some other fruits in that cell division occurs during the entire growth of the fruit. In some cases cell enlargement stops when the fruit reaches 50% of its full mature size and cell division accounts for the continued growth.

The outstanding feature of avocados is their high lipid content. About 4–25% of the fresh weight is fat, depending on the race. Protein content is also high, about 1.5% compared with less than 1% for most other fruits. Despite the high fat content the evidence does not support the idea of lipid utilization as respiratory substrates during the course of the climacteric. Sugars disappear with ripening and so do insoluble pectins but it is not known if these account for all of the energy requirements during respiration. The trends in O_2 consumption and CO_2 production are similar to those shown by other climacteric fruit. Since avocados are tropical fruit they are subject to chilling injury and can not be stored below 5°C. Temperatures above 25°C also prevent normal ripening. Similar to other fruits, the storage life can be extended with low O_2 (5%) and high CO_2 (10%) levels.

Biale and Young (30) and others (190) have examined the role of phosphorylation during the climacteric. They have found that oxidative phosphorylation progresses actively and improves during the rise in respiration. In addition, the structure and function of ATP-forming machinery remains fully intact and operative during the climacteric. The basic reason for the rise in respiration remains unknown, although Biale and Young (30) have shown that thiamine pyrophosphate (TPP) may play a significant role. The oxidation of malic acid and α-oxoglutaric acid was increased with TPP in preclimacteric mitochondria but not in climacteric ones. They suggested that respiration may be limited by concentrations of TPP and changes of TPP may occur during the climacteric.

Ethylene functions to accelerate the rate of softening, respiration, and flavor changes and skin coloring. Although the enzymology behind these changes is unknown, it has been shown that protein, presumably enzyme synthesis, increases during ripening with the greatest rate of protein synthesis occurring during the rise in respiration. Associated with the increase in respiration, protein synthesis, and softening were a decrease in pectin esterification and protopectin and an increase in soluble pectin.

2. *Banana*

This tropical fruit represents the second largest fruit crop produced in the world, the grape being the largest (172). Most edible bananas are derived from *Musa acuminata* and *M. balbisiana*. While there are many varieties produced, the export trade is limited to Gros Michel and Cavendish varieties. The Cavendish have recently replaced Gros Michel as the principal export type, largely because Cavendish cultivars are resistant to Panama disease caused by *Fusarium*. Bananas represent the ovary of a single flower developed parthenocarpically. Development of the edible portion of the fruit is associated with the growth of the pulp from inside the peel.

Similar to other fruits, removal of the banana from the tree results in accelerated ripening. Burg and Burg (39) reported that apple bananas at the light full three-quarter stage remain green on the tree for 7 weeks while those that were removed ripened in 1 or 2 weeks. Bananas for export are harvested green and generally treated with ethylene by local distributors before they are sold. The time between cutting and endogenous ripening is influenced by a number of factors including the conditions under which the plants are grown, the time of harvest, and the variety grown. A 90-day-old Cocos (a Gros Michel sport) ripened in 17 days, while a 120-day fruit ripened in 9 days. Corresponding figures for Valery (a Cavendish variety) were 21 days and 14 days.

Internal levels of ethylene in harvested fruit are around 0.1 ppm and concentrations equal to or greater than this must be applied to the fruit to promote ripening. The time required to degreen bananas varies from 16 days with 111-day fruit treated with 0.1 ppm ethylene to 2 days for those treated with 5 ppm ethylene. Gas storage (high CO_2, low O_2) has been used to delay banana ripening. In addition, storage in polyethylene bags, bags filled with ethylene absorbers, and hypobaric storage conditions have also been employed.

Bananas are climacteric fruit and show increased respiration during ripening. The response to ethylene may be rapid and occur within less than an hour (183). Removal of ethylene can result in a return to the basal rate of respiration and reapplication of ethylene can result in additional bursts of CO_2 production (183). Little is known about the control mechanisms that result in the rise of CO_2 production and O_2 uptake. Brady *et al.* (35) examined the role of cellular compartmentalization during the respiratory rise. They found that ethylene increased the rate of respiration in whole fruit and peel segments although leakage did not increase until some time later. Their data discount the suggestion that increased respiration was due to decreased decompartmentalization of the tissue.

The conversion of starch into sugar occurs during ripening. The pulp of fresh green fruit is 20–25% starch and after a week of ripening it is almost completely hydrolyzed into sugar. The total carbohydrate content of the fruit decreased 2–5% during ripening and is presumably utilized during respiration. Starch in the peel is also converted into sugar during the ripening process. Other changes include the solubilization of pectin and a decrease in cellulose and hemicellulose. Degreening of the peel begins shortly after the climacteric peak and is completed 3–7 days afterward. Volatile production is complex and involves over 200 components. The production of these compounds starts approximately 24 hours after the climacteric. The biochemistry of these compounds is complex, though hydrocarbon amino acids such as leucine, isoleucine, and valine appear to be rapidly incorporated into the volatiles. This observation makes one wonder if there is any relationship between methionine incorporation into ethylene and the incorporation of other amino acids into other hydrocarbon gases.

The reduction of tannins also accompanies ripening. Tannins are phenolic substances responsible for the sensation of astringency through a reaction with protein in the mouth. They occur primarily in the latex vessels of the pulp and skin and are polymerized during ripening. During ripening the pH of the tissue drops from 5.4 to 4.5. This is caused by increased levels of a variety of organic acids, including a large number of oxo acids such as pyruvic, glyoxylic, and oxalacetic acids.

3. *Citrus Fruits*

The citrus of commerce may be classified into four groups, all of which belong to the genus *Citrus* (212); *C. sinensis*, sweet orange, of which 100 varieties including Valencia, Pineapple, Hamlin, Parson Brown, Jaffa, and Shamouti are grown commercially; *C. aurantium*, the bitter or sour orange, used as rootstock; *C. reticulata*, the mandarin orange, grown for marmalade; and others, *C. grandis*, grapefruit, *C. limonia*, lemon, and *C. acida*, lime.

The citrus fruit can be subdivided into flavedo, albedo, and carpel segments. The flavedo is the pigmented outer portion of the skin and contains oil glands from which essential oils are obtained. The yield of oil varies from 1.8 to 9.7 pounds per ton of peel residue at the cannery. The albedo is the white spongy portion of the peel. The edible portion consists of the carpel segments separated by carpel walls. Within the carpel are many juice vesicles developed from hairlike papillae on the segment membrane. These vesicles enlarge as the fruit develops and are attached to the membrane with threadlike stalks through which water and other solutes are translocated from other parts of the plant.

Citrus differ from typical climacteric fruit in that an increase in respiration does not normally accompany other changes in fruit color and quality after the fruit has been removed from the tree. Because of this citrus are generally recognized as nonclimacteric fruit. However, ethylene can induce an increase in respiration in citrus at various stages of maturity and an increase in both respiration and ethylene production have been noted when young oranges were removed from the tree (70). The mechanism of ethylene-induced respiration is atypical of many ethylene effects because CO_2 has been shown to mimic, not block, ethylene action on respiratory changes in lemons (230).

Changes in pigmentation do not necessarily accompany ripening in terms of edibility, and in fact certain oranges such as Valencias can regreen after they have turned orange. Biale (28) pointed out that regreened fruit are more resistant to ethylene than nongreened fruits. Ethylene has no effect on the characteristics of the edible portion of citrus fruit, and its action as a ripening agent seems to be limited to controlling the rate of chlorophyll degradation and volatile production (169).

During development of oranges the sugar content of the juice increases until 75% or more of the total soluble solids consist of sugar, primarily sucrose, glucose, and fructose in a 2:1:1 ratio. The acid content of most citrus fruit is citric acid, which results in a juice pH of 2 for lemons and a pH of 3 or more for ripe oranges. Flavenoids play an important role in determining the taste or bitterness of citrus. The bitter taste of bitter oranges is naringin, which can be detected orally at a concentration of 20 ppm. The total flavenoid content of citrus increases during early stages of development and then remains constant. The relative concentration of these flavonoids decreases during maturation due to dilution as the size of the fruit increases.

Unlike climacteric fruit, citrus do not undergo any rapid physical or chemical changes after they are removed from the tree except to lose moisture during storage. Because of this, fruit is generally left on the tree unless they are losing moisture due to a water stress situation. The commercial use of ethylene is limited primarily to degreening the peel and making the fruit more acceptable for market.

4. Grape

Grapes represent the single most important fruit grown by man in terms of economic value (173). About 25 million acres of land are devoted to its culture of which 40% are used for wine making, 59% for fresh fruit, and 1% for raisins. The important species, *Vitis vinifera*, consists of many white and black varieties.

The respiratory curve of grapes is high during the initial stages of development, then decreases, and increases again during the change from small green berries into larger, colored, softer fruit. Respiration then levels off as grapes continue to swell, store carbohydrates, and develop their characteristic odor. Grapes have been classified as nonclimacteric because no further change in respiration occurs after the fruit has been removed from the vine. However, the classification may be inappropriate if one takes the point of view that all of the significant changes have taken place before the respiratory studies have been made. Studies on the role of ethylene in grape development are few in number. Hale *et al.* (91) reported that ethylene hastened the ripening of Doradillo grapes when it was applied for 10 days starting midway through the slow growth phase. Ethrel applied to Shiraz grapes had the same effect. Ethrel has also been used (223) to accelerate flower and berry abscission.

5. *Mango*

The mango (*Mangifera indica* L.) is an important fruit in tropical portions of the world (114). Botanically, the fruit is a drupe. The edible portion is the mesocarp and the endocarp modified into a tough leathery covering of the seed.

The mango is a climacteric fruit and an increase in respiration immediately precedes full maturity. Associated with the increase in respiration is the conversion of starch into sucrose and a fall in acidity. In addition to changes in metabolic reserves, the skin color changes from green to yellow and the fruit softens.

Burg (38) has reported that mangoes contain 1.87 ppm ethylene while on the tree and that this concentration drops to 0.084 ppm after harvest. The fact that mangoes do not ripen in the presence of high concentrations of ethylene on the tree is explained by endogenous factors produced by the tree which prevent endogenous ethylene from acting and delaying ripening.

6. *Melon*

The muskmelons or cantaloupes share the genus *Cucumis* with cucumbers and other nondessert fruit (177). Melons (*Cucumis melo* L.) and watermelons (*Citrullus vulgaris*) are generally monoecious vines but may also be andromonoecious. They are classified as pepos or inferior berries; they have the edible flesh derived from the placentas as in the watermelon or have a central cavity as in the melons and have the flesh derived from the pericarp.

Melons are climacteric fruit. An increase in respiration and ethylene production occurs after harvest and ethylene is capable of inducing these increases. The development of the climacteric in melons is very closely associated with the time that has lapsed following the pollination of the fruit. If fruit are harvested at various times following pollination and held at a constant temperature, they all show increased respiration and ethylene production at the same time. Apparently, factors that control the sensitivity of melons to ethylene or increase the rate of ethylene production are controlled by some precise internal chronometer.

Pigment changes during ripening are due to a decrease in chlorophyll and an increase in carotenoid synthesis. These pigment changes start taking place 10 days before the climacteric and accelerate as the rate of ethylene production increases. The predominant carotene is β-carotene.

Ripening of melons is associated with an increase in sugar content of the flesh. Melons are among the sweetest of fruits and as much as 16% of the juice may be sugars. At the same time the flesh softens and insoluble pectins are solubilized.

The flavor of melons is determined by the time the fruit are harvested from the vine. Cantaloupes harvested before the abscission layer is fully developed will never have the same flavor as those left for the full period.

7. Olive

The fruit of the olive (*Olea europaea*) is a drupe and is grown primarily for its oil (72). Oleuropein is responsible for the bitter taste of fresh olives and must be removed with NaOH before the fruit can be eaten. For table use fruits are generally treated with 2% NaOH, washed with water, and then stored in a NaCl brine, during which time they ferment.

Maxie *et al.* (148) reported that detached fruit do not show a climacteric rise in respiration although they may undergo one while still on the tree. Ethylene had little or no effect on respiration or ripening as measured by an increased softening or development of red color. However, it did increase the destruction of chlorophyll.

8. Persimmon

Persimmons belong to the genus *Diospyros* (117). The commercially important species is *Diospyros kaki*, or the Japanese persimmon, and over 1000 varieties are known. The varieties are subdivided into classes based on astringency and change in coloration during pollination. The astringency of the persimmon is due to water-soluble tannins present in tannin cells. The astringency is simply and rapidly detected by pressing freshly cut persimmon fruits against dry filter paper previously dipped in 5% ferric chloride solution. The size and density of the tannin cells differ in different varieties. Generally the nonastringent varieties have smaller and a fewer number of of tannin cells. The main component of the tannin is called diospyrin, a leucodelphinidin-3-glucoside. Removal of astringency is associated with coagulation of the tannin and subsequent water insolubility. Other changes during growth and maturation of the fruit include increases in sugar content and carotenoids.

A number of workers (31, 69, but see 59) have reported that respiratory changes during the maturation of persimmons was of the climacteric type. Davis and Church (59) and others (66, 69, 171) indicated that ethylene increased respiration in addition to promoting softening and color changes in the fruit. They also found that ethylene removed astringency of the fruit but it is doubtful that the effect is a typical ethylene-regulated phenomenon because a number of workers (171, 200) have shown that CO_2 will also remove astringency but leave the fruit firm and uncolored. This suggests that ripening is a normal ethylene effect while removal of astringency is not. The latter idea is confirmed by the observation that unrelated compounds such as alcohol and ethylene chlorohydrin (171) also remove astringency.

9. Pineapple

The pineapple (*Ananas comosus*) is different from most other fruits because it is a composite or collective fruit and represents a fusion of a number of smaller fruits (67). The pineapple is harvested 18–22 months after the plants are set out and each plant produces one fruit, called the plant crop fruit. Approximately 12 months later a second smaller crop, called the ratoon crop, is harvested.

The pineapple is classified as a nonclimacteric fruit because the rise in respiration associated with ripening does not show a maximum but instead increases gradually during softening and yellowing and finally senescence. However, ethylene increases respiration before the fruit ripens naturally and in addition appears to accelerate the degreening of the tissue slightly (68). In addition, other workers (66, 107) found that it did increase flavor and sweetness. Although the function of the enzyme is unknown, protease activity is usually absent from the flesh of the fruit following flowering but increases subsequently and remains at a high level until just before fruit harvest (89). Methionine showed an interesting trend during development. Only insignificant amounts of the amino acid were present until fruit ripening was initiated. Methionine then rose to a high level, becoming one of the major free amino acids in the juice (89). Since this amino acid is the precursor of ethylene, and ethylene production rises near the end of the functional life of fruits, it would be interesting to learn if a similar rise in methionine content occurs in other fruits.

10. Pome Fruits

Pome fruits are produced by members of the Rosaceae and include the apple, pear, quince, and medlar (115). Only the apple (*Malus sylvestris*) and pear (*Pyrus communis*) will be discussed here.

Apples and pears will ripen while still attached to the tree and optimal eating quality in apples may be obtained this way. However, commercial practice is aimed at harvesting fruit before ripening and storing the fruit for a period of time before sale.

Kidd and West (125) in 1925 first reported the characteristic rise in the rate of respiration, measured as CO_2 production, when apples were detached from the tree and stored at normal ripening temperatures. This rise was found to occur for fruit detached or left on the tree and is therefore a normal part of fruit maturation. They later (126) showed that ethylene increased respiration and that ripening apples produced additional quantities of ethylene, thus establishing the autocatalytic nature of ethylene production in ripening. The rate of ethylene production varies from variety to variety and is closely associated with the keeping qualities of apples (166).

Apple mitochondria show an active coupling between O_2 uptake and inorganic phosphate esterification, giving P/O ratios close to theoretical values. No uncoupling of oxidative phosphorylation occurs as the fruit ripens except when the fruit becomes overripe, so it is unlikely that the climacteric is due to uncoupling of oxidative phosphorylation. However, there is some evidence that respiratory control by ADP might be changing in mitochondria prepared at successive stages during the climacteric. The addition of glucose and hexokinase (which would cause a regeneration of ADP from ATP) had an increasingly stimulatory effect on the activity of the mitochondria as the climacteric progresses. The increase in respiration may also be due to an increase in the numbers of mitochondria in ripening tissue. However, evidence in favor of this idea has not been obtained. Other proposed regulatory mechanisms include an increase in critical cofactors such as NAD and ADP. Again no positive evidence for these interpretations has been obtained.

Changes in metabolic activity are also due to changes in enzyme levels in apple tissue during senescence. Among the enzymes showing increased activity are NADP-malic enzyme, pyruvate decarboxylase, lipoxidase, chlorophyllase, acid phosphatase, and ribonuclease.

During the climacteric the malate and pyruvate decarboxylating systems increase in activity and levels of malic acid decrease in both peel and pulp tissue. These enzyme systems contribute both reduced pyridine nucleotides required for synthetic functions as well as CO_2 which accounts for the high respiratory quotient in ripening fruit.

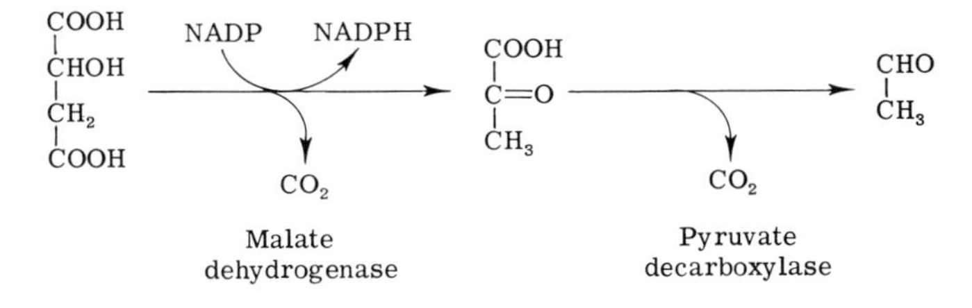

Other changes during ripening include a conversion of starch into sugar, an increase in the waxy surface of the skin, and the production of over 120 volatile products. Among the products is farnesene, and it is now thought that it or its metabolic products contribute to storage disorders called superficial scale.

Pigment changes include the breakdown of chlorophyll and buildup of carotenoids. Anthocyanins, primarily cyanidin-3-galactoside, account for the red color. Red light is required for its formation and postharvest light treatments have been used to increase reddening of apples.

The texture of apples is derived from pectins, hemicelluloses, cellulose, pentosans, and hexosans. The shortening of chains of polygalacturonic acid of which pectin is composed does not seem to be a part of the ripening process since pectinase appears to be absent from the apple. According to Hulme and Rhodes (115), softening is due to enzymatic breakdown of protopectin. Protopectin consists of chains of polygalacturonic acid cross-linked in various ways with Ca^{2+} and Mg^{2+} ions, hydrogen bonding between hydroxyl groups, and possibly by methylene bridges. However, a protopectinase has not yet been identified from apples.

11. Soft Fruits

These fruits are grouped together primarily on the basis of their eating qualities as opposed to any botanical considerations. This group consists of the berry fruits of the genus *Rubus* (blackberries, raspberries, loganberries, boysenberries) and *Morus* (mulberry). The currants consist of members of *Ribes* (gooseberry and currants), and *Vaccinium* (blueberry) and strawberries (*Fragaria*) are also included. Data on the role or effect of ethylene on growth and development of members of the berry fruits are absent from the readily available literature.

The Chinese gooseberry (*Actinidia chinensis*) is considered to be a nonclimacteric fruit (229). Ethylene has been found to increase ripening and respiration of these fruits (99, 229) and ethylene production appears to regulate the normal maturation of fruit in storage (209). Ripening in the Chinese gooseberry consists primarily of a decrease in starch and an increase in sugar. No color changes are apparently involved in maturation.

The three important species of *Vaccinium* are *V. angustifolium* (lowbush blueberry), *V. corymbosum* (highbush blueberry), and *V. macrocarpon* (cranberry). During the development of these three species the greatest level of ethylene production occurred immediately after pollination. This agrees with known stimulating effects of pollination on ethylene production. However, following the low production of ethylene in the early green stage of all three species, there is a rise in production of the gas long before the fruit is ripe. This would be in agreement with the present concept of ethylene as a ripening hormone but would also mean that initial effects of ethylene on ripening in these species is at a rather early stage of fruit development.

However, even though there was a rise in ethylene production and maturation of color, no increase in respiration was observed (79). Similarly, other workers have reported no increase in respiration when cranberries (65, 81) were treated with ethylene. All members of the blueberry group are therefore considered nonclimacteric.

The change in cranberry color is a complex process involving degreening and anthocyanin synthesis. Similar to apples, ethylene can degreen the fruit in the dark (81) but has little or no effect on anthocyanin formation unless light is present (52). Ethylene has no other effect on fruit maturation, i.e., sugar or acid content.

Ethylene has been reported to have no effect on the ripening of strawberries (87, 175). On the contrary, CO_2 has been reported to increase color development of these fruits (71). Strawberries are also considered to be nonclimacteric fruit.

12. Stone Fruits

Stone fruits are grouped together because they are all members of the genus *Prunus* and all have a stony endocarp. The group consists of cherry (*P. avium*, sweet cherry, and *P. cerasus*, sour cherry), plum (*P. domestica*), apricot (*P. armeniaca*), and peach (*P. persica*).

All members of this group are characterized by a double sigmoid growth curve. Phase I consists of the enlargement of all parts of the ovary with the exception of the endosperm and embryo. Lignification takes place during phase II and growth is confined principally to the endosperm and embryo. It is during phase III that expansion of the mesocarp (edible portion) is resumed. All members of this group are considered climacteric.

Reports on the effects of ethylene on cherries are few and superficial. Chace and Sorber (46) reported that ethylene accelerated the softening of cherries and Pinelle (175) claimed that ethylene increased the acid content of the fruit.

Plums both produce ethylene (198, 213) and respond to the gas (198, 199, 213). Uota (213) reported that a correlation existed between the ability of fruit to evolve ethylene and the speed at which they ripened. The Beauty variety evolved relatively large amounts of the gas and ripened normally; Santa Rosa produced an intermediate amount of ethylene and the fruit ripened, but without full color. Duarte and Kelsey, on the other hand, produced practically no ethylene and did not ripen normally. Ethylene treatment prior to holding at high temperature resulted in an improved ripening response and increased the production of ethylene by the fruit.

Apricot and peach are both considered to be climacteric fruit and ethylene has been shown to promote ripening in both cases (31, 46, 96, 99).

13. Tomato

The tomato (*Lycopersicon esculentum*) is a berry and all parts are edible, including the jelly-like parenchyma arising from the placenta (109). While not a fruit from a culinary point of view, it shares with other fruits the metabolic changes associated with ripening, including a climacteric rise in respiration, a decrease in chlorophyll, an increase in carotene, a conversion of starch into reducing sugars, a softening of the tissues, and an increased production of ethylene and other hydrocarbon volatiles. The acid content of the fruit, primarily citric and malic, may or may not decrease during ripening depending on the variety of fruit.

The main constituents of the cell walls are pectic substances, hemicelluloses, cellulose, and some protein. The progressive loss of firmness with ripening is the gradual result of solubilization of protopectin in the cell walls to form pectin and other products. Concomitantly, either erosion of the cell wall or stretching of the wall due to enlargement of the cell, or both, may be responsible for the thinness of the cell walls. Even though pectic substances are known to undergo transformation and degradation by enzymatic mechanisms during ripening, the specific enzymes involved do not appear to have been identified. Although a cellulase-like enzyme is present in tomato fruit, evidence at the moment points toward its having a very limited effect on cellulose in the primary and secondary walls.

Similar to studies with other fruits such as the avocado and apple, the evidence suggests that the capacity for phosphorylation and synthesis remains intact during the climacteric. It isn't until the fruit is overripe that the efficiency of oxidative phosphorylation decreases. The basic control mechanism that accounts for the rise in respiration is not clearly defined, although changes in the levels of substrate or cofactors have been postulated.

14. Other Fruits

The effects of ethylene on the maturation or ripening of other fruits has been observed. A climacteric and an increase in ethylene production has been reported for figs (145), passion fruit (18, 27), vanilla beans (20), cherimoya (31), sapote (31), papaya (31, 107), and guava (122, 123, 165). While data on respiratory changes were not reported, ethylene was shown to accelerate ripening of pepper (47, 197) and jujubes (107). Data on dates are contradictory, some workers reporting a promotion of ripening with ethylene (66, 107) while others failed to observe any effect (45). Studies on the development of bean pods, a dry fruit, are interesting in that they showed an increase in respiration during maturation but no increase in ethylene production (222). The role of ethylene in fig ripening has also been discussed in Chap. 5.

E. Methods for Ripening and Storing Fruit

The methods used to apply ethylene and control ripening of fruit have been described in detail in a series of government (90, 228) and other (19, 92) documents. Details for successful storage of fresh fruits, vegetables, flowers, and other nursery stock have also appeared (141). The idea of using high CO_2 (5–10%) and low O_2 (5–10%) levels to prolong the storage life of fruits arose from the research of Kidd and West in the 1920's. The principle behind this practice stems from the fact that CO_2 acts as an antagonist of ethylene action and low O_2 levels reduce the rate of ethylene production and overall metabolism of the tissue. A considerable body of literature on this topic has appeared and the reader is referred to a review by Kidd and West (127) for details.

Storage life of plant materials can also be extended by absorbing or removing ethylene from the gas phase. Potassium permanganate adsorbed on various supports has been used effectively to protect fruits and flowers during shipping and storage (71, 78, 195, 209). The use of other techniques such as ozone (86) and silver sulfate traps (73) has also been described.

The use of hypobaric techniques by Burg and Burg (40–42) promises to be a new and effective way of storing fruit and other produce. The strategy behind this technique is to speed up the escape of ethylene produced by fruits by reducing the atmospheric pressure around them. If the rate of ripening or senescence is limited by the level of ethylene present, then the life of the tissue will be extended to the point that food reserves can maintain the integrity of the tissue and ethylene levels remain below the threshold for an effect. The basic details of this technique are to place fruit in a chamber and reduce the atmospheric pressure to 150 mm Hg. Oxygen is bled into the chamber so that respiratory activity is normal. Desiccation of the tissue is reduced by saturating the oxygen with water. Burg and Burg have shown that bananas were still green after 3 months and would ripen normally after exposure to ethylene. This technique also worked with other fruits including tomato, avocado, mango, sweet cherry, lime, and guava.

III. Abscission

A. Introduction

Early reports on the effect of ethylene on abscission are associated with literature on the effects of leaking illuminating gas on street trees (129, 206, 225), leaking gas systems in laboratories (64, 77, 98), and tobacco smoke on plants (77, 159). The active constituent was first shown to be ethylene by Harvey in 1913 (105); this fact was confirmed later by Doubt (64) and Goodspeed *et al.* (88). Other components of gas and smoke such as CO had also been examined (88, 185) but it was clear that the most effective compound was ethylene. Doubt found that 0.1 ppm ethylene caused abscission of *Mimosa pudica*, while more than 50 ppm CO was required for the same effect.

The introduction of the abscission zone explant by Kendall (124) in 1918 and later again by Livingston (139) in 1950 was another significant advance in abscission research. The explant technique facilitated the study of abscission physiology because it made available large samples for both statistical and biochemical work under controlled and hence reproducible conditions. While whole plant studies appear to be more natural than isolated sections, the removal of abscission zones permitted the study of cell separation isolated from the effects of the rest of the plant. However, normal cell separation at the abscission zone is controlled by auxin produced by the leaf blade and this fact has to be kept in mind during an analysis of experiments utilizing explants.

The demonstration by Fitting (76) and Morita (162) that a watery extract of pollen prevented the abscission of orchid flowers set the stage for the idea that a chemical factor played a regulatory role in cell separation and controlled senescence and juvenility in plants. Laibach (131) reported that orchid pollen delayed the abscission of debladed *Coleus* petioles and then subsequently reported (132) a similar retardation when auxin was used.

Two other contributions that played an important role in abscission research were the discovery that the enzyme responsible for cell separation was cellulase (111) and the introduction of quantitative techniques for measuring abscission in terms of a reduction of breakstrength (53, 61, 163). For a more detailed review of abscission research in general, the following papers can be consulted: 5, 14, 16, 17, 60.

B. Physiology of Leaf Abscission

The physiology of cell separation during leaf, fruit, and floral abscission and fruit dehiscence is probably the same for all processes. For example, in addition to promoting leaf abscission, ethylene is known to promote fruit (25) and flower (160) abscission and the dehiscence of pecan (137) and walnut (201–203) fruits. Carbon dioxide, which inhibits leaf abscission, also retards floral abscission (75) and dehiscence of walnuts (203).

The relationship between physiological age and susceptibility to ethylene is the basis of the idea that abscission is not controlled by ethylene alone. The following observations indicate that the ability of ethylene to initiate or accelerate cell separation ultimately depends on the levels of auxin at or near the cell separation layer. First of all, it is a common observation that older leaves abscise sooner than younger ones. Second, older leaves are usually more susceptible to ethylene than younger ones (179, 231, 232). However, in the case of *Ilex vomitoria*, *Bougainvillia* (232), and cotton (160), the younger leaves are more sensitive than the older ones. Similarly, concentrations that are too low to cause leaf abscission of cotton leaves can cause abscission of young flower buds and fruit (160). Correlated with these observations is the fact that many investigators have observed a reduction in auxin content of leaves as they age (34) (for a review on this topic, see 60). In other words, leaves that are insensitive to externally supplied ethylene have a high auxin content.

Finally, if the supply of auxin to separation layer is cut off by excising the abscission zone or deblading the petiole, abscission can be prevented if auxin is applied soon after excision. A promotion of abscission can also be observed in intact leaves if the petiole is ringed with an inhibitor of auxin transport (161). If the application of auxin to tissue is delayed for a number of hours following excision, the residual endogenous supply falls off and the tissue becomes sensitive to ethylene generated by the addition of auxin (8, 9, 11, 12). This period of ethylene insensitivity followed by a period of ethylene sensitivity has been summarized in a two-stage theory of abscission. It, like others before it (3), were useful as models, and has been modified subsequently as more has been learned about the overall processes. It is clear from the above discussion that ethylene alone does not control abscission. Auxin and perhaps other factors are thought to control the sensitivity of the separation cells to ethylene and the levels of these hormones must drop before the gas can promote cellulase synthesis. This relatively simple idea of auxin-controlled ethylene sensitivity is complicated by two facts. First, ethylene itself can promote aging or senescence (8), and second, other chemical factors can play a role in the juvenility of susceptibility of tissue to ethylene. As an example, it has been shown that branches supply factors controlling abscission of citrus (25). Another way of emphasizing that the ability of ethylene to promote abscission depends on internal factors is to compare the relative effectiveness of ethylene as a defoliant for various species. For example, a 2-ppm 18-hour treatment of ethylene defoliated huisache but had no effect up to 30 hours on mesquite (23). Funk *et al.* (82) exposed a variety of plants to apple volatiles and found that within 2 days *Mimosa pudica* was defoliated; within 1 week, *Ilex aquifolia*, *Ligustrum vulgare*, *Sparmannia africana*, and *Coleus blumei;* within 2 weeks, *Gneista anglica*, *Beloperone guttata*, *Acacia linifolia*, *Camellia japonica*, and *Gisellinia macrophylla;* and within 3 weeks, *Azalea indica* and *Veronica diosmaefolia*. Some plants such as *Quercus robur*, *Fagus silvatica*, *Pinus silvestris, and Hedera helix* did not lose their leaves. Additional data on the relative rate of defoliation of 114 species of plants have been presented by Heck and Pires (108).

Relative sensitivity among varieties of a single species has also been observed. Beans (eighteen varieties) were exposed to 10 ppm ethylene for 30 hours. A sensitive variety Top Crop showed a 61% reduction of breakstrength; Plentiful, a moderately sensitive variety, 28%; and Slenderwhite, 16% (53). Haney (97) surveyed snapdragon varieties for tolerance to ethylene in terms of "shatter resistance." He listed twenty-five varieties he considered to be resistant. McMillan and Cope (152) found differences in the degree of defoliation by geographical variants of *Acacia farnesiana*. Depending on the variety, 2% CO resulted in a variation of abscission from 83% in responsive varieties to 7% in the more resistant ones. The adaptive significance of these observations was unknown. Milbrath and Hartman (154) found that the time to achieve full defoliation of roses varied from one variety to another. While most varieties were defoliated in 4 days, a resistant variety required 6 days.

Anatomic studies of abscission have suggested that separation was due to both cell separation and cell wall breakdown (224). Hydrolysis occurs in living parenchyma cells and the separation of vascular tissue appears to take place by cell wall breakage. The wall breakage commonly observed in the parenchyma is probably due to the fragility of the thin walls of hydrolyzed cells which makes them susceptible to damage during the preparation for microscopic examination. Davenport and Marinos (57) have shown that all of cell separation during abscission was due to separation and rounding up of the cell walls (see Fig. 7-3). An electron microscopic examination of separating walls showed the initial level of attack occurs at the middle lamella (see Fig. 7-4). These figures confirm the observation made by Webster (224) that ruthenium red staining materials (pectic substances) are lost during the cell wall hydrolysis. Morré (163) examined the levels of various sugars during abscission and reported a loss of galactose in the separation layer. However, other sugars such as arabinose, xylose, mannose, and glucose were also degraded. Stösser (207) has also reported the degradation of pectic materials and the loss of Ca and Mg from the middle lamella during abscission. While the role of these metals is not completely known, they are thought to function in a binding capacity between pectic acid polymers.

The enzyme responsible for cell wall hydrolysis was shown by Horton and Osborne (111) to be cellulase. Although other carbohydrate polymer degrading enzymes (pectinase, hemicellulase, etc.) have been tested, no support for the role of other enzymes has been unequivocally established. The criteria for enzymes responsible for cell separation include localization, in the separation layer, an increase in activity prior to the loss of breakstrength, and a dependence on ethylene levels, that is, an addition of ethylene increases the rate of enzyme synthesis, while a removal of ethylene or the addition of CO_2 causes a decrease in enzyme synthesis.

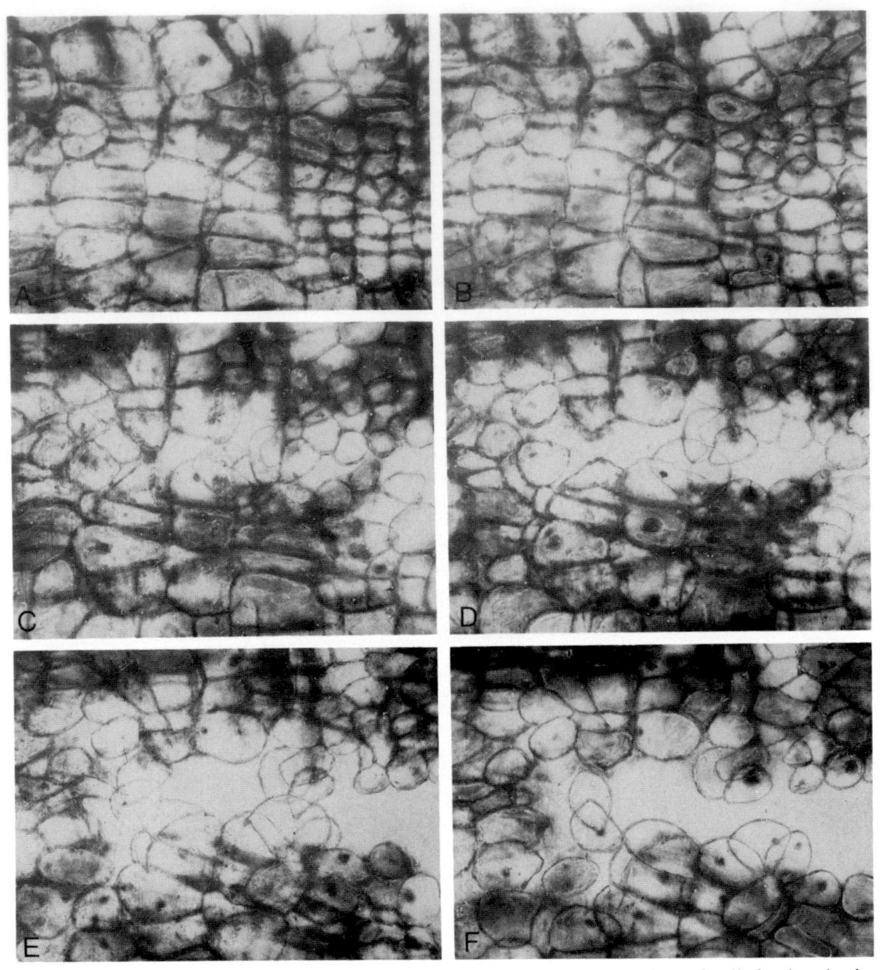

Fig. 7-3. Stages, selected from a time-lapse sequence, in the separation of cells in the abscission zone of *Coleus blumei*. The elapsed time in hours from setting up the culture are as follows: A, 0; B, 60; C, 65; D, 67; E, 69; F, 90. The separation process starts at the ventral side of the petiole tissue. [Courtesy of Davenport and Marinos (57).]

Horton and Osborne (111) and others (6, 53, 135) have shown that the increase in cellulase is localized in the separation layer of abscission zones. Cellulase occurs elsewhere in the plant in small quantities; however, ethylene is not able to increase its activity there (186).

Techniques that decrease abscission (auxin, CO_2, inhibitors of RNA and protein synthesis, cytokinins) or increase it (ethylene, abscisic acid) cause a

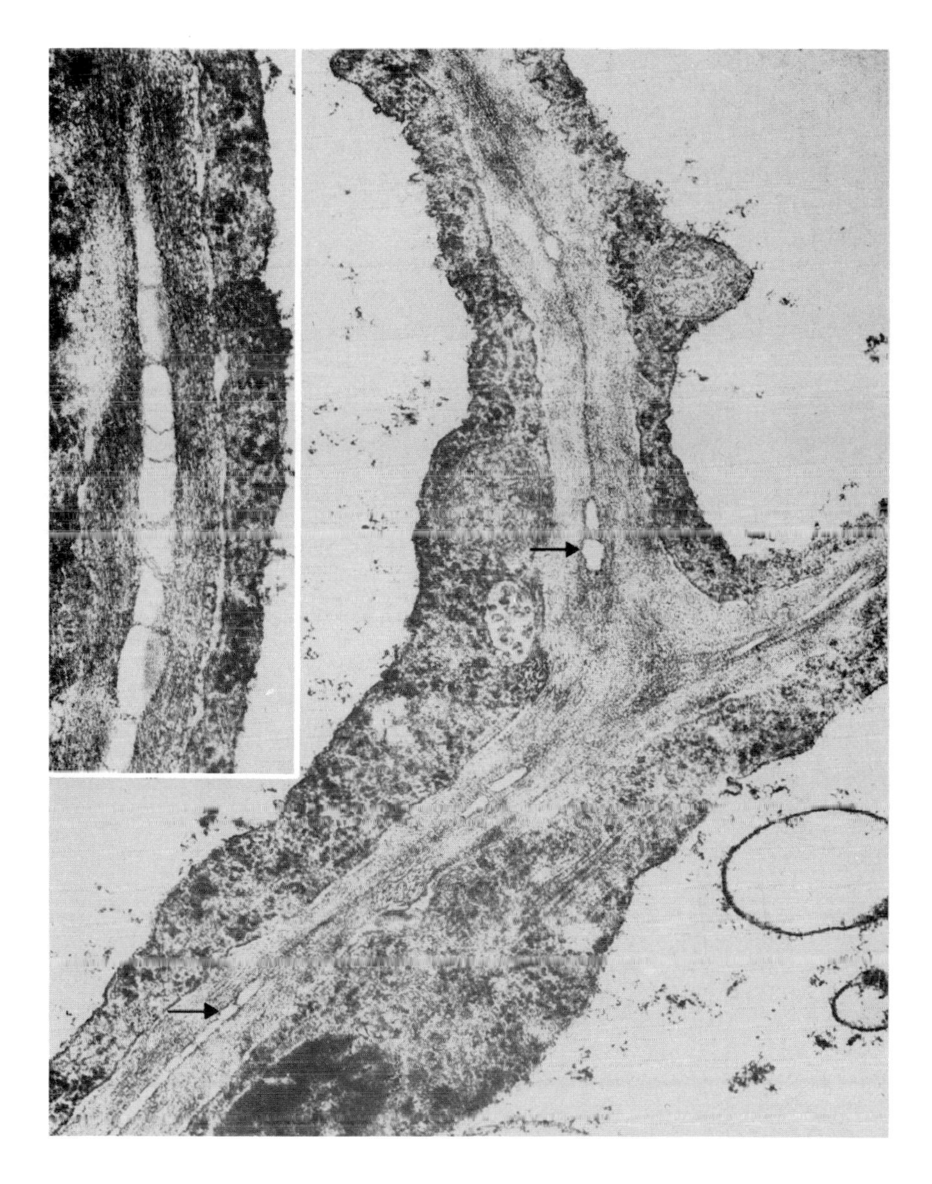

Fig. 7-4. Typical appearance of cell walls in the abscission zone 20 hours after excision of the explant. The arrows indicate regions where the middle lamella begins to dissolve. × 37,200. The insert shows a higher magnification (× 48,600) of such a region. [Courtesy of Davenport and Marinos (57).]

concomitant decrease or increase in cellulase activity (6, 53, 111, 135, 170, 181, 182). Figure 7-5 shows time-course data which show an increase in cellulase activity before the reduction in breakstrength. Similar data have been obtained by others (61). This line of evidence has been taken as proof for the idea that cellulase plays a central role in hydrolyzing the cell walls of separation layer cells. It is worth remembering, however, that this enzyme is called cellulase because of its ability to hydrolyze the β-1,4 linkages of carboxymethyl cellulose in test tube assays. This enzyme assay substrate is not identical to cell wall material and it is likely that abscission cellulase may be able to attack other kinds of glucose polymers with β-1,4 linkages. The only other enzyme that has been implied as having a role in abscission has been pectinase (163). However, it might be more accurate to describe the enzyme as a cell macerating enzyme since the substrate used was cucumber fruit disks and the assay consisted of recording the loss of weight during the enzyme incubation.

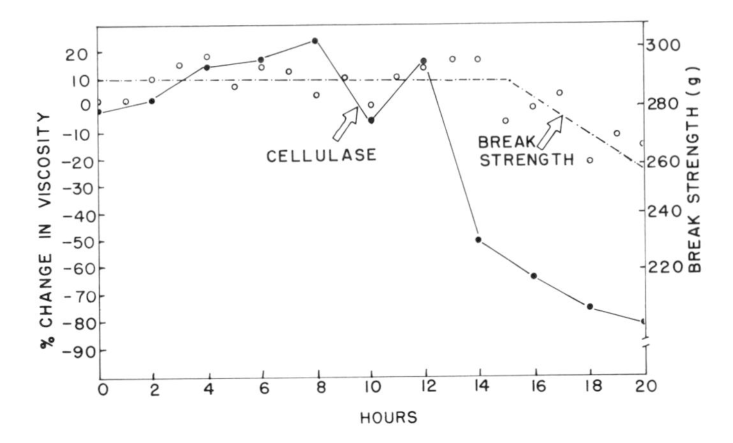

Fig. 7-5. Change in break strength and cellulase content of bean abscission-zone explants. Data on the left ordinate represent the loss of viscosity of a solution of carboxymethyl cellulose. The larger the reduction in viscosity, the greater the amount of enzyme activity. [Courtesy of Craker and Abeles (54).]

The increase in cellulase activity during cell separation is probably due to the synthesis of new protein, which in turn is regulated by the synthesis of messenger RNA and subsidiary RNA such as soluble and ribosomal RNA. Cellulase is probably synthesized de novo, because inhibitors of protein synthesis, such as cycloheximide, block abscission (10, 61), addition of D_2O to explants during abscission results in the formation of a cellulase with increased density (135), and labeled leucine is incorporated in separation layer cells of explants (170, 208, 224) during abscission (Fig. 7-6). A similar argument can be advanced for the role of RNA in abscission. Inhibitors of RNA

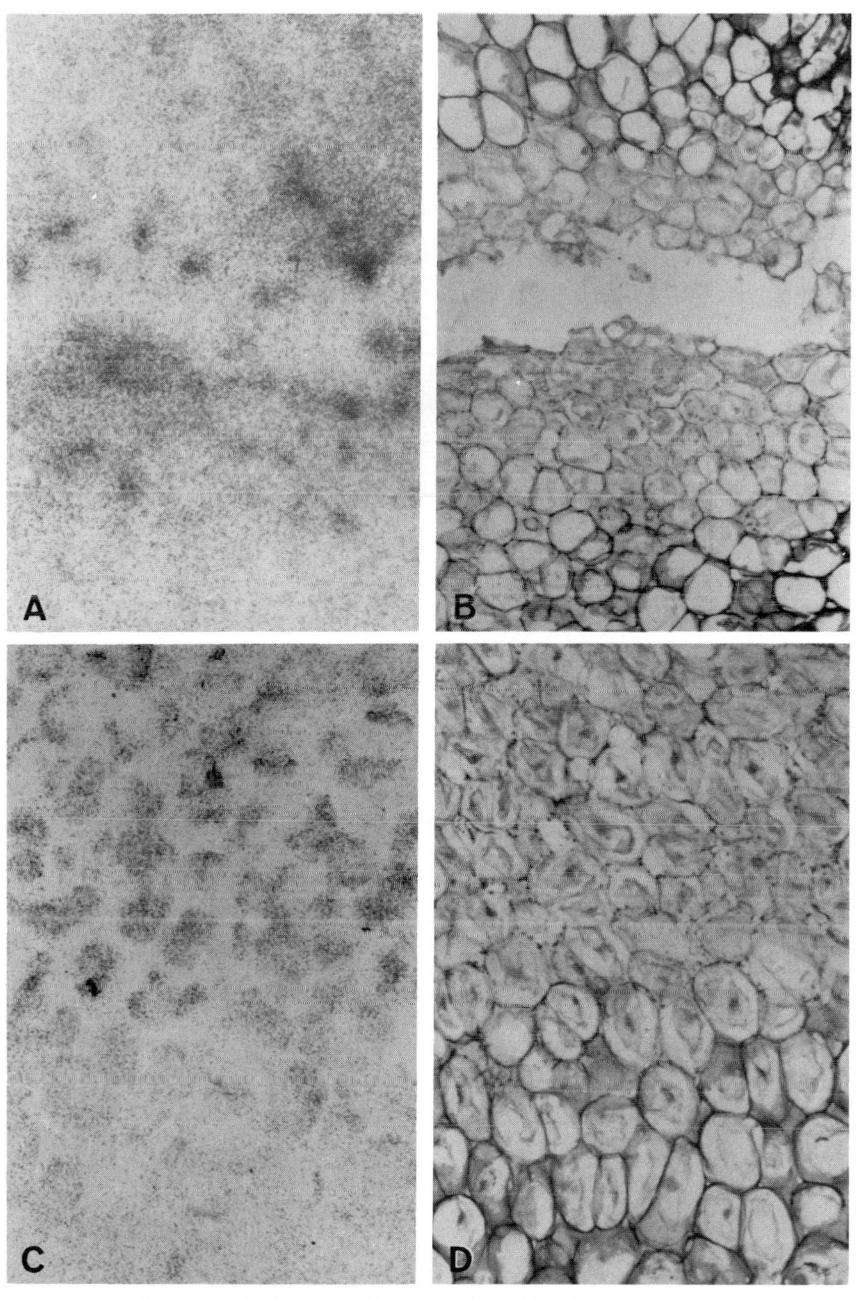

Fig. 7-6. Radioautograph illustrating incorporation of labeled L-leucine in the cells of the abscission layer of *Prunus cerasus*. A. [^{14}C]Leucine. B. Corresponding histological section; $\times$ 176. C. [^{3}H]Leucine. D. Corresponding histological section; $\times$ 240. [Courtesy of Stösser (208).]

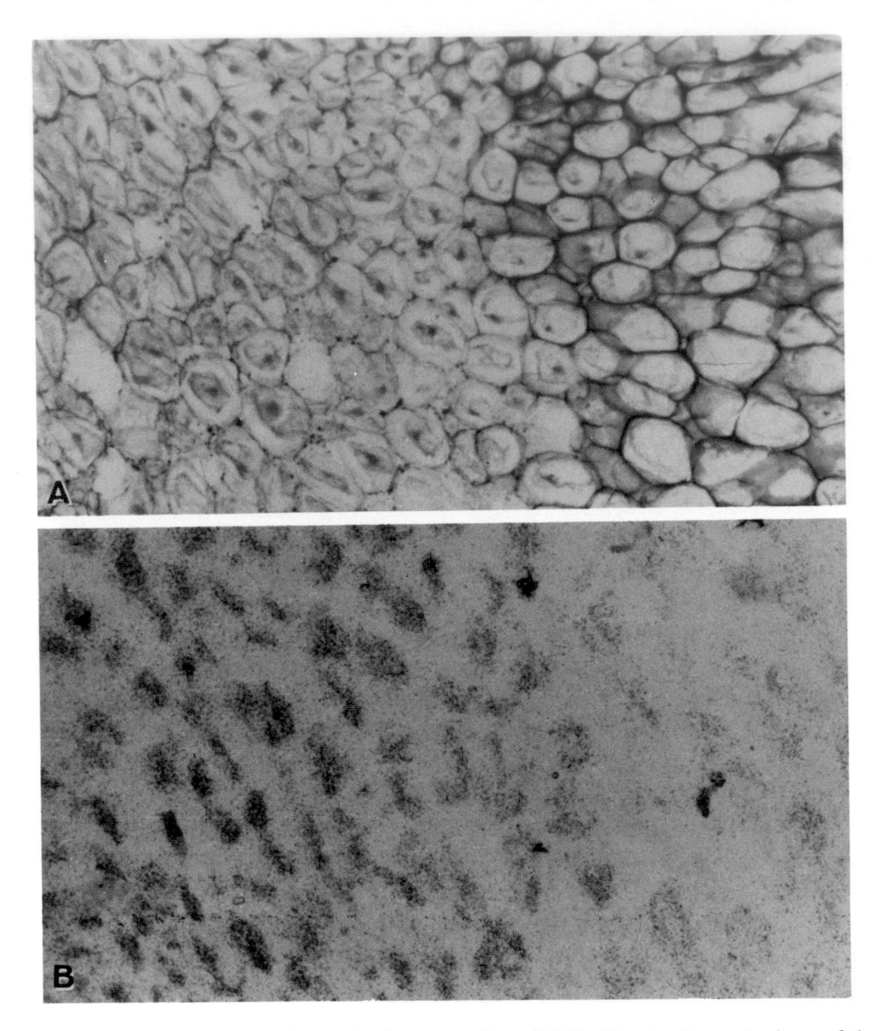

Fig. 7-7. Radioautograph illustrating incorporation of [^{3}H]uridine in the protoplasm of the cells of the abscission layer (B), and the corresponding tissue section (A). × 270. [Courtesy of Stösser (208).]

synthesis such as actinomycin D block abscission (9, 10, 110) and labeled nucleic acids or ^{32}P appear to be localized in the separation layer (110, 170, 208, 224). Figure 7-7 shows a radioautograph illustrating the incorporation of [^{3}H]uridine into the protoplasm of abscission layer cells. Kinetic studies also support the idea of a role of RNA synthesis in abscission. Enhancement of ^{32}P incorporation into RNA following ethylene treatment occurred after 1 hour while the lag for leucine incorporation into protein was 2 hours (9).

These incorporation studies support the idea of hormonal regulation of nucleic acid metabolism followed by the synthesis of enzymes that perform the physiological action. The details of facts that support this interpretation have been reviewed in Chap. 10 and elsewhere (5, 7, 14).

It now appears that cell wall loosening is not automatic once cellulase has been synthesized. Recent data show that cellulase secretion from the cytoplasm (the site of protein synthesis) to the cell wall (the site of cellulase action) is important and appears to be regulated by ethylene (13). It has also been shown that in addition to cell wall degradation, the separation layer contains a repair mechanism if abscission has proceeded only partway. In other words, abscission is reversible in certain cases. Ben-Yehoshua and Biggs (24) and others (14) have reported that the breakstrength of abscission zone from intact plants can return to normal if ethylene is removed before abscission has gone to completion.

To summarize, abscission involves (*1*) aging, (*2*) RNA synthesis, (*3*) cellulase synthesis, (*4*) cellulase secretion, and, to some extent, (*5*) cell wall repair. The data have shown that ethylene can influence all of these processes, which indicates that in even a relatively simple process, such as abscission, the action of ethylene is complex and varied.

C. Hormonal Regulation of Abscission

1. Auxin

Auxin plays a central role in abscission. As described above, the application of auxin to debladed leaves or explants prevents abscission. This effect is not limited to IAA but also occurs with other compounds with auxin activity such as NAA and 2,4-D. The chemical inhibition of abscission has been taken advantage of in situations where premature defoliation or flower or fruit drop represents an economic loss or production problem. For example, Milbrath and Hartman (155, 156) recommended the use of NAA to prevent leaf drop in cut holly wreaths. Other examples of auxin-controlled abscission have been reviewed by Cooper *et al.* (49).

There is good reason to believe that internal auxin levels in leaves play a central role in controlling the rate of abscission. In general, most workers report that young leaves yield more auxin than older senescent ones (34, 60). The ability of auxin or cytokinins to delay or prevent abscission is probably associated with their juvenility or antisenescent activity as opposed to their growth-regulating capacity. Using the loss of chlorophyll, RNA, and protein from the pulvinus of bean explants as a criterion of senescence, Abeles *et al.* (11) demonstrated that the ability of auxin and other similar compounds to prevent abscission was associated with their ability to delay the degradation of these vital cellular components. Senescence of the pulvinus can also be controlled by the way the tissue is excised. Separation of bean abscission zone explants into a pulvinal and petiolar portion delays the senescence of the pulvinus. This observation suggests that the senescence of pulvinal tissue is modified or controlled by a sink-to-source relationship as by chemicals.

Since auxin delays or prevents abscission, any treatment that reduces auxin levels should also increase abscission. Auxin levels in the separation layer can be reduced a number of ways. These include inhibition of synthesis, inhibition of transport, oxidation, decarboxylation, and conjugation. All of these processes have been examined as sites for ethylene action during the promotion of abscission. Valdovinos *et al.* (215) have shown that ethylene reduced auxin levels in coleus and peas. According to their findings, ethylene decreased the activity of the enzyme system which converted tryptophan to auxin.

According to Beyer and Morgan (26), the inhibitor of auxin transport from leaf blade to the base of the petiole is the major site of ethylene action during abscission. They found that levels of ethylene required to promote abscission of intact cotton plants also reduced the capacity of the auxin transport system to function. They also noted that there was a decline in the auxin transport capacity of petioles as they aged.

The role of auxin oxidase or peroxidase in abscission has also been examined. Hall and Morgan (93) reported that the level of IAA oxidase in leaves increased after ethylene treatment. Schwertner and Morgan (194) supported this idea with the observation that auxin oxidase cofactors (monophenols) speeded abscission. However, Gahagan *et al.* (83) obtained evidence which suggested that this concept was incorrect. They failed to find an obvious correlation between abscission and peroxidase or between the activity of phenolic substances and the rate of abscission.

Valdovinos *et al.* (215) reported that decarboxylation of auxin took place in leaf tissue. However, they failed to observe any effect of ethylene on this process, which suggested that decarboxylation was not an important control mechanism.

Craker *et al.* (55) have shown that conjugation of auxin with aspartic acid is an important means of decreasing auxin levels in explants. They reported that a 1-hour application of auxin to explants delayed abscission for 8 hours and during that time it was converted into indoleacetyl-aspartate and other conjugates. If auxin was reapplied at a later time, abscission could be reinhibited. On the other hand, 2,4-D, which is not destroyed or conjugated by this system, did not lose its ability to inhibit abscission. They concluded that IAA destruction via conjugation is one of the processes involved in the aging stage of abscission.

One anomaly of auxin is that it inhibits or promotes abscission depending on the concentration used or on the time of application. Any confusion over the dual action of auxin can be readily resolved if the fact that auxin promotes ethylene production is kept in mind. The promotive effects of auxin on abscission can be invariably explained (15, 191) by the presence of increased levels of ethylene under conditions where auxin is unable to get to the separation layer in sufficient time or concentration to delay senescence, then ethylene is without effect.

2. Ethylene

There are two distinct points of view with regard to how ethylene controls abscission. One assigns ethylene a passive role while the other assumes that it takes an active part in regulating abscission. In a passive role, the control of abscission depends on an increase in the sensitivity of the tissue to ethylene that is already being produced. In other words, ethylene production proceeds in more or less constant fashion and remains in a position to control abscission when the levels of juvenility factors have fallen due to the aging of the tissue. The basic idea is that ultimate control of abscission depends upon the availability of juvenility factors to the separation layer and as long as supplies of these hormones are high, ethylene is without effect.

The active role for ethylene assumes that senescence and subsequent abscission take place as a result of increased rates of ethylene production during the terminal life of the organ. Under these conditions, ethylene is thought to be the primary factor in regulating the levels of juvenility factors and subsequent senescence of the tissue.

Apparently both points of view are correct for some plants under certain conditions and neither mechanism serves as the sole explanation for abscission. On one hand, the observation that some plants are readily defoliated by ethylene, or others less so, and still others not at all, suggests that in some plants ultimate control depends on factors other than ethylene. This is also seen when Ethrel is used as a defoliant. In some cases, Ethrel will not act as a defoliant unless some other compound, primarily a herbicide, is added to the formulation. The herbicides apparently traumatize the tissue, accelerate senescence, and make the tissue sensitive to ethylene.

The fact that ethylene production increases during the maturation of fruits and at the end of the physiological life of leaves (26, 119, 134, 170) provides the evidence that ethylene plays an active role in inducing abscission. Many species are predisposed to respond to ethylene and it is easy to see where an increase in ethylene production would be a rate-limiting step in the control of abscission.

3. Abscisic Acid

Abscisic acid has been implicated as playing a central role in abscission and has been shown to increase abscission of both explants (33, 54) and leaves (49, 170). In addition, Böttger (34) has shown that levels of abscisic acid increase during leaf senescence. An excellent review on the physiology of abscisic acid has been written recently by Addicott (17), one of the discoverers of this hormone.

The action of abscisic acid appears to be twofold. Abscisic acid is able to promote ethylene production (4, 54), a nonspecific effect shared by other compounds such as amino acids, KI (4), cycloheximide (10), ascorbic acid, iodoacetic acid (48), and defoliants (94, 118). However, the increase in ethylene production does not account for all of abscisic acid's activity because the presence of a saturating dose of ethylene did not completely mask the ability of abscisic acid to increase abscission. However, this effect is not specific for abscisic acid. Morré *et al.* (164) have shown that galactose was able to promote abscission even though a saturating level of ethylene was present in the gas phase. Craker and Abeles (54) subsequently examined the effect of abscisic acid on two other aspects of abscission, aging and cellulase synthesis. In agreement with Osborne (170), they found little or no effect of abscisic acid on senescence or aging of leaf tissue. However, abscisic acid increased the rate of cellulase synthesis, suggesting that the mode of action of this hormone was directed toward effects on RNA or protein synthesis.

References

1. Abdel-Kader, A. S., Morris, L. L., and Maxie, E. C. (1968). *Proc. Amer. Soc. Hort. Sci.* **92,** 553.
2. Abdel-Kader, A. S., Morris, L. L., and Maxie, E. C. (1968). *Proc. Amer. Soc. Hort. Sci.* **93,** 831.
3. Abeles, F. B. (1966). *Plant Physiol.* **41,** 585.
4. Abeles, F. B. (1967). *Physiol. Plant.* **20,** 442.
5. Abeles, F. B. (1968). *Plant Physiol.* **43,** 1577.
6. Abeles, F. B. (1969). *Plant Physiol.* **44,** 447.
7. Abeles, F. B. (1972). *Annu. Rev. Plant Physiol.* **23,** 259.
8. Abeles, F. B., Craker, L. E., and Leather, G. R. (1971). *Plant Physiol.* **47,** 7.
9. Abeles, F. B., and Holm, R. E. (1966). *Plant Physiol.* **41,** 1337.
10. Abeles, F. B., and Holm, R. E. (1967). *Ann. N. Y. Acad. Sci.* **144,** 367.
11. Abeles, F. B., Holm, R. E., and Gahagan, H. E. (1967). *Plant Physiol.* **42,** 1351.
12. Abeles, F. B., Holm, R. E., and Gahagan, H. E. (1967). *In* "Biochemistry and Physiology of Plant Growth Substances" (F. Wightman and G. Setterfield, eds.), p. 1515. Runge Press, Ottawa.
13. Abeles, F. B., and Leather, G. R. (1971). *Planta* **97,** 87.

14. Abeles, F. B., Leather, G. R., Forrence, L. E., and Craker, L. E. (1971). *HortScience* **6,** 371.
15. Abeles, F. B., and Rubinstein, B. (1964). *Plant Physiol.* **39,** 963.
16. Addicott, F. T. (1968). *Plant Physiol.* **43,** 1471.
17. Addicott, F. T. (1970). *Biol. Rev.* **45,** 485.
18. Akamine, E. K., Young, R. E., and Biale, J. B. (1957). *Proc. Amer. Soc. Hort. Sci.* **69,** 221.
19. American Society of Heating, Refrigeration and Air-conditioning Engineers, Inc. (1964). "Guide and Data Book."
20. Arana, F. E. (1944). *U. S. Dep. Agr., Fed. Exp. Sta., Puerto Rico, Bull.* No. 42.
21. Asmaev, P. G. (1937). *Proc. Agr. Inst. Krasnodar* **6,** 49.
22. Banerjee, H. K., and Kar, B. K. (1939). *Trans. Bose Res. Inst., Calcutta* **14,** 171.
23. Baur, J. R., and Morgan, P. W. (1969). *Plant Physiol.* **44,** 831.
24. Ben-Yehoshua, S., and Biggs, R. H. (1970). *Plant Physiol.* **45,** 604.
25. Ben-Yehoshua, S., and Eaks, I. L. (1970). *Bot. Gaz. (Chicago)* **13,** 144.
26. Beyer, E. M., and Morgan, P. W. (1971). *Plant Physiol.* **48,** 208.
27. Biale, J. B. (1960). *Advan. Food Res.* **10,** 293.
28. Biale, J. B. (1961). *In* "The Orange, Its Biochemistry and Physiology." (W. B. Sinclair, ed.), p. 96. Univ. of California Press, Berkeley.
29. Biale, J. B., and Young, R. E. (1962). *Endeavour* **21,** 164.
30. Biale, J. B., and Young, R. E. (1971). *In* "The Biochemistry of Fruits and Their Products" (A. C. Hulme, ed.), Vol. 2, p. 2. Academic Press, New York.
31. Biale, J. B., Young, R. E., and Olmstead, A. J. (1954). *Plant Physiol.* **29,** 168.
32. Borgström, G. (1945). *Sver. Pomol. Foeren. Arsskr.* **46,** 202.
33. Bornman, C. H. (1967). *S. Afr. J. Agr. Sci.* **10,** 143.
34. Böttger, M. (1970). *Planta* **93,** 205.
35. Brady, C. J., O'Connell, P. B. H., Smydzuk, J., and Wade, N. L. (1971). *Aust. J. Biol. Sci.* **23,** 1143.
36. Brady, C. J., Palmer, J. K., O'Connell, P. B. H., and Smillie, R. M. (1970). *Phytochemistry* **9,** 1037.
37. Bramlage, W. J., and Couey, H. M. (1965). *U. S. Dep. Agr., Marketing Res. Rep.* **717.**
38. Burg, S. P. (1964). *In* "Régulateurs naturels de la croissance végétale" (J. P. Nitsch, ed.), p. 719. CNRS, Paris.
39. Burg, S. P., and Burg, E. A. (1965). *Bot. Gaz. (Chicago)* **126,** 200.
40. Burg, S. P., and Burg, E. A. (1965). *Science* **148,** 1190.
41. Burg, S. P., and Burg, E. A. (1966). *Science* **153,** 314.
42. Burg, S. P., and Burg, E. A. (1969). *Qual. Plant Mater. Veg.* **19,** 185.
43. Burg, S. P., and Dijkman, M. J. (1967). *Plant Physiol.* **42,** 1648.
44. Chace, E. M. (1934). *Amer. J. Pub. Health Nat. Health* **24,** 1152.
45. Chace, E. M., and Church, C. G. (1927). *Ind. Eng. Chem.* **19,** 1135.
46. Chace, E. M., and Sorber, D. G. (1930). *Canning Age* **11,** 391.
47. Chmelar, F. (1941). *Sb. Cesk. Akad. Zem.* **16,** 2.
48. Cooper, W. C., Henry, W. H., Rasmussen, G. K., and Hearn, C. J. (1969). *Proc. Fla. State Hort. Soc.* **82,** 99.
49. Cooper, W. C., Rasmussen, G. K., Rogers, B. J., Reece, P. C., and Henry, W. H. (1968). *Plant Physiol.* **43,** 1560.
50. Cottrell, G. G. (1968). *Florist Nursery Exch.* **148,** 5.
51. Cousins, H. H. (1910). *Annu. Rep. Dep. Agr. Jamaica.*
52. Craker, L. E. (1971). *HortScience* **6,** 137.
53. Craker, L. E., and Abeles, F. B. (1969). *Plant Physiol.* **44,** 1139.
54. Craker, L. E., and Abeles, F. B. (1969). *Plant Physiol.* **44,** 1144.

55. Craker, L. E., Chadwick, A. V., and Leather, G. R. (1970). *Plant Physiol.* **45,** 790.
56. Crocker, W. C., and Knight, L. I. (1908). *Bot. Gaz.* **46,** 259.
57. Davenport, T. I., and Marinos, N. G. (1971). *Aust. J. Biol. Sci.* **24,** 709.
58. Davidson, O. W. (1949). *Proc. Amer. Hort. Sci.* **53,** 440.
59. Davis, W. B., and Church, C. G. (1931). *J. Agr. Res.* **42,** 165.
60. De la Fuente, R. K., and Leopold, A. C. (1968). *Plant Physiol.* **43,** 1486.
61. De la Fuente, R. K., and Leopold, A. C. (1969). *Plant Physiol.* **44,** 251.
62. Dijkman, M. J., and Burg, S. P. (1970). *Amer. Orchid. Soc., Bull.* [N. S.] Sept., p. 799.
63. Dostal, H. C., and Leopold, A. C. (1967). *Science* **158,** 1579.
64. Doubt, S. L. (1917). *Bot. Gaz.* (*Chicago*) **63,** 209.
65. Doughty, C. C., Patterson, M. E., and Shawa, A. Y. (1967). *Proc. Amer. Soc. Hort. Sci.* **91**, 192.
66. Dufrenoy, J. (1929(. *Rev. Bot. Appl.* **95,** 441.
67. Dull, G. G. (1971). *In* "The Biochemistry of Fruits and Their Products" (A. C. Hulme, ed.), Vol. 2, p. 303. Academic Press, New York.
68. Dull, G. G., Young, R. E., and Biale, J. B. (1967). *Physiol. Plant* **20,** 1059.
69. Eaks, I. L. (1967). *Proc. Amer. Soc. Hort. Sci.* **91,** 868.
70. Eaks, I. L. (1970). *Plant Physiol.* **45,** 334.
71. Eaves, C. A., Forsyth, F. R., and Lockhart, C. L. (1969). *Can. Inst. Food Technol. J.* **2,** 46.
72. Fernandez-Diaz, M. J. (1971). *In* "The Biochemistry of Fruits and Their Products" (A. C. Hulme, ed.), Vol. 2, p. 255. Academic Press, New York.
73. Fidler, J. C. (1948). *J. Hort. Sci.* **24,** 178.
74. Fischer, C. W. (1949). *N. Y. State Flower Growers , Bull.* **52,** 5.
75. Fischer, C. W. (1950). *N. Y. State Flower Growers, Bull.* **61,** 1.
76. Fitting, H. (1909). *Z. Bot.* **1,** 1.
77. Fitting H. (1911). *Jahrb. Wiss. Bot.* **49,** 187.
78. Forsyth, F. R., Eaves, C. A., and Lightfoot, H. J. (1969). *Can. J. Plant Sci.* **49,** 567.
79. Forsyth, F. R., and Hall, I. V. (1969). *Natur. Can.* **96,** 257.
80. Frenkel, C., Klein, I., and Dilley, D. R. (1968), *Plant Physiol.* **43,** 1146.
81. Fudge, B. R. (1930). *N. J. Agr. Exp. Sta., Bull.* **504**.
82. Funke, G. L., de Coeyer, F., de Decker, A., and Maton, J. (1938). *Biol. Jaarb.* **5,** 335.
83. Gahagan, H. E., Holm, R. E., and Abeles, F. B. (1968). *Physiol. Plant.* **21,** 1270.
84. Galliard, T., Rhodes, M. J. C., Wooltorton, L. S. C., and Hulme, A. C. (1968). *Phytochemistry* **7,** 1453.
85. Gane, R. (1935). *Gt. Brit., Dep. Sci. Ind. Res., Food Invest. Bd., Rep. 1934* p. 122.
86. Gane, R. (1937). *New Phytol.* **36,** 170.
87. Gerhart, A. R. (1930). *Bot. Gaz.* (*Chicago*) **89,** 40.
88. Goodspeed, T. H., McGee, J. M., and Hodgson, R. W. (1918). *Univ. Calif., Berkeley, Publ. Bot.* **5,** 439.
89. Gortner, W. A., and Singleton, V. L. (1965). *J. Food Sci.* **30,** 24.
90. Grierson, W., and Newhall, W. F. (1960). *Fla., Agr. Exp. Sta., Bull.* **620**.
91. Hale, C. R., Coombe, B. G., and Hawker, J. S. (1970). *Plant Physiol.* **45,** 620.
92. Hall, E. G. (1940). *Agr. Gaz. N. S. Wa.* **51,** 98.
93. Hall, W. C., and Morgan, P. W. (1964). *In* "Régulateurs naturels de la croissance végétale" (J. P. Nitsch, ed.), p. 727. CNRS, Paris.
94. Hall, W. C., Truchelut, G. B., Leinweber, C. L., and Herrero, F. A. (1957). *Physiol. Plant.* **10,** 306.
95. Hallaway, M., and Osborne, D. J. (1969). *Science* **163,** 1067.

96. Haller, M. H. (1952). *U. S. Dep. Agr., Biblio. Bull.* **21**.
97. Haney, W. J. (1958). *Florists Exch., Hort. Trade World* **130,** 17.
98. Hannig, E. (1913). *Z. Bot.* **5,** 417.
99. Hansen, E. (1939). *Proc. Amer. Soc. Hort. Sci.* **36,** 427.
100. Hansen, E. (1939). *Plant Physiol.* **14,** 145.
101. Hansen, E. (1946). *Plant Physiol.* **21,** 588.
102. Hansen, E. (1966). *Annu. Rev. Plant Physiol.* **17,** 459.
103. Hansen, E., and Blanpied, G. D. (1968). *Proc. Am. Soc. Hort. Sci.* **93,** 807.
104. Hansen, E., and Hartman, H. (1937). *Plant Physiol.* **12,** 441.
105. Harvey, E. M. (1913). *Bot. Gaz. (Chicago)* **56,** 439.
106. Harvey, R. B. (1925). *Minn., Agr. Exp. Sta., Bull.* **222**.
107. Harvey, R. B. (1928). *Minn., Agr. Exp. Sta., Bull.* **247**.
108. Heck, W. W., and Pires, E. G. (1962). *Tex., Agr. Exp. Sta., Misc. Publ.* **MP-603**.
109. Hobson, G. E., and Davies, J. N. (1971). *In* "The Biochemistry of Fruits and Their Products" (A. C. Hulme, ed.), Vol. 2, p. 439. Academic Press, New York.
110. Holm, R. E., and Abeles, F. B. (1967). *Plant Physiol.* **42,** 1094.
111. Horton, R. F., and Osborne, D. J. (1967). *Nature (London)* **214,** 1086.
112. Hulme, A. C. (1954). *J. Exp. Bot.* **5,** 159.
113. Hulme, A. C., ed. (1970). "The Biochemistry of Fruits and Their Products," Vol. 1. Academic Press, New York.
113a. Hulme, A. C., ed. (1971). "The Biochemistry of Fruits and Their Products," Vol. 2. Academic Press, New York.
114. Hulme, A. C., ed. (1971). *In* "The Biochemistry of Fruits and Their Products," Vol. 2, p. 233. Academic Press, New York.
115. Hulme, A. C., ed. (1971). *In* "The Biochemistry of Fruits and Their Products," Vol. 2, p. 333. Academic Press, New York.
116. Hulme, A. C., Rhodes, M. J. C., and Wooltorton, L. S. C. (1971). *Phytochemistry* **10,** 749.
117. Ito, S. (1971). *In* "The Biochemistry of Fruits and Their Products" (A. C. Hulme, ed.), Vol. 2, p. 281. Academic Press, New York.
118. Jackson, J. M. (1952). *Arkansas Acad. Sci. Proc.* **5,** 73.
119. Jackson, M. B., and Osborne, D. J. (1970). *Nature (London)* **225,** 1019.
120. Jones, J. D., Hulme, A. C., and Wooltorton, L. S. C. (1965). *New Phytol.* **64,** 158.
121. Kaltaler, R. E. L., and Boodley, J. W. (1970). *HortScience* **5,** Sect. 2, 67th Meet. Abstr., p. 356.
122. Kar, B. K., and Banerjee, H. K. (1939–1941). *Trans. Bose Res. Inst., Calcutta* **14,** 91.
123. Kar, B. K., and Banerjee, H. K. (1940). *Curr. Sci.* **9,** 321.
124. Kendall, J. N. (1918). *Univ. Calif., Berkeley, Publ. Bot.* **5,** 347.
125. Kidd, F., and West, C. (1925). *Gt. Brit., Dep. Sci. Ind. Res., Food Invest. Bd., Rep. 1924* p. 27.
126. Kidd, F., and West, C. (1933). *Gt. Brit., Dep. Sci. Ind. Res., Food Invest. Bd., Rep. 1932* p. 55.
127. Kidd, F., and West, C. (1937). *J. Pomol. Hort. Sci.* **14,** 299.
128. Kidd, F., West, C., and Hulme, A. C. (1939). *Gt. Brit., Dep. Sci. Ind. Res., Food Invest. Bd., Rep. 1938* p. 119.
129. Kny, L. (1871). *Bot. Ztg.* **29,** 852.
130. Kraynev, S. I. (1937). *Proc. Agr. Inst. Krasnodar* **6,** 101.
131. Laibach, F. (1933). *Ber. Deut. Bot. Ges.* **51,** 336.
132. Laibach, F. (1933). *Ber. Deut. Bot. Ges.* **51,** 386.

133. La Rotonda, C., Rossi, U., and Petrosini, G. (1943). *Z. Unters. Lenbenmittel.* **85,** 64.
134. Lewis, L. N., Palmer, R. L., and Hield, H. Z. (1968). *In* "Biochemistry and Physiology of Plant Growth Substances" (F. Wightman and G. Setterfield, eds.), p. 1303. Runge Press, Ottawa.
135. Lewis, L. N., and Varner, J. E. (1970). *Plant Physiol.* **46,** 194.
136. Linder, R. C. (1946). *Hawaii, Agr. Exp. Sta., Sta. Progr. Notes* **49**.
137. Lipe, J. A., and Morgan, P. W. (1970). *HortScience* **5,** 266.
138. Lipton, W. J. (1961). *Proc. Amer. Soc. Hort. Sci.* **78,** 367.
139. Livingston, G. A. (1950). *Plant Physiol.* **25,** 711.
140. Looney, N. E., and Patterson, M. E. (1967). *Nature* (*London*) **214,** 1245.
141. Lutz, J. M., and Hardenburg, R. E. (1968). *U. S., Dep. Agr., Agr. Handb.* **66**.
142. Mack, W. B. (1927). *Plant Physiol.* **2,** 103.
143. Mack, W. B., and Livingston, B. E. (1933). *Bot. Gaz.* (*Chicago*) **94,** 625.
144. Mapson, L. W. (1970). *Endeavour* **29,** 29.
145. Marei, N., and Crane, J. C. (1971). *Plant Physiol.* **48,** 249.
146. Marei, N., and Romani, R. (1971). *Plant Physiol.* **48,** 806.
147. Marth, P. C., and Mitchell, J. W. (1949). *Bot. Gaz.* (*Chicago*) **110,** 514.
148. Maxie, E. C., Catlin, P. B., and Hartmann, H. T. (1960). *Proc. Amer. Soc. Hort. Sci.* **75,** 275.
149. Maxie, E. C., Sommer, N. F., Muller, C. J., and Rae, H. L. (1966). *Plant Physiol.* **41,** 437.
150. McCready, R. M., and McComb, E. A. (1954). *Food Res.* **19,** 530.
151. McGlasson, W. B., Palmer, J. K., Vendrell, M., and Brady, C. J. (1971). *Aust. J. Biol. Sci.* **24,** 1103.
152. McMillan, C., and Cope, J. M. (1969). *Amer. J. Bot.* **56,** 600.
153. Meigh, D. F., Jones, J. D., and Hulme, A. C. (1967). *Phytochemistry* **6,** 1507.
154. Milbrath, J. A., and Hartman, H. (1940). *Oreg., Agr. Exp. Sta., Bull.* **385**.
155. Milbrath, J. A., and Hartman, H. (1940). *Science* **92,** 401.
156. Milbrath, J. A., and Hartman, H. (1942). *Oreg., Agr. Exp. Sta., Bull.* **413**.
157. Miller, E. V. (1946). *Bot. Rev.* **12,** 393.
158. Miller, E. V., Winston, J. R., and Schomer, H. A. (1940). *J. Agr. Res.* **60,** 259.
159. Molisch, H. (1911). *Sitzungsberg. Kaiserl. Akad. Wiss. Wien* **120,** 813.
160. Morgan, P. W. (1969). *Plant Physiol.* **44,** 337.
161. Morgan, P. W., and Durham, J. I. (1972). *Plant Physiol.* **50,** 313.
162. Morita, K. (1918). *Bot. Mag.* **32,** 39.
163. Morré, D. J. (1968). *Plant Physiol.* **43,** 1545.
164. Morré, D. J., Rau, B., Vieira, R., Stanceu, T., and Dion, T. (1969). *Proc. Indiana Acad. Sci.* **78,** 146.
165. Mukherjee, S. K., and Dutta, M. N. (1967). *Curr. Sci.* **36,** 674.
166. Nelson, R. C. (1938). *Proc. Minn. Acad. Sci.* **6,** 37.
167. Nichols, R. (1966). *J. Hort. Sci.* **41,** 279.
168. Nichols, R. (1968). *J. Hort. Sci.* **43,** 335.
169. Norman, S., and Craft, C. C. (1968). *HortScience* **3,** 66.
170. Osborne, D. J. (1968). *In* "Biochemistry and Physiology of Plant Growth Substances" (F. Wightman and G. Setterfield, eds.), p. 815. Runge Press, Ottawa.
171. Overholser, E. L. (1927). *Proc. Amer. Soc. Hort. Sci.* **24,** 256.
172. Palmer, J. K. (1971). *In* "The Biochemistry of Fruits and Their Products" (A. C. Hulme, ed.), Vol. 2, p. 65. Academic Press, New York.
173. Peynaud, E., and Ribéreau-Gayon, P. (1971). *In* "The Biochemistry of Fruits and Their Products" (A. C. Hulme, ed.), Vol. 2, p. 171. Academic Press, New York.

174. Pfützer, G., and Losch, H. (1935). *Umschau* **39,** 202.
175. Pinelle, J. (1930). *Bull. Mens. Soc. Nat. Hort. Fr.* [5] **3,** 235.
176. Pratt, H. K. (1961). *Recent Advan. Bot.* p. 1160.
177. Pratt, H. K. (1971). *In* "The Biochemistry of Fruits and Their Products" (A. C. Hulme, ed.), Vol. 2, 207. Academic Press, New York.
178. Pratt, H. K., and Goeschl, J. D. (1968). *In* "Biochemistry and Physiology of Plant Growth Substances" (F. Wightman and G. Setterfield, eds.), p. 1295. Runge Press, Ottawa.
179. Pridham, A. M. S., and Hsu, R. (1954). *Proc. Northeast Weed Contr. Conf.* **21, 221.**
180. Rakitin, Iu. V. (1946). *Biokhimiya* **11,** 1.
181. Rasmussen, G. K., and Jones, J. W. (1971). *HortScience* **6,** 402.
182. Ratner, A., Goren, R., and Monselise, S. P. (1969). *Plant Physiol.* **44,** 1717.
183. Regeimbal, L. O., Vacha, G. A., and Harvey, R. B. (1927). *Plant Physiol.* **2,** 357.
184. Rhodes, M. J. C., Gaillard, T., Woolterton, L. S. C., and Hulme, A. C. (1968). *Phytochemistry* **7,** 405.
185. Richards, H. M., and MacDougal, D. T. (1904). *Bull Torrey Bot. Club* **3,** 57.
186. Ridge, I., and Osborne, D. J. (1969). *Nature* (*London*) **223,** 318.
187. Rood, D. H., Cook, H. T., and Redit, W. H. (1951). *U. S., Dep. Agr., Biblio. Bull.* 13.
188. Rood, P. (1956). *Proc. Amer. Soc. Hort. Sci.* **68,** 296.
189. Rossi, U. (1934). *Boll. Tec. Regio Ist. Sper. Coltiv. Tab. "Leonardo Angeloni"* **30,** 221.
190. Rowan, K. S., Pratt, H. K., and Robertson, R. N. (1958). *Aust. J. Biol. Sci.* **2,** 329.
191. Rubinstein, B., and Abeles, F. B. (1965). *Bot. Gaz.* (*Chicago*) **126,** 255.
192. Ryerson, N. (1927). *Calif., Agr. Exp. Sta., Bull.* **416.**
193. Sacher, J. A., and Salminen, S. O. (1969). *Plant Physiol.* **44,** 1371.
194. Schwertner, H. A., and Morgan, P. W. (1966). *Plant Physiol.* **41,** 1513.
195. Scott, K. J., McGlasson, W. B., and Roberts, E. A. (1970). *Aust. J. Exp. Agr. Anim. Husb.* **10,** 237.
196. Sievers, A. F., and True, R. H. (1912). *U. S., Dep. Agr., Bur. Plant Ind., Bull.* **232.**
197. Sims, W. L., Collins, H. B. and Gledhill, B. L. (1970). *Calif. Agr.* **24,** 4.
198. Smith, W. H. (1939). *Gt. Brit., Dep. Sci. Ind. Res., Food Invest. Bd., Rep. 1938* p. 165.
199. Smith, W. H. (1940). *J. Pomol. Hort. Sci.* **18,** 74.
200. Soldatenkov, S. V. (1935). *Dokl. Akad. Nauk SSSR* **2,** 313.
201. Sorber, D. G. (1934). *Calif. Cultiv.* **81,** 125.
202. Sorber, D. G. (1934). *Diamond Walnut News* June, p. 3.
203. Sorber, D. G., and Kimball, M. H. (1950). *U. S., Dep. Agr., Tech. Bull.* **996**.
204. Spencer, M. (1965). *In* "Plant Biochemistry" (J. Bonner and J. E. Varner, eds.), 2nd ed., p. 793. Academic Press, New York.
205. Steffens, G. L., Alphin, J. G., and Ford, Z. T. (1970). *Beitr. Tabakforsch.* **5,** 262.
206. Stone, G. E. (1913). *Mass. Agr. Exp. Sta., Bull.* **31,** 45.
207. Stösser, R. (1970). *Planta* **90,** 299.
208. Stösser, R. (1971). *Z. Pflanzenphysiol.* **64,** 328.
209. Strachan, G. (1970). *Orchardist, N. Z.* **43,** 32.
210. Tager, J. M., and Biale, J. B. (1957). *Physiol. Plant.* **10,** 79.
211. Thompson, W. W. (1969). *Proc. Int. Citrus Symp. 1st,* Vol. 3. p. 1163.
212. Ting, S. V., and Attaway, J. A. (1971). *In* "The Biochemistry of Fruits and Their Products" (A. C. Hulme, ed.), Vol. 2, p. 107. Academic Press, New York.
213. Uota, M. (1955). *Proc. Amer. Soc. Hort. Sci.* **65,** 231.
214. Uota, M. (1969). *J. Amer. Soc. Hort. Sci.* **94,** 598.
215. Valdovinos, J. G., Ernest, L. C., and Henry, E. W. (1967). *Plant Physiol.* **42,** 1803.
216. Vendrell, M. (1969). *Aust. J. Biol. Sci.* **22,** 601.

217. Vendrell, M. (1970). *Aust. J. Biol. Sci.* **23,** 553.
218. Vendrell, M. (1970). *Aust. J. Biol. Sci.* **23,** 1133.
219. von Loesecke, H. W. (1950). "Bananas." Wiley (Interscience), New York.
220. Wade, N. L., and Brady, C. J. (1971). *Aust. J. Biol. Sci.* **24,** 165.
221. Wang, C. Y., Mellenthin, W. M., and Hansen, E. (1972). *J. Amer. Soc. Hort. Sci.* **97,** 9.
222. Watada, A., and Morris, L. L. (1967). *Plant Physiol.* **42,** 757.
223. Weaver, R. J., and Pool, R. M. (1969). *J. Amer. Soc. Hort. Sci.* **94,** 474.
224. Webster, B. D. (1968). *Plant Physiol.* **43,** 1512.
225. Wehmer, C. (1900). *Z. Pflanzenkr.* **10,** 267.
226. Wilkins, H. F. (1965). Ph.D. Thesis, University of Illinois, Urbana.
227. Wilkinson, B. G. (1963). *Nature (London)* **199,** 715.
228. Winston, J. R. (1955). *U. S., Dep. Agr. Circ.* **961**.
229. Wright, H. B., and Heatherbell, D. A. (1967). *N. Z. J. Agr. Res.* **10,** 405.
230. Young, R. E., Romani, R. J., and Biale, J. B. (1962). *Plant Physiol.* **37,** 416.
231. Zimmerman, P. W., Crocker, W. C., and Hitchcock, A. E. (1930). *Proc. Amer. Soc. Hort. Sci.* **27,** 53.
232. Zimmerman, P. W., Crocker, W. C., and Hitchcock, A. E. (1933). *Contrib. Boyce Thompson Inst.* **5,** 195.
233. Zimmerman, P. W., Hitchcock, A. E., and Crocker, W. (1931). *Contrib. Boyce Thompson Inst.* **3,** 459.

Chapter 8

Regulation of Metabolic and Physiological Systems by Ethylene

I. Role of Ethylene in Disease and Disease Resistance

Increased ethylene production is often associated with plant pathogens or diseased tissue. As discussed more fully in Chap. 5, increased ethylene production has been observed in diseases caused by viruses, bacteria, and fungi. In some cases, the host is the source of ethylene, while in others, the parasite produces ethylene. Does the increase in ethylene production play a role in pathogenesis or is it merely symptomatic of stressed and damaged tissue? A number of potential roles for ethylene exist. They include aiding the progress of the pathogen by accelerated senescence and destruction of host tissue, accelerating abscission of diseased tissue so as to reduce further disease damage, or acting as a regulator to induce disease resistance. In the latter case, ethylene may diffuse from diseased tissue into surrounding healthy tissue which in turn induces the activation of a disease-resistance mechanism. Evidence in favor of all three concepts has been reported and the following discussion summarizes the status of current research in this area.

A. Increased Disease Susceptibility

Most investigators have observed that fruits or plants treated with ethylene decay or rot faster than untreated controls. This effect has been reported for plums (9, 163), celery (122), vanilla beans (11), and citrus (15, 20, 26,

66–68, 98, 119, 181). The acceleration of disease in fruits may represent accelerated development of the pathogen because of improved growth conditions in ripe tissue as opposed to any specific effect of ethylene itself. Also, the environment under which ethylene is applied may be more favorable for fungal growth. In support of the former idea, Winston and Roberts (181) reported that washed and processed oranges were as sensitive to disease whether they were treated with ethylene or not.

Ethylene promoted disease development in tomato and wheat plants by converting disease-resistant varieties into disease-susceptible ones. Collins and Scheffer (39) reported that the Jefferson variety of tomatoes was resistant to *Fusarium oxysporum* while Bonny Best was not. Ethylene was applied to the resistant Jefferson variety as a water-saturated solution. Inoculated resistant varieties treated with ethylene developed no symptoms. In comparison, susceptible Bonny Best normally developed disease symptoms in 3 days. *Fusarium* was recovered from all parts of ethylene-treated resistant tissue but was found only in the basal part of inoculated untreated resistant cuttings. Ethylene also increased the rate of disease development in the susceptible Bonny Best variety.

Wheat plants infected with rust (*Puccinia graminis*) were found to produce more ethylene than healthy controls (45), and greater rates of ethylene production were observed with plants normally considered susceptible. However, the conversion of resistant plants to susceptibility was not specific for ethylene. The same effect was observed by raising the temperature from 20° to 25°C.

B. Decreased Disease Susceptibility

In a few cases, ethylene has been found to decrease disease development. Lockhart *et al.* (117) reported that increasing ethylene levels decreased rot development in apples by *Gloeosporium album*. In their experiments they found that ethylene reduced fungal growth in culture and postulated that the action of the gas was directed against the pathogen as opposed to increasing the resistivity of the host. Less mold was observed on walnuts dehulled with an ethylene treatment (166). However, this could be due to the fact that rapid dehulling resulted in less opportunity for fungal development.

Reduced disease development has also been reported after citrus was degreened with Ethrel (184). When Ethrel was applied as a preharvest spray it accelerated degreening of Robinson and Lee tangerines and reduced the incidence of decay from 25% for controls to 4% for treated fruit.

C. Antibiotic Systems

Stahmann and co-workers have developed the concept that peroxidase may play a role in disease development in sweet potato roots (168). They showed that varieties of sweet potatoes resistant to black rot (*Ceratocystis fimbriata*) had higher levels of peroxidase than susceptible ones. This observation has been confirmed by others (61). Sweet potato roots treated with ethylene developed higher levels of peroxidase and were found to be more resistant to pathogenic strains of *Ceratocystis*. However, Chalutz and DeVay (36) were unable to confirm these observations. They reported that ethylene had no effect on the rate of disease development in sweet potato roots challenged with spores of *Ceratocystis*.

In addition to peroxidase, other enzymes and metabolic products of diseased tissue have been found to increase following ethylene treatment. Abeles *et al.* (3) found that ethylene increased β-1,3-glucanase and chitinase in bean leaves, and, once formed, remained stable for at least 3 days. The function of these enzymes in plants not known since, except for phloem callose, mature vegetative structures of plants contain little β-1,3-linked glucose and no chitin. However, ethylene has been shown to decrease phloem callose (4, 158) though the significance of these observations is unclear. Fungi on the other hand contained both β-1,3-linked glucans and chitin in their cell walls. Because of this it was suggested that β-1,3-glucanase and chitinase served as defense mechanisms against pathogens because they could attack and degrade fungal cell walls but have no effect on wall material of the host. Fungi are known to operate in much the same fashion. In the progress of invasion of the host tissue they produce cellulases and pectin lyases which attack the host's cell wall but do not alter the pathogen's cell wall which does not contain pectin or cellulose. The role of β-1,3-glucanase and chitinase was tested by treating bean plants with ethylene prior to exposure to bean rust (*Uromyces phaseoli*). Abeles *et al.* (8) found that while ethylene did not decrease the total number of pustules ultimately formed as a result of a challenge with *Uromyces* it slowed disease development.

The role of peroxidase has been discussed above in conjunction with disease development in sweet potato. Little is really known concerning the true function of this enzyme though it has been established that most of the peroxidase activity is localized in the cell wall. In this location it may play a role in either cell wall development or as a first line of resistance to pathogenic invasion by forming antibiotic oxidation products of phenols or related compounds. However, Daly *et al.* (45) claimed that they found no correlation between disease resistance of wheat to rust infections and levels of peroxidase in the leaves.

Ethylene has been shown to induce a series of compounds called phytoalexins. These compounds are thought to have antifungal activity and are supposed to be produced by the host in response to invasion by a pathogen. Two such compounds are pisatin (I) and isocoumarin (II) (see Fig. 8-1). Chalutz and Stahmann (38) reported that ethylene increased the concen-

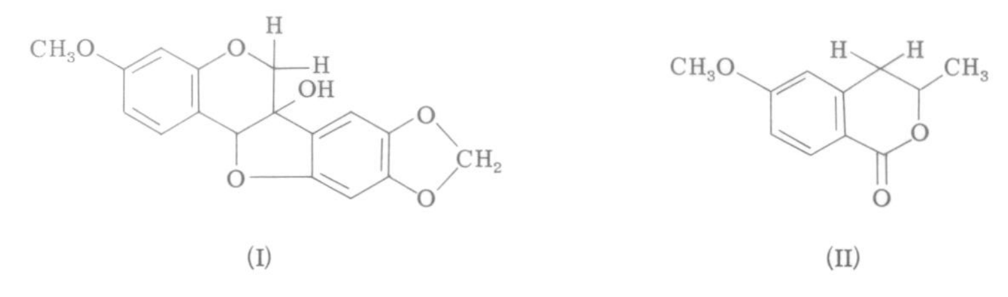

Fig. 8-1. Structure of pisatin (I) and isocoumarin (II).

tration of pisatin in pea pods. Pisatin was also induced by the fungus *Monilinia fructicola*. However, in contrast to many reports on the effect of fungi on ethylene production by the host, *Monilinia* did not cause an increase in ethylene production. Apparently, pisatin production can be regulated by more than one mechanism.

Isocoumarin was identified by Sondheimer (164, 165) and others (35) as the bitter component of carrots stored in the presence of apples and ethylene. Sondheimer (165) noted that similar compounds occurred in *Aspergillus melleus* and that isocoumarin did not have any antibiotic properties against *Micrococcus pyogenes*, *Bacillus subtilis*, and *Serratia marcescens*. Chalutz *et al.* (37) noted that the ability of a variety of isolates of *Ceratocystis fimbriata* and *Helminthosporium carbonum* to induce isocoumarin was directly correlated with their ability to increase ethylene production. However, they also doubted that these compounds had any antibiotic activity. They observed that there was no correlation between the sensitivity of the fungi to isocoumarin and the rate the organisms induced it.

Other compositional changes in plants following ethylene treatment include increases in chlorogenic acid (94) and decreases in phenolics (72), sulfhydryl groups (80), and ascorbic acid (170, 171). The role of these compounds in disease resistance or susceptibility is unknown.

At the present time we know that an increase in ethylene production is often associated with plant disease and plant pathogens. However, we do not know the significance of disease-related ethylene production in the course of disease development.

II. Role of Ethylene in Enzyme Synthesis

A. Acid Phosphatase

An increase in acid phosphatase following ethylene fumigation of cotton was reported by Herrero and Hall (80). Madeikyte and Turkova (123) found that adenosine triphosphatase in tomato and sunflower plants increased after they were sprayed with ethylene-saturated water. During a study showing increases in RNA and decreases in DNA content of tomato plants, Turkova *et al.* (172) reported that ethylene doubled ATPase activity of leaf and stem tissue. However, Stahmann *et al.* (168), studying the role of ethylene in disease resistance, reported that ethylene had no effect on acid or alkaline phosphatase content of sweet potato.

B. α-Amylase

Increase in α-amylase activity from cotton leaves (80), rice seedlings (107), and sweet potato roots (168) following the application of ethylene has been observed. Similarly, increases in starch hydrolysis and α-amylase activity in tobacco have been reported (107, 161) though prolonged exposures were found to decrease enzyme activity. Results with barley seeds have been variable. Nord and Weichherz (138) claimed that ethylene increased both germination and α-amylase formation while Scott and Leopold (157) found that it inhibited α-amylase formation. Similarly, high concentrations of CO retarded α-amylase development in barley (143).

C. Catalase

Potato sprouts (56), cotton leaves (80), pears, oranges, cucumbers (96), and mangoes (13, 14) have been found to have greater quantities of catalase following ethylene treatment than controls. In the case of mangoes, the increase was thought to be due to the removal of a heat-stable inhibitor of catalase activity (125, 126).

D. Cellulase

The original observation of Horton and Osborne (85) that ethylene increased, and auxin decreased, cellulase activity in separation layer cells has been confirmed by a number of other workers (2, 7, 42, 116, 147). However, Morré raised the question that additional enzymes may be involved (136). The induction of cellulase in abscission was shown to represent de novo synthesis by Lewis and Varner (116). Recently Abeles and Leather (7) suggested that ethylene was required for the release or secretion of cellulase in addition to its synthesis.

E. Chitinase

Abeles *et al.* (3) reported that as much as 4% of the soluble leaf protein of ethylene-treated leaves was chitinase. A 3-day fumigation with 10 ppm ethylene resulted in a 50-fold increase in enzyme activity. The function of this enzyme is unknown, though it was suggested that it served an antibiotic function since plants do not contain chitin while insects and fungi do.

F. Chlorophyllase

Numerous workers have shown that ethylene causes the removal of chlorophyll from fruits and leaves. However, the biochemistry of chloroplast degradation remains almost completely unknown. Looney and Patterson (118) presented evidence that chlorophyllase activity increased with the climacteric and ripening of apples and bananas and suggested that this enzyme played some role in chloroplast destruction.

G. Phenylalanine Ammonia-Lyase (PAL) and Cinnamic Acid 4-Hydroxylase (CAH)

These enzymes convert phenylalanine into *p*-coumaric acid with *trans*-cinnamic acid as an intermediate. Ethylene has been shown to increase levels of PAL in sweet potato (92, 94), citrus flavedo (155), and excised sections of cucumber seedlings (57). However, ethylene fumigation of intact cucumber seedlings causes a reduction in PAL. Unlike the usual dose-response curve, high concentrations of ethylene had little effect on PAL in either isolated sections or intact seedlings. This is in contrast to the results with sweet potato, which yielded more characteristic dose-response curves. Riov *et al.* (155) found that continuous ethylene treatment was required to maintain high PAL levels. The PAL content decreased rapidly once ethylene was removed. They also found that protein synthesis was probably required for the increase in PAL since the effect of ethylene was blocked by cycloheximide.

The product of PAL activity is *trans*-cinnamic acid. CAH converts this molecule into *p*-coumaric acid. Hyodo and Yang (90) reported that ethylene increased CAH activity in etiolated pea seedlings and that cycloheximide prevented the appearance of the enzyme.

H. Cytochrome c Reductase

In a study on enzymatic changes in ripening apples, Jones *et al.* (99) reported that levels of cytochrome c reductase increased.

I. Diaphorase

Levels of diaphorase were reported to rise in ripening apples (99).

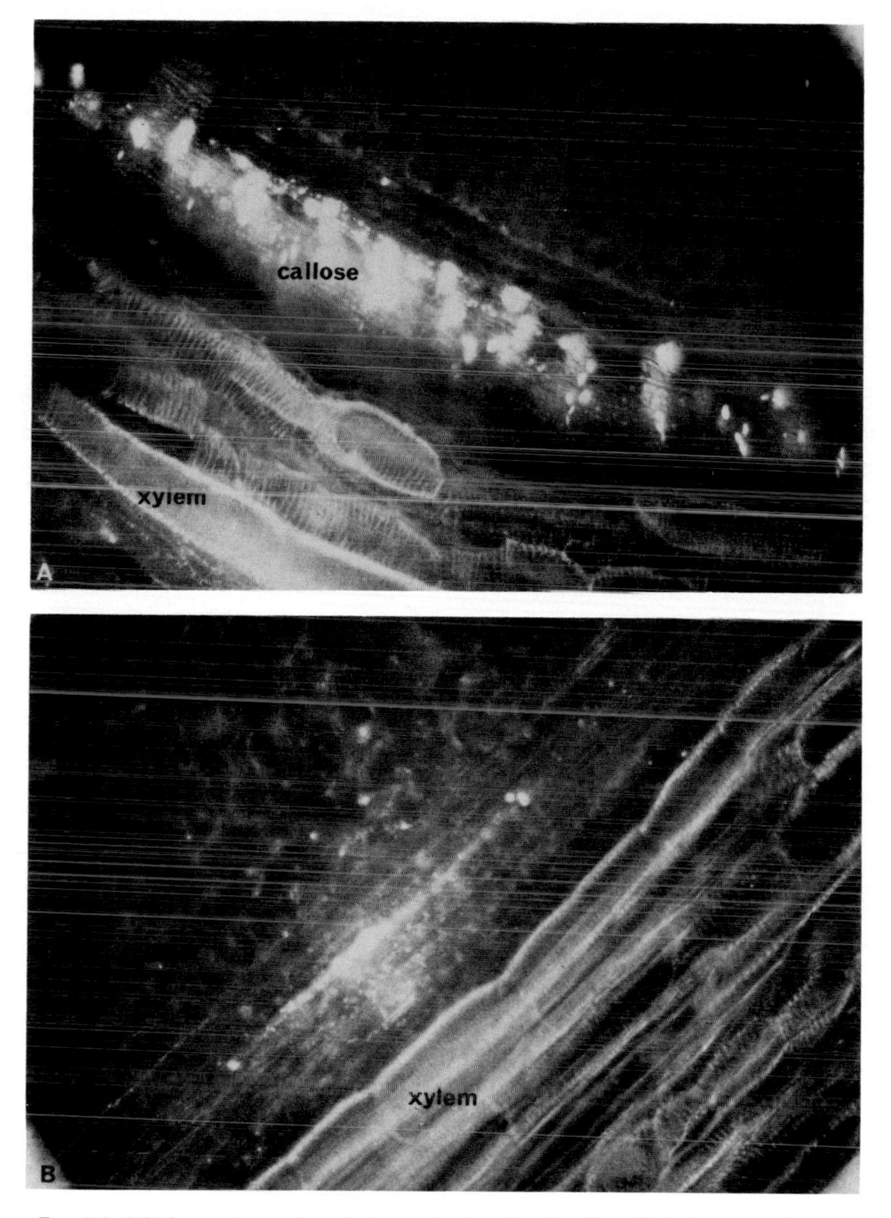

Fig. 8-2. UV fluorescence microphotographs showing the effect of ethylene on phloem callose deposits in bean petiole tissue. Illuminated spots above the xylem represent fluorescence by accumulated aniline blue absorbed by phloem callose. A, Vascular tissue before ethylene treatment. B, Reduction in callose deposits after 3-day 10-ppm ethylene treatment.

J. β-1,3-Glucanase

Abeles and Forrence found that β-1,3-glucanase activity increased in excised sections of bean leaves or in ethylene-treated plants. Induction was blocked by cycloheximide but not actinomycin D. The molecular weight of the enzyme was 12,000 and had an isoelectric point of pH 11 (3). The function of the enzyme is unknown, but it has been suggested that it controlled callose levels in the phloem (see Fig. 8-2A and 2B) or served as an antibiotic (3).

K. Invertase

Ethylene has been reported to increase levels of invertase in tomatoes, oranges, tangerines (112), apples, persimmons (96), cucumbers (95, 96), pineapples (148), tobacco (160, 161), and yeast (159). Invertase levels of lemons were not affected by ethylene (112). Scott and Leopold (157) found that ethylene reduced invertase activity in sugar beet but that gibberellic acid overcame the inhibition.

L. Lipase

Padoa (143) fumigated castor bean seeds with CO and found an increase in lipase activity. However, high concentrations were used (50%) and it is possible that the effect represents a change in respiratory patterns.

M. Malic Enzyme

Hulme and co-workers (88, 99) have found that the increase in malic enzyme coincided with the climacteric of apples. This enzyme converts malic acid into CO_2 and pyruvic acid and is thought to play a role in the production of CO_2 during enhanced respiration associated with ripening. Studies with isolated peel disks showed that the malic enzyme increased after a 6-hour lag and that cycloheximide and actinomycin D blocked the formation of the enzyme (152). The lag period was reduced when the peel disks were treated with ethylene.

N. Pectin Esterase

Herrero and Hall (80) reported that a 15-hour, 100-ppm ethylene treatment increased levels of pectin esterase by 138% in cotton leaves.

O. Peroxidases

While the true functions for these enzymes have not been established, they are easy to measure and continually attract the interest of plant scientists. Investigators have found that ethylene will increase peroxidase activity in tobacco (12), cotton (72, 80), sweet potato (61, 91, 92, 94), potato, burdock, radish (91), cucumbers (95), persimmons (96), and mangoes (125, 126). However, it has little or no effect on peroxidase activity in parsnips, carrot, radish (168), or bean (61). Studies on peroxidase have been hampered or complicated by the fact that tissues often contain dialysable inhibitors which can mask peroxidase activity (72, 125, 126). Induction of certain isozymes of peroxidase in sweet potato roots occurs within 4 hours and was blocked by cycloheximide. Actinomycin D, however, had only a partial effect (61). Similar results with pea seedlings have been obtained by Osborne and coworkers (141, 153).

Peroxidase is unusual in that it contains hydroxyproline, and as such, accounts for some of the hydroxyproline associated with plant cell walls. Ridge and Osborne (154) found that ethylene increased hydroxyproline in the cell walls of pea seedlings and at the same time, levels of peroxidase. Most plant peroxidase is located in the cell wall region and is either soluble, ionically bound, or covalently attached to the wall. Ridge and Osborne (154) reported that the largest increase in peroxidase was in the ionically bound fraction.

Hall and Morgan (72) have proposed that the peroxidase represents an auxin oxidase and postulated that the reduction of auxin in tissue after ethylene treatment was due to the activity of this enzyme.

Stahmann *et al.* (168) have proposed that ethylene-induced peroxidase may play a role in disease resistance. Generally speaking, varieties of sweet potato resistant to black rot (*Ceratocystis fimbriata*) have higher peroxidase levels than susceptible ones (61, 168). Stahmann *et al.* (168) found that when ethylene induced resistance to black rot, it also caused an increase in peroxidase activity.

P. Polygalacturonase

McCready and McComb (127) found that polygalacturonase increased during the ripening of pears and avocados. They postulated that it was responsible for the soft consistency of ripening fruit.

Q. Polyphenol Oxidase

Polyphenol oxidase is a copper-containing enzyme which catalyzes the oxidation of phenols such as tyrosine and is responsible for the blackening of cut raw potatoes on exposure to air. Knapp *et al.* (106) found that peaches treated with Ethrel in an immature stage did not darken as rapidly as controls. They also found that Ethrel-treated fruit had a lower content of polyphenol oxidase. However, the effect was not specific for ethylene, since Alar and gibberellic acid also reduced darkening.

Most workers have reported that ethylene increased levels of polyphenol oxidase. Increases have been observed in cotton (80), sweet potato (92, 94), tobacco (160), potato, and parsnip (168). However, ethylene did not affect polyphenol oxidase activity in carrot and radish (168).

R. Protease

Reports on the induction of protease are limited to observations with pineapple (77, 148). However, the fact that destruction of protein is a characteristic feature of senescence and this process is accelerated by ethylene suggests that induction of protease may be more widely prevalent.

S. Pyruvic Carboxylase

Hulme and co-workers (88, 89, 99) have shown that an increase in CO_2 production followed the addition of ethylene to apple peels. Ethylene also promoted RNA and protein synthesis in this tissue. On the basis of these observations they postulated that pyruvic carboxylase is a part of accelerated respiratory activity in ripening apples and that increases in pyruvic carboxylase reflects de novo synthesis.

III. Effect on Respiration

The ability of ethylene to increase respiration was first described by Denny in 1924 (47, 48). He reported that the active component of kerosene fumes which promoted the degreening of citrus was ethylene, and in addition to hastening ripening, increased the rate of respiration. Later Harvey (77) found that ethylene also enhanced the respiration of bananas. Kidd and West demonstrated that the respiration of apples normally increased during ripening. By 1933 (105) they were aware that apples produced ethylene and that ethylene induced the climacteric rise of CO_2 production. Since that time, numerous investigators have shown that ethylene promotes or increases respiration from a variety of fruits and other plant tissues.

A. Fungi

Shaw (159) reported that ethylene promoted CO_2 production by yeast. The significance of his finding was difficult to interpret because the response was obtained after the yeast cells were added to sucrose which was given a prior treatment with ethylene. Vesselov (177) reported that 1000 ppm ethylene stimulated the respiration of *Aspergillus* by 10%. In spite of these reports, it is doubtful that ethylene has a significant effect on the respiration of fungi.

B. Vegetative Tissue

Almost without exception, ethylene has no effect on the respiration of leaf (5, 79) and seedling (27, 31, 63, 156, 162) tissue. However, Hale *et al.* (70) claimed ethylene increased the respiration of wheat seedlings (cf. 156) and Shcherbakov (160) reported that ethylene increased the rate of respiration of tobacco leaves.

C. Storage Organs

A number of investigators have shown that ethylene promotes respiration of potato tuber (34, 86, 87, 150) and swede and parsnip root tissue (151).

The ability of ethylene to increase the respiration of potato tubers varies according to the age of the tissues. Huelin (86) and Reid and Pratt (150) reported that ethylene increased tuber respiration shortly after harvest but had little or no effect on tissue stored for long periods of time. On the other hand, Huelin and Barker (87) reported that the response to ethylene increased with time after harvest. They observed no effect between 2 and 34 days, and a 29% to 53% increase in respiration after 63 and 98 days, respectively. Ethylene increased respiration without increasing sugar content of young tubers. Tubers which had been stored for a period of time showed both an increase in respiration and sugar content (86, 87). Treatment of potato tubers with ethylene caused a rise in respiration after an 8-hour lag reaching 5–10 times the rate of controls after 30 hours of treatment, and then falling slowly (86, 150). The concentration required for an effect was high compared to other ethylene-mediated processes. A concentration of 1 ppm gave a 28% response, 10 ppm, 68%, and 100 ppm, 96% (86). The effect of ethylene was reversed by the addition of CO_2 to the gas phase (34).

D. Flowers

The respiration of roses (101) and carnations (137) falls after harvest and then increases again as the flowers undergo senescence. Ethylene enhances both the increase in respiration and premature fading or senescence of the blossoms.

E. Fruits

Except for cranberries (60), ethylene has been shown to increase the respiration of all fruits tested. The list of sensitive fruits includes apple (105), avocado (146), banana (62), citrus (47), cucumber (96), fig (124), mango (104), melon (128), pineapple (54), passion fruit (10), persimmon (46), pear (74), and tomato (21). The response time varies in the case of mangoes (29) and banana (77, 149), to an hour or more in the case of cantaloupes (128), and a day or more in the case of tomato (21) and pear (75). While the increase in respiration is often associated with other aspects of ripening, such as softening and changes in color, these processes are not necessarily linked together in an obligatory fashion. For example, pears can undergo an acceleration of softening by ethylene without undergoing an acceleration of respiration (179). Similarly, ethylene can induce respiration changes in young pineapple fruit without any increase in softening or change in flavor (54). Bananas treated with 2,4-D can still exhibit an increase in respiration after ethylene treatment but do not undergo changes in skin pigmentation (175).

Even though the ability of ethylene to increase fruit respiration is well known, little is understood concerning its mode of action or what aspect of respiration is being affected. Ku and Leopold (110) have shown that mitochondria are not the site of ethylene action in pea and cauliflower plants contrary to claims by others (121, 139, 140). Peas and cauliflower are not normally responsive to ethylene and mitochondria from fruit have not been carefully studied. Hulme and co-workers (88) have shown that levels of pyruvic and malic decarboxylase increase in ripening apples and these changes occur parallel to increases in CO_2 production. The concept that protein synthesis takes place during the increase induced by ethylene is supported by the observation that cycloheximide prevented the climacteric and ripening in pears (58) and bananas (24). It is possible that the action of ethylene is directed more toward the regulation of enzyme synthesis during respiration than a direct control of the Krebs cycle, glycolysis, ATP formation, or electron transport.

IV. Effects on Pigmentation

The degreening of leaves and fruits is a well-known effect of ethylene. In addition to decreasing chlorophyll levels, ethylene has also been reported to hasten or delay the formation of carotenoids and anthocyanins. The significance of these pigment changes is not clear. Degreening is usually associated with abscission and ripening and represents one facet of senescence. Increases in carotenoids and anthocyanins also occur in fruit development or maturation.

A. Chlorophyll

The destruction of chlorophyll is an important part of the blanching of celery (76, 82). The removal of chlorophyll is associated with the disappearance of the acrid bitter taste of unblanched celery. In self-blanching varieties the chlorophyll was removed in 6 days while in the green winter varieties about 10 days were required. Only chlorophyll was destroyed by the ethylene treatment. No effect on carotenoids or anthocyanin was observed in the varieties which contained these pigments (76).

Degreening of fruits is a well-known effect of ethylene and is used commercially in the treatment of citrus and bananas prior to sale in the retail market. As discussed more fully in Chap. 7, the loss of green color is associated with the destruction of chloroplasts.

The removal of chlorophyll is also an important part of tobacco curing. Steffens *et al.* (169) have described the use of Ethrel to accelerate maturation and harvesting of tobacco. Heck and Pires (78) surveyed the effect of ethylene on a wide variety of plants. Enhanced yellowing followed by abscission was observed in many varieties and was a common feature of ethylene damage. However, they noticed that new leaves developed in the presence of ethylene, which, though smaller, were darker than controls. This observation suggests that ethylene action is directed primarily toward the acceleration of chlorophyll breakdown as opposed to an effect on chlorophyll synthesis. Chlorophyll breakdown appears to require protein synthesis since the addition of cycloheximide to leaves (6) and fruits (58) delays the degreening processes. The exact enzymology of chlorophyll degradation is unknown though an increase in chlorophyllase activity has been reported to occur during the maturation of apples (118). Based on these preliminary observations, it appears that degreening of tissue is due to the destruction of chloroplasts by degradative enzymes induced by ethylene where the gas is acting in the capacity of an aging hormone.

B. Carotene

Maturation of tomatoes is associated with a change in pigmentation from green to red. The red color is due to an increase in β-carotene and lycopene (21, 51). Treatment of tomatoes with ethylene induces an increase in the concentration of these pigments after a 3-day lag. Respiration changes occur faster and occur after a 1-day lag.

Kang and Burg (102) have shown that ethylene prevented the accumulation of carotenoids in the shoot apex of the etiolated pea. The inhibition is rapid and occurs within 1 or 2 hours. Ethylene had no effect on carotenoid breakdown because levels of carotene remained constant in the presence of the gas. Carotene synthesis in the tissue was under the control of phytochrome since red light increased the concentration of carotene and far-red light reversed its effect. Kang and Burg (102) postulated that red light regulated carotene levels by controlling endogenous ethylene production. They observed that ethylene production decreased following irradiation with red light and that the addition of ethylene blocked the formation of carotene in the presence of red light. In addition, CO_2 and the growth of the tissue under hypobaric conditions also reduced the levels of carotene in the tissue.

C. Anthocyanin

A number of investigators have studied the effect of ethylene on cranberry development. Fudge (60) noted that ethylene had no effect on anthocyanin levels in cranberries but destroyed chlorophyll. Later Devlin and co-workers (25, 49) reported that cranberries treated with Ethrel developed more anthocyanin than controls. In addition, the vines of sprayed plants also developed more anthocyanin. The difference between these results was explained by the observation that light was required for ethylene-enhanced anthocyanin production by cranberries (41). The experiments by Fudge were performed in a dark incubation chamber while the work with Ethrel had been done in the field.

Hale *et al.* (69) reported that ethylene hastened the ripening of grapes. In addition to changes in composition and texture, maturation of grapes was also associated with the increase in anthocyanin pigments.

A number of workers have shown that ethylene or ethylene analogs inhibit the formation of anthocyanin in seedlings of *Vicia faba* (65, 109, 132), corn (132), bean (135), *Strobilanthus dyerianus* (131), and sorghum (44). The effect of ethylene on anthocyanin biosynthesis in sorghum was found to vary according to the time the tissue was exposed to the gas. Craker *et al.* (44) found that plants receiving ethylene during the early lag phase of pigment synthesis had a higher anthocyanin content at 24 hours than control plants receiving no ethylene treatment. Plants exposed to ethylene after the lag phase had a lower anthocyanin content at 24 hours than controls which had no ethylene treatment.

Promotion of anthocyanin synthesis has also been observed in radish (78) and cotton plants (71, 73) treated with ethylene.

V. Regulation of Hormone Biochemistry and Physiology

A. Synthesis

Changes in auxin levels in plant tissue can be attributed to changes in synthesis, transport, or degradation. Valdovinos *et al.* (173) studied the effect of ethylene on the conversion of tryptophan to auxin in *Coleus*. They found that tryptophan conversion was inhibited by treating plants with ethylene. By measuring the efficiency by which cell-free preparations would decarboxylate tryptophan they showed that cell-free preparations extracted from ethylene-treated plants were less active than controls. However, ethylene had no effect on the preparations themselves. They concluded that ethylene regulated auxin levels by controlling the activity of auxin biosynthesis.

B. Effect on Auxin Levels

Auxin effects are often opposite to those of ethylene. For example, auxin promotes elongation while ethylene inhibits it, and auxin delays abscission while ethylene accelerates it. Because of this, it seems reasonable to postulate that a mechanism of ethylene action includes control of auxin levels in plant tissue. In other words, ethylene might slow growth and promote abscission by reducing the levels of auxin in plant tissue. While a number of investigators have explored this possibility, the evidence is fragmentary, due for the most part to the fact that auxin assays are difficult to perform and are always subject to criticism with regard to specificity and precision. A review of the available data shows that ethylene reduces auxin levels in oat and pea tissue (22, 28, 113, 173, 178). Burg *et al.* (28) have examined this phenomenon more closely and have found that the effect of ethylene varies according to the portion of the seedling under consideration. The largest reduction of auxin occurs in the subapex of pea seedlings, less in the hook region, and an increase in the apical region. Increases in auxin content after ethylene treatment have been observed by a number of other investigators. Dostal (52) reported increases in pea epicotyls following treatment with illuminating gas. He also (53) found that treating potato seed pieces with ethylene gave rise to plants that had higher levels of auxin in their tubers. Turkova (171) reported that ethylene increased the amount of auxin in the base of tomato petioles. Increases in the auxin content of leaf bases may explain the increase in cell size associated with leaf epinasty. The ethylene homolog CO has also been reported to increase auxin levels in leaves of *Mercurialis ambigua* (81).

C. Auxin Transport

Auxin transport is thought to play a role in abscission, growth, and epinasty. It follows that one possible mechanism of ethylene action would center around the ability to reduce auxin movement from the site of production to the site of action. For example, if auxin did not reach the separation layer, abscission would occur, or if auxin failed to reach the zone of elongation, no growth would occur. Evidence has been obtained that ethylene inhibits auxin transport and that this plays a role in the various phenomena described. However, it is important to distinguish between a direct effect on the machinery which moves auxin in plant tissue and an effect on the general metabolism resulting in reduced transport. The way to demonstrate a direct effect of ethylene on auxin transport is to conduct experiments with isolated sections of plants. All of the investigators who have done such experiments have failed to observe an effect of ethylene on auxin transport (1, 30, 32, 113, 142) in stem or petiole tissue.

However, ethylene blocks auxin transport if it is measured in sections isolated from plants pretreated with ethylene (16–19, 32, 133, 134, 142, 144, 178). The response is rapid and can be seen as soon as 1.5 hours after fumigation (17). Auxin transport can be regulated by the amount of auxin moved or by the rate of movement. Burg and Burg (32) have shown that the point of control is at the level of capacity and not velocity of movement. Beyer and Morgan (17) confirmed this finding, though a close examination of the data they presented reveals a slight reduction in velocity as well. Reduced auxin transport after ethylene treatment was shown in a variety of species. Morgan *et al.* (133) found that cotton and okra were highly susceptible, cowpea and English pea were intermediate in sensitivity, while tomato and sunflower showed little response after 15 hours. However, Palmer and Halsall (144) reported that they observed an inhibition of auxin transport in ethylene-fumigated tomato plants.

The reduced capacity for transport can be due to either an effect on uptake or degradation. A number of investigators (16, 18, 30, 32, 178) have shown that ethylene does not influence the uptake of auxin through the cut surface of plant tissue. Similarly, degradation of auxin via decarboxylation does not appear to regulate transport. Burg and Burg (32) and Valdovinos *et al.* (173) found that ethylene had no effect on the formation of $^{14}CO_2$ from [1-^{14}C]IAA. On the other hand, Beyer and Morgan (16, 133) reported an increase in $^{14}CO_2$ production from [1-^{14}C]IAA following ethylene treatment. However, in the same experiments, decarboxylation of [1-^{14}C]NAA was not influenced by ethylene even though transport of this IAA analog was reduced. Hall and Morgan (72) have suggested that an increase in auxin oxidase occurred in cotton leaves following ethylene treatment. However, Gowing and Leeper (64) failed to find an increase in auxin oxidase in pineapples after ethylene treatment.

Beyer and Morgan (17) have found that auxin was readily converted into a number of metabolites, including indoleacetyl aspartate, and that significantly more [^{14}C]IAA was recovered from ethylene-pretreated sections. However, they pointed out that accumulation of auxin metabolites might be the result, instead of the cause, of disrupted auxin transport.

Phototropic and geotropic curvature of certain plants can be blocked by ethylene. Tropistic phenomena involve, among other things, rapid lateral auxin transport, and Burg *et al.* (28) suggested that the cause of ethylene action may be an inhibition of lateral transport. There is good reason to believe that their interpretation is correct. Dostál (52) reported that illuminating gas prevented the lateral transport of auxin in horizontal pea tissue. This was confirmed by Burg and Burg (32), who showed that the effect was very rapid in pea tissue but did not occur to a great extent in oat coleoptiles. Burg *et al.* (28) suggested that their results agree with observations on the effects of ethylene on geotropic curvature of intact plants, since geotropic curvature of peas was effectively blocked by ethylene while it was only partially inhibited in monocot tissue. They also pointed out that ethylene does not prevent phototropic curvature of the fungus *Phycomyces*, which apparently has the same light receptor as higher plants but does not utilize auxin transport to mediate the preception to light. They also proposed that blockage of lateral auxin transport may explain why ethylene prevents nutation and the rapid curvature of isolated pea sections when they are placed in an aqueous media.

Epinasty is thought to involve a growth phenomenon resulting from an accumulation of auxin in the upper side of the petiole. Lyon (120) postulated that this accumulation of auxin resulted from a disruption of lateral auxin transport in the petiole. He found that ethylene altered lateral auxin transport in the petiole and that in the presence of ethylene, auxin accumulated on the upper side of the petiole.

D. Enhancement of Auxin Action

Van der Laan (113) observed that ethylene altered the ability of oat coleoptile sections to respond to auxin. After a 2-hour fumigation, ethylene increased the activity of auxin, and after 6 hours, it decreased auxin activity. Similar experiments were performed by Michener (129). He found that ethylene increased the response of both oat and pea seedlings to auxin. These phenomena have been confirmed by Yamaki (182) and Burg *et al.* (28). Increased sensitivity to auxin may explain the ability of ethylene to promote the growth of rice seedlings (12, 107, 111). Imaseki and Pjon (93) reported that ethylene enhanced the sensitivity of rice coleoptiles to auxin, though ethylene itself had no effect on isolated rice coleoptile sections. A possible explanation of ethylene action on the growth of rice seedlings may be that it enhances the action of endogenous auxin, and it has no separate effect of inhibiting elongation. In general, ethylene is less effective in inhibiting the growth of monocots than dicots.

Another example where ethylene action can be enhanced by auxin is in rooting. Krishnamoorthy (108) reported that Ethrel-induced rooting of mung bean hypocotyls was improved by the addition of IAA.

E. Reversal of Ethylene Action

1. Auxin

The ability of auxin to block abscission in the presence of large quantities of ethylene is a well-known example of the reversal of ethylene action by other plant hormones (23, 55, 115, 130, 180). In a similar fashion, auxin blocked ethylene action in the promotion of ripening (175). The mechanism of action appears to involve the maintenance of a state of juvenility by auxin. As long as the tissue remains in this juvenile stage, ethylene is without effect. The action of auxin is not competitive in the sense that the addition of auxin to an ongoing process will stop ethylene action in much the same fashion that CO_2 acts as a competitive inhibitor. In addition to effects on abscission and ripening, other ethylene-mediated processes can be blocked by auxin. The list includes the induction of β-1,3-glucanase (4), induction of flowering in pineapples (40, 64), promotion of stem growth (83), and maintenance of auxin transport (142).

The inhibition of ethylene action by auxin in the promotion of flowering was originally described by Cooper (40). He reported that pineapple plants treated with NAA and ethylene simultaneously did not flower. When NAA was applied 3 days after ethylene treatment, a collar of undeveloped flowers developed midway on the fruit. The capacity of bean petiole segments to transport [^{14}C]IAA decreases by 90% within 10 hours of excision. Osborne and Mullins (142) reported that ethylene enhanced this loss of auxin transport capacity but a pretreatment with auxin protected the transport system from natural and ethylene-induced aging.

2. Gibberellin

Gibberellic acid has a number of opposing effects to ethylene. The list includes: the induction of β-1,3-glucanase (4), α-amylase (114, 157), and invertase from sugar beet (157), promotion of the growth of soybean (84), lettuce (157), and *Poa pratensis* (174), promotion of hook opening (103, 145), maintenance of auxin levels (173), regulation of female flowers in cucurbits (97, 167), and ripening (176). Application of Ethrel to pumpkins (167) resulted in the development of female flowers. Gibberellin on the other hand caused maleness and when applied together with Ethrel, the plants had longer internodes and a greater number of male flowers than ethylene-treated controls.

3. Cytokinins

Ethylene inhibits the growth of pea buds. Kinetin was shown (33) to reverse this action of ethylene and in turn promote bud growth in the presence of ethylene. Kinetin also blocked the ability of ethylene to induce β-1,3-glucanase (4).

F. Promotion of Ethylene Action

In a few cases, plant hormones have been found to enhance ethylene action. For example, abscisic acid was found to enhance ethylene-induced production of cellulase (43). Jones (100) reported that ethylene had an inhibitory effect on the total amount of α-amylase synthesized by barley halfseeds. When gibberellic acid was added, the release of α-amylase was promoted. Finally, gibberellin has been found to promote abscission in the presence of saturating concentrations of ethylene (115).

VI. Cold Hardiness

A number of workers have explored the possibility that ethylene may regulate the development of cold hardiness in plants. Dollwet and Kumamoto (50) reported that ethylpropylphosphonate, an ethylene-releasing agent, retarded growth and increased cold hardiness in a number of woody plants. Ethrel, on the other hand, had no effect on inducing cold hardiness in citrus (183). However, other chemicals such as malic hydrazine increased cold hardiness while abscisic acid reduced cold hardiness. The effect of ethylene alone on cold hardiness has also been measured. Concentrations of ethylene capable of causing 50% abscission on Red-osier dogwood plants had no effect on the induction of cold hardiness (59). These results suggest that the effect of ethylpropylphosphonate may be due to the chemical itself as opposed to any release of ethylene.

References

1. Abeles, F. B. (1966). *Plant Physiol.* **41,** 946.
2. Abeles, F. B. (1969). *Plant Physiol.* **44,** 447.
3. Abeles, F. B., Bosshart, R. P., Forrence, L. E., and Habig, W. H. (1971). *Plant Physiol.* **47,** 129.
4. Abeles, F. B., and Forrence, L. E. (1970). *Plant Physiol.* **45,** 395.
5. Abeles, F. B., and Gahagan, H. E. (1968). *Plant Physiol.* **43,** 1255.
6. Abeles, F. B., Holm, R. E., and Gahagan, H. E. (1967). *In* "Biochemistry and Physiology of Plant Growth Substances" (F. Wightman and G. Setterfield, eds.), p. 1515. Runge Press, Ottawa.

7. Abeles, F. B., and Leather, G. R. (1971). *Planta* **97,** 87.
8. Abeles, F. B., Leather, G. R., Forrence, L. E., and Craker, L. E. (1971). *HortScience* **6,** 371.
9. Adam, W. B., and Gillespy, T. G. (1942). *Annu. Rep. Fruit Veg. Preserv. Campden Res. Sta.* p. 42.
10. Akamine, E. K., Young, R. E., and Biale, J. B. (1957). *Proc. Amer. Soc. Hort. Sci.* **69,** 221.
11. Arana, F. E. (1944). *U. S., Dep. Agr., Fed. Exp. Sta., Puerto Rico, Bull.* **42.**
12. Asmaev, P. G. (1937). *Proc. Agr. Inst. Krasnodar* **6,** 49.
13. Banerjee, H. K., and Kar, B. K. (1939). *Trans. Bose Res. Inst., Calcutta* **14,** 171.
14. Banerjee, H. K., and Kar, B. K. (1941). *Curr. Sci.* **10,** 289.
15. Ben-Yehoshua, S., and Eaks, I. L. (1969). *J. Amer. Soc. Hort. Sci.* **94,** 292.
16. Beyer, E. M., and Morgan, P. W. (1969). *Plant Cell Physiol.* **10,** 787.
17. Beyer, E. M., and Morgan, P. W. (1969). *Plant Physiol.* **44,** 1690.
18. Beyer, E. M., and Morgan, P. W. (1970). *Plant Physiol.* **46,** 157.
19. Beyer, E. M., and Morgan, P. W. (1971). *Plant Physiol.* **48,** 208.
20. Biale, J. B. (1948). *Proc. Amer. Soc. Hort. Sci.* **52,** 187.
21. Boe, A. A., and Salunkhe, D. K. (1967). *Econ. Bot.* **21,** 312.
22. Botjes, J. O. (1942). *Proc. Kon. Ned. Akad. Wetensch.* **45,** 999.
23. Bradley, M. V., Marei, N., and Crane, J. C. (1969). *J. Amer. Soc. Hort. Sci.* **94,** 316.
24. Brady, C. J., Palmer, J. K., O'Connell, P. B. H., and Smillie, R. M. (1970). *Phytochemistry* **9,** 1037.
25. Bramlage, W. J., and Devlin, R. M. (1970). *HortScience* **5,** Sect. 2, Abstr. 67th Meet., p. 359.
26. Brooks, C. (1944). *J. Agr. Res.* **68,** 363.
27. Burg, S. P. (1964). *In* "Regulateurs naturels de la croissance végétale" (J. P. Nitsch, ed.), p. 719. CNRS, Paris.
28. Burg, S. P., Apelbaum, A., Eisinger, W., and Kang, B. G. (1971). *HortScience* **6,** 359.
29. Burg, S. P., and Burg, E. A. (1962). *Plant Physiol.* **37,** 179.
30. Burg, S. P., and Burg, E. A. (1966). *Proc. Nat. Acad. Sci. U. S.* **55,** 262.
31. Burg, S. P., and Burg, E. A. (1967). *Plant Physiol.* **42,** 144.
32. Burg, S. P., and Burg, E. A. (1967). *Plant Physiol.* **42,** 1224.
33. Burg, S. P., and Burg, E. A. (1968). *Plant Physiol.* **43,** 1069.
34. Burton, W. G. (1952). *New Phytol.* **51,** 154.
35. Carlton, B. C., Peterson, C. E., and Tolbert, N. E. (1961). *Plant Physiol.* **36,** 550.
36. Chalutz, E., and DeVay, J. E. (1969). *Phytopathology* **59,** 750.
37. Chalutz, E., DeVay, J. E., and Maxie, E. C. (1969). *Plant Physiol.* **44,** 235.
38. Chalutz, E., and Stahmann, M. A. (1969). *Phytopathology* **59,** 1972.
39. Collins, R. P., and Scheffer, R. P. (1958). *Phytopathology* **48,** 349.
40. Cooper, W. C. (1942). *Proc. Amer. Soc. Hort. Sci.* **41,** 93.
41. Craker, L. E. (1971). *HortScience* **6,** 137.
42. Craker, L. E., and Abeles, F. B. (1969). *Plant Physiol.* **44,** 1139.
43. Craker, L. E., and Abeles, F. B. (1969). *Plant Physiol.* **44,** 1144.
44. Craker, L. E., Standley, L. A., and Starbuck, M. J. (1971). *Plant Physiol.* **48,** 349.
45. Daly, J. M., Seevers, P. M., and Ludden, P. (1970). *Phytopathology* **60,** 1648.
46. Davis, W. B., and Church, C. G. (1931). *J. Agr. Res.* **42,** 165.
47. Denny, F. E. (1924). *J. Agr. Res.* **27,** 757.
48. Denny, F. E. (1924). *Bot. Gaz.* (*Chicago*) **7,** 322.
49. Devlin, R. M., and Demoranville, I. E. (1970). *Physiol. Plant* **23,** 1139.
50. Dollwet, H. H. A., and Kumamoto, J. (1970). *Plant Physiol.* **46,** 786.
51. Dostal, H. C., and Leopold, A. C. (1967). *Science* **158,** 1579.
52. Dostal, R. (1942). *Jahrb. Wiss. Bot.* **90,** 199.

53. Dostál, R. (1944). *Bodenk. Pflanzenernaehr.* **33,** 215.
54. Dull, G. G., Young, R. E., and Biale, J. B. (1967). *Physiol. Plant.* **20,** 1059.
55. Edgerton, L. J. (1968). *Proc. N. Y. State Hort. Soc.* **113,** 99.
56. Elmer, O. H. (1936). *J. Agr. Res.* **52,** 609.
57. Engelsma, G., and Van Bruggen, J. M. H. (1971). *Plant Physiol.* **48,** 94.
58. Frenkel, C., Klein, I., and Dilley, D. R. (1968). *Plant Physiol.* **43,** 1146.
59. Fuchigami, L. H., Weiser, C. J., and Evert, D. R. (1971). *Plant Physiol.* **47,** 98.
60. Fudge, B. R. (1930). *N. J., Agr. Exp. Sta., Bull.* **504.**
61. Gahagan, H. E., Holm, R. E., and Abeles, F. B. (1968). *Physiol. Plant.* **21,** 1270.
62. Gane, R. (1936). *Gt. Brit., Dep. Sci. Ind. Res., Food Invest. Bd., Annu. Rep. 1935* p. 123.
63. Goeschl, J. D., Rappaport, L., and Pratt, H. K. (1966). *Plant Physiol.* **41,** 877.
64. Gowing, D. P., and Leeper, R. W. (1961). *Bot. Gaz. (Chicago)* **123,** 34.
65. Grafe, V., and Richter, O. (1912). *Sitzungsber. Kaiserl. Akad. Wiss. Wien* **120,** 1187.
66. Grierson, W. (1956). *Proc. Fla. State Hort. Soc.* **39,** 165.
67. Grierson, W., and Newhall, W. F. (1955). *Proc. Amer. Soc. Hort. Sci.* **65,** 244.
68. Grierson, W., and Newhall, W. F. (1960). *Fla., Agr. Exp. Sta., Bull.* **620.**
69. Hale, C. R., Coombe, B. G., and Hawker, J. S. (1970). *Plant Physiol.* **45,** 620.
70. Hale, W. S., Schwimmer, S., and Bayfield, E. G. (1943). *Cereal Chem.* **20,** 224
71. Hall, W. C. (1952). *Bot. Gaz. (Chicago)* **113,** 310.
72. Hall, W. C., and Morgan, P. W. (1964). *In* "Regulateurs naturels de la croissance végétale" (J. P. Nitsch, ed.), p. 727. CNRS, Paris.
73. Hall, W. C., Truchelut, G. B., Leinweber, C. L., and Herrero, F. A. (1957). *Physiol. Plant.* **10,** 306.
74. Hansen, E. (1939). *Plant Physiol.* **14,** 145.
75. Hansen, E. (1943). *Proc. Amer. Soc. Hort. Sci.* **43,** 69.
76. Harvey, R. B. (1925). *Minn. Hort.* **53,** 41.
77. Harvey, R. B. (1928). *Minn., Agr. Exp. Sta., Bull.* **247.**
78. Heck, W., and Pires, E. (1962). *Tex., Agr. Exp. Sta., Misc. Publ.* **MP-613.**
79. Heck, W., Pires, E., and Hall, W. C. (1961). *J. Air Pollut. Contr. Asso.* **11,** 549.
80. Herrero, F., and Hall, W. C. (1960). *Physiol. Plant.* **13,** 736.
81. Heslop-Harrison, J., and Heslop-Harrison, Y. (1957). *New Phytol* **56,** 352.
82. Hibbard, R. P. (1930). *Mich., Agr. Exp. Sta., Tech. Bull.* **104.**
83. Holm, R. E., and Abeles, F. B. (1967). *Planta* **78,** 293.
84. Holm, R. E., and Key, J. L. (1969). *Plant Physiol.* **44,** 1259.
85. Horton, R. F., and Osborne, D. J. (1967). *Nature (London)* **214,** 1086.
86. Huelin, F. E. (1933). *Gt. Brit., Dep. Sci. Ind. Res., Food Invest. Bd., Rep. 1932* p. 53.
87. Huelin, F. E., and Barker, T. (1939). *New Phytol.* **38,** 85.
88. Hulme, A. C., Jones, J. D., and Wooltorton, L. S. C. (1963). *Proc. Roy. Soc., Ser. B* **158,** 514.
89. Hulme, A. C., Rhodes, M. J. C., and Wooltorton, L. S. C. (1971). *Phytochemistry* **10,** 749.
90. Hyodo, H., and Yang, S. F. (1971). *Arch. Biochem. Biophys.* **143,** 338.
91. Imaseki, H. (1970). *Plant Physiol.* **46,** 170.
92. Imaseki, H., Asahi, T., and Uritani, I. (1968). *In* "Biochemical Regulation in Diseased Plants or Injury, p. 189. Phytopath. Soc. Jap.
93. Imaseki, H., and Pjon, C. J. (1970). *Plant Cell Physiol.* **11,** 827.
94. Imaseki, H., Uchiyama, M., and Uritani, I. (1968). *Agr. Biol. Chem.* **32,** 387.
95. Ivanoff, N. N. (1932). *Biochem. Z.* **254,** 71.
96. Ivanoff, N. N., Prokoshev, S. M., and Gabunya, M. K. (1930–1931). *Bull. Appl. Genet. Plant Breed.* **25,** 262.
97. Iwahori, S., Lyons, J. M., and Smith, O. E. (1970). *Plant Physiol.* **46,** 412.
98. Jahn, O. L., Chace, W. G., and Cubbedge, R. H. (1969). *J. Amer. Soc. Hort. Sci.* **94,** 123.

99. Jones, J. D., Hulme, A. C., and Wooltorton, L. S. (1965). *New Phytol.* **64,** 158.
100. Jones, R. L. (1968). *Plant Physiol.* **43,** 442.
101. Kaltaler, R. E. L., and Boodley, J. W. (1970). *Hort. Science* **5,** Sect. 2, Abstr., 67th Meet., p. 356.
102. Kang, B. G., and Burg, S. P. (1972). *Plant Physiol.* **49,** 631.
103. Kang, B. G., and Ray, P. M. (1969). *Planta* **87,** 193.
104. Kar, B. K., and Banerjee, H. K. (1939). *Nature (London)* **144,** 597.
105. Kidd, F., and West, C. (1933). *Gt. Brit., Dep. Sci. Ind. Res., Food Invest. Bd., Rep. 1932* p. 55.
106. Knapp, F. W., Hall, C. B., Buchanan, D. W., and Biggs, R. H. (1970). *Phytochemistry* **9,** 1453.
107. Kraynev, S. I. (1937). *Proc. Agr. Inst. Krasnodar* **6,** 101.
108. Krishnamoorthy, H. N. (1970). *Plant Cell Physiol.* **11,** 979.
109. Kropfitsch, M. (1951). *Phyton* **3,** 108.
110. Ku, H. S., and Leopold, A. C. (1970). *Plant Physiol.* **46,** 842.
111. Ku, H. S., Suge, H., Rappaport, L., and Pratt, H. K. (1969). *Planta* **90,** 333.
112. Kursanov, A., and Krukova, N. (1938). *Biokhimiya* **3,** 202.
113. Laan, P. A. van der. (1934). *Rec. Trav. Bot. Neer.* **31,** 691.
114. Leopold, A. C. (1967). *Symp. Soc. Exp. Biol.* **21,** 507.
115. Lewis, L. N., Palmer, R. L., and Hield, H. Z. (1968). *In* "Biochemistry and Physiology of Plant Growth Substances" (F. Wightman and G Setterfield, eds.), p. 1303. Runge Press, Ottawa.
116. Lewis, L. N., and Varner, J. E. (1970). *Plant Physiol.* **46,** 194.
117. Lockhart, C. L., Forsyth, F. R., and Eaves, C. A. (1968). *Can. J. Plant Sci.* **48,** 557.
118. Looney, N. E., and Patterson, M. E. (1967). *Nature (London)* **214,** 1245.
119. Loucks, E. W., and Hopkins, E. F. (1946). *Phytopathology* **36,** 750.
120. Lyon, C. J. (1970). *Plant Physiol.* **45,** 644.
121. Lyons, J. M., and Pratt, H. K. (1964). *Arch. Biochem. Biophys.* **104,** 318.
122. Mack, W. B. (1927). *Plant Physiol.* **2,** 103.
123. Madeikyte, E., and Turkova, N. S. (1965). *Liet. TSR Mokslu Akad. Darb., Ser. C* No. 2, p. 37.
124. Marei, N., and Crane, J. C. (1971). *Plant Physiol.* **48,** 249.
125. Matoo, A. K., and Modi, V. V. (1969). *Plant Physiol.* **44,** 308.
126. Matoo, A. K., Modi, V. V., and Reddy, V. V. R. (1968). *Indian J. Biochem.* **5,** 111.
127. McCready, R. M., and McComb, E. A. (1954). *Food Res.* **19,** 530.
128. McGlasson, W. B., and Pratt, H. K. (1964). *Plant Physiol.* **39,** 120.
129. Michener, H. D. (1938). *Amer. J. Bot.* **25,** 711.
130. Milbrath, J. A., and Hartman, A. (1940). *Science* **92,** 401.
131. Molisch, H. (1911). *Sitzungsber. Kaiserl. Akad. Wiss. Wien* **120,** 813.
132. Molisch, H. (1937). "The Influence of One Plant on Another. Allelopathy." Fischer, Jena.
133. Morgan, P. W., Beyer, E., and Gausman, H. W. (1968). *In* "Biochemistry and Physiology of Plant Growth Substances" (F. Wightman and G. Setterfield, eds.), p. 1255. Runge Press, Ottawa.
134. Morgan, P. W., and Gausman, H. W. (1966). *Plant Physiol.* **41,** 45.
135. Morgan, P. W., and Powell, R. D. (1970). *Plant Physiol.* **45,** 553.
136. Morré, D. J. (1968). *Plant Physiol.* **43,** 1545.
137. Nichols, R. (1968). *J. Hort. Sci.* **43,** 335.
138. Nord, F. F., and Weichherz, J. (1929). *Hoppe-Seyler's Z. Physiol. Chem.* **183,** 218.
139. Olson, A. O., and Spencer, M. (1968). *Can. J. Biochem.* **46,** 277.
140. Olson, A. O., and Spencer, M. (1968). *Can. J. Biochem.* **46,** 283.
141. Osborne, D. J. (1968). *Sci. (Soc. Chem. Ind., London) Monogr.* **31,** 236.
142. Osborne, D. J., and Mullins, M. G. (1969). *New Phytol.* **68,** 977.

143. Padoa, M. (1932). *Nature (London)* **129,** 686.
144. Palmer, O. H., and Halsall, D. M. (1969). *Physiol. Plant* **22,** 59.
145. Powell, R. D., and Morgan, P. W. (1970). *Plant Physiol.* **45,** 548.
146. Pratt, H. K., and Biale, J. B. (1944). *Plant Physiol.* **19,** 519.
147. Ratner, A., Goren, R., and Monselise, S. P. (1969). *Plant Physiol.* **44,** 1717.
148. Regeimbal, L. O., and Harvey, R. B. (1927). *J. Amer. Chem. Soc.* **49,** 1117.
149. Regeimbal, L. O., Vacha, G. A., and Harvey, R. B. (1927). *Plant Physiol.* **2,** 357.
150. Reid, M. S., and Pratt, H. K. (1972). *Plant Physiol.* **49,** 252.
151. Rhodes, M. J. C., and Wooltorton, L. S. C. (1971). *Phytochemistry* **10,** 1989.
152. Rhodes, M. J. C., Wooltorton, L. S. C., Galliard, T., and Hulme, A. C. (1968). *Phytochemistry* **7,** 1439.
153. Ridge, I., and Osborne, D. J. (1970). *J. Exp. Bot.* **21,** 720.
154. Ridge, I., and Osborne, D. J. (1970). *J. Exp. Bot.* **21,** 843.
155. Riov, J., Monselise, S. P., and Kahan, R. S. (1969). *Plant Physiol.* **44,** 631.
156. Roberts, D. W. A. (1951). *Can. J. Bot.* **29,** 10.
157. Scott, P. C., and Leopold, A. C. (1967). *Plant Physiol.* **42,** 1021.
158. Scott, P. C., Miller, L. W., Webster, B. D., and Leopold, A. C. (1967). *Amer. J. Bot.* **54,** 730.
159. Shaw, F. H. (1935). *Aust. J. Exp. Biol.* **13,** 95.
160. Shcherbakov, A. P. (1939). *Izv. Akad. Nauk SSSR, Ser. Biol.* No. 6, p. 975.
161. Smirnov, A. I., and Krainev, S. I. (1940). *Izv. Akad. Nauk SSSR, Ser. Biol.* No. 4, p. 577.
162. Smith, A. J. M., and Gane, R. (1933). *Gt. Brit., Dep. Sci. Ind. Res., Food Invest. Bd., Rep. 1932* p. 156.
163. Smith, W. H. (1939). *Gt. Brit., Dep. Sci. Ind. Res., Food Invest. Bd., Rep. 1938* p. 165.
164. Sondheimer, E. (1957). *Food Res.* **22,** 296.
165. Sondheimer, E. (1957). *J. Amer. Chem. Soc.* **79,** 5036.
166. Sorber, D. G., and Kimball, M. H. (1950). *U. S., Dep. Agr., Tech. Bull.* **996.**
167. Splittstoesser, W. E. (1970). *Physiol. Plant* **23,** 762.
168. Stahmann, M. A., Clare, B. G., and Woodbury, W. (1966). *Plant Physiol.* **41,** 1505.
169. Steffens, G. L., Alphin, J. G., and Ford, Z. T. (1970). *Beit. Tabakforsch.* **5,** 262.
170. Thornton, N. C. (1946). *Contrib. Boyce Thompson Inst.* **2,** 535.
171. Turkova, N. (1942). *Bull. Acad. Sci. USSR* **6,** 391.
172. Turkova, N. S., Vasileva, L. N., and Cheremukhina, L. F. (1965). *Sov. Plant Physiol.* **12,** 721.
173. Valdovinos, J. G., Ernest, L. C., and Henry, E. W. (1967). *Plant Physiol.* **42,** 1803.
174 van Andel, O. M. (1970). *Naturwissenschaften* **57,** 396.
175. Vendrell, M. (1969). *Aust. J. Biol. Sci.* **22,** 601.
176. Vendrell, M. (1970). *Aust. J. Biol. Sci.* **23,** 553.
177. Vesselov, I. J. (1937). *Mikrobiologiya* **6,** 510.
178. von Guttenberg, H., and Steinmetz, E. (1947). *Pharmazie* **2,** 17.
179. Wang, C. Y., and Hansen, E. (1970). *J. Amer. Soc. Hort. Sci.* **95,** 314.
180. Weaver, R. J., and Pool, R. M. (1969). *J. Amer. Soc. Hort. Sci.* **94,** 474.
181. Winston, J. R., and Roberts, G. L. (1944). *Proc. Fla. State Hort. Soc.* p. 140.
182. Yamaki, T. (1947). *Proc. Jap. Acad.* **23,** 53.
183. Young, R. (1970). *HortScience* **5,** *Sect.* 2, Abstr., 67th Meet., p. 306.
184. Young, R., Jahn, O., Cooper, W. C., and Smoot, J. J. (1970). *HortScience* **5,** 268.

Chapter 9

Ethylene Analogs and Antagonists

I. Ethylene Analogs

The activity of ethylene relative to other hydrocarbon analogs is shown in Table 9-1 for growth inhibition and abscission. Although other processes have not been as carefully analyzed, similarities between various systems exist. These data are summarized in Table 9-2. Except for mitochondrial swelling and perhaps promotion of rice growth, a number of generalizations are possible. Burg and Burg (26) observed that for a molecule to be an effective substitute for ethylene it must have the following characteristics:

1. Only unsaturated compounds are effective. The double bond confers more activity than a triple bond, and single bond compounds are inactive.
2. Activity decreases with increasing chain length. In the ethylene series the amounts needed for equivalent effective activity for ethylene, propylene, and 1-butene are 1 : 130 : 140,000. In the acetylenic series the ratio is 1 : 3.7 : 61 for acetylene, methyl acetylene, and ethyl acetylene. Burg and Burg reasoned that the acetylene series is probably more effective because the alkyl substitution is held in line with the triple bond causing a minimum of steric hindrance compared with the ethylenic series where the substitution is positioned off to one side with respect to the carbon-to-carbon double bond.
3. Substitutions that lower the bond order of the unsaturated position by causing electron delocalization reduce biological activity. On the basis of size the biological activity should be

$$FCH{=}CH_2 > ClCH{=}CH_2 > CH_3{-}CH{=}CH_2 > BrCH{=}CH_2$$

Instead, propylene is more active than the halides and among the halides the order observed is Cl > F > Br.

Table 9-1

BIOLOGICAL ACTIVITY OF ETHYLENE AND OTHER UNSATURATED COMPOUNDS AS DETERMINED BY THE PEA STRAIGHT GROWTH TEST AND ABSCISSION

		ppm in gas phase for ½ maximum activity	
Compound[a]	Structure	Pea growth[b]	Abscission[c]
Ethylene	$CH_2{=}CH_2$	0.1	0.1
Deuterated ethylene	$CD_2{=}CD_2$	0.1[d]	0.1[e]
Propylene	$CH_3{-}CH{=}CH_2$	10	6
Vinyl chloride	$ClCH{=}CH_2$	140	—
Carbon monoxide	$C{=}O$	270	175
Acetylene	$CH{\equiv}CH$	280	125
Vinyl fluoride	$FCH{=}CH_2$	430	250
Methyl acetylene	$CH_3{-}C{\equiv}CH$	800	—
Vinyl bromide	$BrCH{=}CH_2$	1600	—
Allene	$CH_2{=}C{=}CH_2$	2900	—
Vinyl methyl ether	$CH_3{-}O{-}CH{=}CH_2$	10,000	—
Ethylene acetylene	$CH_3{-}CH_2{-}C{\equiv}CH$	11,000	—
1-Butene	$CH_3{-}CH_2{-}CH{=}CH_2$	27,000	10,000
Vinyl ethyl ether	$CH_3{-}CH_2{-}O{-}CH{=}CH_2$	30,000	—
1,1-Difluoroethylene	$F_2C{=}CH_2$	350,000	—
Butadiene	$CH_2{=}CH{-}CH{=}CH_2$	500,000	10,000

[a] Not effective at 300,000 ppm: ethane, *trans*-2-butene, *cis*-2-butene, isobutene, and nitrous oxide. The following are inactive at concentrations below toxic levels indicated in parentheses: dichloromethane, *cis*-dichloroethylene, *trans*-dichloroethylene, trichloroethylene, tetrachloroethylene, H_2S, ethylene oxide, allyl chloride (all approximately 1%), HCN (0.03 *M*), acrylonitrile (0.17 m*M*), H_2O_2 (0.01 *M*), KN_3 (4 m*M*), allyl alcohol (1 m*M*), and HCHO (0.1 m*M*).

[b] Burg and Burg (26).

[c] Abeles and Gahagan (5).

[d] Beyer (14a).

[e] Abeles *et al.* (8a).

The observed order is thought to represent a compromise between steric hindrance due to size and lowering of electron density in the double bond because of the electrophilic nature of the halide substitution. Vinyl halides undergo resonance to form compounds with reduced electron density between the carbon atoms.

$$XCH{=}CH_2 \longleftrightarrow {}^{\oplus}X{=}CH{-}CH_2^{\ominus}$$

Table 9-2

RELATIVE ORDER OF EFFECT OF ETHYLENE AND ITS ANALOGS ON VARIOUS PROCESSES

Process	Active	Inactive	Reference number
Abscission	Ethylene, propene, CO, acetylene, vinyl fluoride, 1-butene, 1,3-butadiene	NH_3	5, 65, 108
Breaking bud dormancy	Ethylene, acetylene		138
Epinasty	Ethylene, acetylene, propylene, CO, butylene, ethyl bromide, ethyl iodide	Ethylene chlorohydrin, saturated olefins, benzine and other ring compounds, NH_3, pyridine, alcohols, aldehydes	39, 47, 104, 119
Floral induction	Ethylene, acetylene, propylene, propyl chloride, acetonitrile, butylene, amylene, mesityle oxide, vinyl acetate, benzol, petroleum ether, H_2S, acetone, NH_3	Acetone, bromine, turpentine, dichloroethylene, trichloroethylene, tetrachloroethylene, acetaldehyde, diethyl ether	34, 39, 47, 60, 104, 119
Growth inhibition and swelling	Ethylene, propylene, CO, acetylene, vinyl fluoride	H_2S, SO_2, CS_2, turpentine, benzene, ethyl ether, chloroform	24, 26, 29, 84
Growth promotion, rice		Methane, ethane, propene, propylene, butane	88
Latex flow promotion	Ethylene, acetylene		41
Mitochondrial swelling	1-Butene, propylene, propane, ethylene, ethane, acetylene		87
Ripening	Ethylene, propylene, CO, acetylene, methyl acetylene, ethyl acetylene, allene	Methylene chloride, ethylene chloride, amylene	26, 63, 64, 85, 123, 139, 142
Rooting	Ethylene, acetylene, propylene		86, 145
Spore release (*Isoetes*)	Ethylene, acetylene, isoamylene, CO, ethyl ether, methyl ether	Methane, ethane, ethylene bromide, ethylene chloride	55

The inductive effect is in the order of F > Cl > Br so that the fluoride derivative has the lowest bond order and the highest dipole even though it has the smallest size.

4. The unsaturated position must be adjacent to a terminal carbon atom. This is indicated by the activity of 1-butene as opposed to the inactivity of 2-butene. Since acetonitrile ($CH_3C{\equiv}N$) is completely inactive while methyl acetylene is active it appears that nitrogen will not substitute for the carbon adjacent to the double bond. Its inactivity may also be explained by extensive hyperconjugation.

5. The terminal carbon atom must be not positively charged. The requirement is suggested to account for the activity of CO and the inactivity of formaldehyde. The resonance form $^{\ominus}C{=}O^{\oplus}$ contributes to the structure of CO overcoming the inherent polarity of the CO bond to yield a molecule in which there is a slight negative charge on the carbon. The polarity of the carbonyl group of formaldehyde causes its carbon to be strongly positive in charge.

Four years after Neljubov discovered that the physiological effects of illuminating gas were due to ethylene, Richards and MacDougal (113) observed that another constituent also had a wide range of physiological effects. They reported that CO retarded seedling growth, altered geotropic and phototropic sensitivity, reduced chlorophyll content, and promoted abscission. Since that time CO has been shown to have the same physiological effects as ethylene. Both CO and ethylene are known to promote abscission (5, 59, 96, 108, 119) and fruit ripening (24, 29), inhibit leaf movement of *Mimosa* (144), cause epinasty (39, 144), promote root initiation (143), stimulate sperm release from *Isoetes* (55), cause intumescence formation (143), and induce female flowers in *Mercurialis* (66), *Cannabis* (67), and cucumbers (99–101).

II. Ethylene Antagonist—CO_2

The observation that CO_2 overcame or blocked the action of ethylene was first reported in 1927. Mack (94) observed that the addition of CO_2 to the gas phase reduced the ability of ethylene to blanch celery. Removing the CO_2 by absorbing it with KOH enhanced the effectiveness of ethylene. In the same year, Wallace (135) reported that CO_2 reduced the formation of intumescences by ethylene. The competitive action of CO_2 was perhaps suggested earlier by Cousins (35). In 1910 he observed that emanations from oranges stimulated the ripening of bananas. Although he was unaware that the active component was ethylene, he noted that CO_2 preserved the fruit.

The idea that CO_2 is a competitive inhibitor of ethylene was first stated by Burg and Burg (24, 26). They pointed out that CO_2 was a close structural analog of allene and CO, compounds which substitute for ethylene.

$$O{=}C{=}O \qquad C{=}O \qquad CH_2{=}C{=}CH_2 \qquad CH_2{=}CH_2$$

Since CO_2 has structural features needed for ethylene action except that it lacks the terminal carbon atom and is negatively charged at both ends, Burg and Burg proposed that it acts as a competitive inhibitor of ethylene. The idea was tested by growing pea stem sections in different concentrations of CO_2 and ethylene and plotting the resulting data in a Lineweaver-Burk plot. While the use of the Lineweaver-Burk plot was originally designed to test models of enzyme action, it also serves as a convenient test for competitive inhibition in those cases where physiological data can be represented by a rate function. As shown in Fig. 9-1, a plot of the reciprocal of the percent inhibition of growth versus the reciprocal of the ethylene concentrations gives a straight line. When CO_2 was added to the system the slope of the line increased and the intercept of the line on the ordinate coincided. This result is identical to that observed when substrate rate curves of enzymes in the presence of inhibitors are plotted, for example, the action of succinic dehydrogenase in the presence of malonic acid. The upper limit for CO_2 in these experiments is in the order of 10%. Higher concentrations of CO_2 ordinarily have secondary effects which are probably due to toxicity. The relative affinity of the site of action for ethylene compared with that for CO_2 appears to be 100,000 to 1. In other words, it usually takes 10% CO_2 to overcome the effect of 1 ppm ethylene.

Table 9-3 summarizes the literature reporting the testing of the action of CO_2 on various ethylene-mediated processes. In most cases CO_2 has

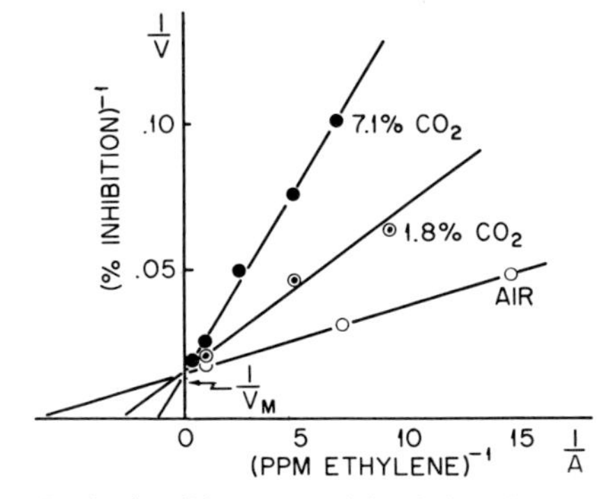

Fig. 9-1. A Lineweaver-Burk plot ($1/V$ versus $1/A$) relating the percent inhibition of growth (V) to the concentration of ethylene (A) at various levels of CO_2. V_M is the maximum effect occurring when the ethylene concentration is infinite ($1/A = 0$). [Courtesy of Burg and Burg (26).]

Table 9-3

COMPETITIVE INHIBITION OF ETHYLENE-MEDIATED PROCESSES BY CO_2

Process	Inhibition (reference number)	No effect or promotion (reference number)
Abscission	1, 5–7, 37, 40, 53, 54, 140	
Loosening of walnut hulls	127	
Growth inhibition, stems	26, 69, 136	2
Growth inhibition, roots	31, 32, 109	
Growth promotion, rice		88
Growth, pollen		122
Growth, slime molds	121	
Hook closure	80	
Enzyme induction		
Glucanase	4	
Phenylalanine ammonium lyase	115	
Peroxidase	56, 74	
Seed germination	50	8, 52, 111, 131
Tendril coiling		111
Celery blanching	94	
Floral senescence	105, 124, 125, 128, 133	
Respiration, potato	30	
Respiration, lemons		17, 18, 141
Carotene biogenesis	79	
Potato tuber sprouting	30, 130	
Isocoumarin formation	33	
Epinasty	40, 46, 119	
Intumescence formation	135	
Fruit ripening		
Avocados	141	
Apples	81, 82, 89	
Bananas	35, 57, 68, 129, 141	
Cucumbers	9	
Oranges	97	
Lemons	97, 141	
Persimmons		106, 118, 126
Peaches	129	

been found to block or retard ethylene action. However, a number of exceptions occur.

In the case of growth promotion in rice, stimulation of seed germination, promotion of lemon respiration, and removal of astringency from persimmons, CO_2 has the same effect as ethylene. Search and Stanley (120) reported that ethylene promoted peach pollen growth and germination. Sfakiotakis *et al.* (122) reported that CO_2 had the same effect; however, in their work they failed to find an effect of ethylene on pollen growth. A number of workers have reported that CO_2 and ethylene both reduced the astringency of persimmons (106, 118, 126). Overholser (106) suggested that removal of astringency was not associated with the ripening process since CO_2 treatment resulted in firm fruit while ethylene promoted softening. On the other hand, Soldatenkov (126) claimed that softening was caused by both CO_2 and ethylene.

The ability of CO_2 to mimic the effect of ethylene in these cases implies that these phenomena are not typical ethylene effects or that CO_2 has secondary effects that circumvent or duplicate ethylene action. An analogous phenomenon has been reported for ethylene anesthesia in animals. Kleindorfer (83) reported that 10% CO_2 had an additive effect on anesthesia produced by subanesthetic concentrations of ethylene, ether, and N_2O.

The fact that CO_2 is able to overcome or block ethylene action has been taken advantage of in a number of cases where workers demonstrated that ethylene played an intermediate role in the action of various chemicals or treatments. For example, CO_2 has inhibited the action of abscisic acid (37) and malformin (40) in inducing abscission and overcoming the ability of IAA and 2,4-D (69) to inhibit growth. Carbon dioxide has also been used to show that ethylene plays an intermediate role in clinostat-induced epinasty (88a) and geotropism of roots (31).

In addition to inhibiting ethylene action, CO_2 has a number of other physiological and biological effects on plants. Carbon dioxide is essential for the growth of plants although seedlings living on their own food resources and saprophytes manage to live and grow without CO_2 fixation. Metabolically, CO_2 is also involved in acid metabolism and lipid synthesis. All plants are exposed to some quantity of CO_2. When plants are illuminated in an enclosed chamber the concentrations do not drop below 100 ppm. Internal levels of CO_2 vary depending on the size of the plant organ and its location. Leaf and stem tissue exposed to air contains levels close to that of ambient air or 300 ppm. Bulky tissues such as tubers and fruit have internal levels in the order of 10–25%. Roots are normally exposed to higher levels of CO_2. At a depth of 2 feet, CO_2 concentrations of 3% were observed in silty clay loam and 1.5% in sandy loam.

Carbon dioxide is required for the growth of certain fungi, such as *Mucor, Aspergillus,* and *Penicillium,* and bacteria. Carbon dioxide also promotes germination or sprouting of potato tubers, seeds such as clover, alfalfa, lettuce, and sorghum, and fungal spores. Sensitive fungi include *Ustilago zeae, Aspergillus niger,* and *Puccinia graminis*. Subsequent growth of seedlings is also sensitive to CO_2. In oats and other grasses, development of long mesocotyls and short coleoptiles is promoted by high levels of CO_2. The sensitivity of seeds and seedlings to high CO_2 levels is of physiological interest since the soil contains high levels of CO_2. Although an increase in CO_2 to about 7% inhibits shoot development in some plants, roots will tolerate concentrations of this order and higher. This is another example of adaption to prevailing CO_2 levels under natural conditions.

Carbon dioxide also has a number of other effects on plant metabolism. It increases the pH of cell sap (instead of decreasing it as anticipated), it increases the size of stomates, stimulates the Hill reaction, and decreases water uptake by roots as evidenced by reduced transpiration and exudation.

III. Dose-Response Relationships

Dose-response relationships for a number of ethylene-mediated processes are summarized in Table 9-4. For a majority of examples a great similarity in the kind of data obtained exists. A concentration of 0.01 ppm is normally required for a threshold effect, 0.1 ppm for a half-maximal effect, and 10 ppm for a saturating dose. A few exceptions have been reported. The threshold for fruit ripening appears to be ten times higher than that observed for other processes. This is probably due to the fact that fruits evolve ethylene at a greater rate than vegetative tissue which in turn results in greater internal levels of the gas. For externally supplied ethylene to have any effect it must be greater than that normally occurring in the tissue. The atypical concentrations of ethylene required to induce mitochondrial swelling are probably due to the fact that this is not a typical ethylene-mediated process. Ku and Leopold (87) reported that other gases such as 1-butene, propylene, and propane were more effective in causing swelling, which suggests that this response is a generalized response to hydrocarbon gases and not a hormonal effect of ethylene.

Generally speaking, after a maximal response to ethylene is observed, higher concentrations have no additional effect. In other words, dose-response curves are asymptotic and do not show reversal or secondary effects at high concentrations. However, a few exceptions have been reported. These occur in ethylene-mediated growth responses of pollen and induction of certain enzymes.

IV. Other Compounds

A. Ethylene Chlorohydrin

Ethylene chlorohydrin (2-chloroethanol, $ClCH_2-CH_2OH$) has been tested as a potential ethylene analog. This idea was strengthened by reports

Table 9-4

DOSE-RESPONSE RELATIONSHIPS FOR ETHYLENE-MEDIATED PHENOMENA[a]

Process	Species	Threshold	½ maximal	Saturation	Additional effects	Reference number
Growth						
Inhibits stem elongation	Pea	0.01	0.1	1	None	25
	Pea	—	0.1	10	None	28
	Pea	—	0.1	1	None	58
	Pea	0.05	0.2	—	None	116
Promotes stem elongation	Rice	0.5	5	100	None	88
Promotes stem swelling	Pea	0.01	1	10	None	137
Inhibits root elongation	Pea	0.01	0.3	100	None	31
	Pea	0.1	1	10	None	32
	Pea	—	0.1	1	None	109
Promotes root curvature	Corn	1	10	100	None	40
Inhibits bud elongation	Pea	—	0.1	10	None	27, 28
Fern gametophytes						
Inhibition of cell division in dark-grown filaments	*Onoclea sensibilis*	0.01	0.1	10	None	98
Promotes filament growth	*Onoclea sensibilis*	0.001	0.005	0.1	Inhibition at 1	98
Growth promotion of slime mold	*Physarum polycephalum*	25%	—	75%	None	121
Promotes pollen germination	Peach	0.01	—	0.01	50% inhibition at 100	22
Promotes pollen tube elongation	Peach	0.01	—	1	100% inhibition at 10	22
Epinasty	Tomato	—	1	100	None	92
	Tomato	0.01	0.1	1	None	Unpublished research
Inhibition of leaf movement	Mesquite	0.1	—	—	None	12
Promotes hook closure	Bean	0.01	0.1	1	None	80

	Bean	0.01	0.1	—	None	102
Intumescence formation	Poplar	0.01	1	100	None	135
Aging						
Promotes abscission	Bean	0.01	0.1	1	None	5
	Citrus	0.01	0.1	10	None	13
	Apple	0.9	—	20	None	110
	Cotton	0.08	—	—	None	15
Promotes fruit ripening	Banana	0.1	—	—	None	19
	Mango	0.1	0.4	40	None	23
	Lemons	0.2	—	100	None	43
	Lemons	0.05	—	—	None	116
	Oranges	0.1	5	100	None	76
	Melons	0.5	5	100	None	107
Promotes floral senescence	Orchids	0.02	—	—	None	42
	Carnations	0.03	—	—	None	133
	Carnations	0.2	—	—	None	105
	Rose	0.2	—	1	None	78
Promotes aging excised tissue	Bean	0.01	0.3	10	None	3
Dormancy						
Stimulates seed germination	Lettuce	—	—	1	None	8
	Witch weed	0.01	—	0.1	None	50
	Witch weed	0.001	0.1	10	None	52
Promotes floral initiation						
6 hours	Pineapple	10	100	1000	None	34
24 hours	*Billbergia*	0.03	0.1	10	None	Unpublished research
Pigmentation						
Chlorophyll degradation	Tobacco	0.1	5	100	None	103
Promotes anthocyanin biosynthesis	Cranberries	0.1	10	—	None	36
Inhibits anthocyanin biosynthesis	Sorghum	0.01	0.1	10	None	38
	Bean	0.01	0.1	1	None	102
Inhibits carotenoid biosynthesis	Pea	0.01	0.1	10	None	79
Metabolism and enzymology						
Promotes isocoumarin formation	Carrot	0.1	1	10	None	33
Increase in cell wall hydroxyproline	Pea	0.1	10	100	None	114

Table 9-4 *(continued)*

Process	Species	Threshold	½ maximal	Saturation	Additional effects	Reference number
Increase in chromatin activity	Soybean	—	—	1	None	70
Promotes respiration	Orange	—	0.1	10	None	16
	Banana	—	0.1	10	None	20
	Cantaloupe	0.1	1	10	None	95
	Potato	1	10	100	None	71, 72
Promotes mitochondrial swelling	Pea	50	—	100	None	87
	Cauliflower	10	100	1000	None	93
Increase in β-1,3-glucanase	Bean	0.01	0.1	10	None	4
Increase in phenylalanine ammonium lyase	Citrus	1	15	100	None	115
	Swede	—	1	4	None	112
	Sweet potato	0.01	0.1	15	Inhibition at 100	75
	Cucumber sections	0.01	0.1	10	Inhibition at 100	51
Decrease in phenylalanine ammonium lyase	Cucumber seedlings	0.01	—	10	Reduced effect at 100	51
Increase in cinnamic acid hydroxylase	Pea	0.1	3	50	Inhibition at 100	73
Increase in α-amylase	Barley	0.04	0.4	4	None	77
Increase in peroxidase	Sweet potato	0.1	1	100	None	74
	Sweet potato	0.01	0.5	5	Inhibition at 100	75

[a] Some of the data shown were extracted from photographs and incomplete graphs and represent extrapolations of available data. Except for the one exception noted, concentrations are given in parts per million.

that it had striking effects on plants, including promotion of sprouting from potatoes (44), formation of adventitious buds from roots (48), promotion of abscission (62), increased seed germination (117), removal of astringency from persimmons (106), breaking dormancy of gladiolus bulbs (45), and increased germination of fungal spores (21). However, there are a number of reasons for believing that ethylene chlorohydrin does not mimic or duplicate ethylene action. First of all, there are a number of processes that are regulated by ethylene that do not respond to ethylene chlorohydrin. These include epinasty (39), flowering of bromeliads (132), and ripening (49, 63). Second, some of the systems that respond positively to ethylene chlorohydrin do not respond to ethylene. These include potato tuber sprouting (44) and shortening the rest period of gladiolus (45). As discussed earlier, removal of astringency from persimmons and promotion of seed germination are unusual ethylene effects because CO_2 fails to act as a competitive inhibitor and in fact duplicates ethylene action. The promotion of spore germination of *Diplodia* and *Phomopsis* can be caused by ethylene. However, as Brooks (21) points out, other compounds such as ethyl acetate and methanol have the same effect. Many compounds are known to promote abscission, and it has been shown that increased ethylene production due to the herbicidal or toxic action of these compounds plays a significant role in their activity. As a number of workers have pointed out, high concentrations of ethylene chlorohydrin can cause damage to plant tissue.

Finally, an examination of ethylene chlorohydrin structure suggests that analog activity is unlikely. As discussed above, saturated hydrocarbons have no ethylene-like activity. Steric hindrance due to the OH and Cl groups also suggests that ethylene activity would be minimal. It is conceivable, however, that some hydrolysis mechanism might convert ethylene chlorohydrin into ethylene in a fashion analogous to the conversion of Ethrel into ethylene.

B. Ethylene Oxide

Ethylene oxide, a cyclic ether, is a highly strained molecule because it is a three-membered ring. Under mild conditions it readily forms ethylene glycol, dioxane, and various polymers. It is a toxic compound and is commonly used as a fumigant for materials sensitive to heat. Plant tissue is also readily killed with this compound. In the 1920's Harvey and co-workers reported that ethylene oxide damaged fruit (63) and at concentrations less than 0.1% in water killed corn seeds (61) and potato tubers (134). In the gas phase, concentrations greater than 2000 ppm cause lesions and damage to fruits (90) and flowers (10). Notwithstanding, Harvey (64) and others (24, 90) have examined the possibility that ethylene oxide can act as an ethylene analog or antagonist. Lieberman and Mapson (91) noted that 7500 ppm ethylene oxide retarded the normal ripening of tomatoes 5–21 days.

The idea that ethylene oxide might be operating in a physiological manner was supported by the observation that the retarding effect of 3000 ppm ethylene oxide was reversed by 50 ppm ethylene. Lieberman and Mapson reported that ethylene oxide was among the volatiles produced by tissue slices and homogenates of fruits (91). However, this observation was not confirmed in a later report (90). However, they did extend their findings on the antagonistic effects of ethylene oxide. For example, roses were subjected to 3500 ppm ethylene oxide and reached full bloom 30 hours later than controls. They indicated that the delay resulted from the absorption of ethylene oxide and lasted as long as ethylene oxide was being released from the tissue. When carnations were exposed to 1 ppm ethylene and 1000 ppm ethylene oxide simultaneously, sleepiness did not occur (cf. 10, 11). Ethylene oxide also reversed the ability of ethylene to induce the triple response of peas. Similar experiments on carnations were performed by Nichols (105), who confirmed the fact that ethylene oxide reversed the effect of ethylene on carnations. However, he noted that the threshold between toxic doses (2000 ppm) and nontoxic ones (1000 ppm), insofar as external appearance of the tissue was concerned at least, was small. Asen and Lieberman (10, 11) also pointed out that there was a narrow limit between concentrations they considered antagonistic to ethylene action (500–1000 ppm) and toxicity (2000 ppm).

Ben-Yehoshua *et al.* (14) on the other hand found that ethylene oxide failed to delay rose maturation without simultaneously damaging the tissue. Burg and Burg (24) also indicated that ethylene oxide was not a competitive inhibitor of ethylene because its effects on peas were not overcome by ethylene. The available data suggest that ethylene oxide is neither an analog nor antagonist of ethylene. Its ability to slow maturation is more likely due to a delay in development caused by tissue damage not readily apparent from external appearances.

References

1. Abeles, F. B. (1967). *Physiol. Plant.* **20,** 442.
2. Abeles, F. B. (1968). *Weed Sci.* **16,** 498.
3. Abeles, F. B., Craker, L. E., and Leather, G. R. (1971). *Plant Physiol.* **47,** 7.
4. Abeles, F. B., and Forrence, L. E. (1970). *Plant Physiol.* **45,** 395.
5. Abeles, F. B., and Gahagan, H. E. (1968). *Plant Physiol.* **43,** 1255.
6. Abeles, F. B., and Holm, R. E. (1966). *Plant Physiol.* **41,** 1337.
7. Abeles, F. B., and Holm, R. E. (1967). *Ann. N. Y. Acad. Sci.* **144,** 367.
8. Abeles, F. B., and Lonski, J. (1969). *Plant Physiol.* **44,** 277.

8a. Abeles, F. B., Ruth, J. M., Forrence, L. E., and Leather, G. R. (1972). *Plant Physiol.* **49,** 669.
9. Apeland, J. (1961). *Bull. Inst. Int. Froid, Annexe* **4,** 45.
10. Asen, S., and Lieberman, M. (1963). *Florists' Rev.* **131,** 27.
11. Asen, S., and Lieberman, M. (1963). *Exch. Flower, Nursery Gard. Cent. Trade* **140,** 23.
12. Baur, J. R., and Morgan, P. W. (1969). *Plant Physiol.* **44,** 831.
13. Ben-Yehoshua, S., and Eaks, I. L. (1970). *Bot. Gaz.* (*Chicago*) **13,** 144.
14. Ben-Yehoshua, S., Juven, B., Fruchter, M., and Halevy, A. H. (1967). *Proc. Amer. Soc. Hort. Sci.* **89,** 677.
14a. Beyer, E. M. (1972). *Plant Physiol.* **49,** 672.
15. Beyer, E. M., and Morgan, P. W. (1971). *Plant Physiol.* **48,** 208.
16. Biale, J. B. (1960). *Advan. Food Res.* **10,** 293.
17. Biale, J. B. (1961). *In* "The Orange, Its Biochemistry and Physiology" (W. B. Sinclair, ed.), p. 96. Univ. of California Press, Berkeley.
18. Biale, J. B. (1962). *Food Preserv. Quart.* **22,** 57.
19. Biale, J. B., and Young, R. E. (1962). *Endeavour* **21,** 164.
20. Brady, C. J., O'Connell, P. B. H., Smydzuk, J., and Wade, N. L. (1971). *Aust. J. Biol. Sci.* **23,** 1143.
21. Brooks, C. (1944). *J. Agr. Res.* **68,** 363.
22. Buchanan, D. W., and Biggs, R. H. (1969). *J. Amer. Soc. Hort. Sci.* **94,** 327.
23. Burg, S. P., and Burg, E. A. (1962). *Plant Physiol.* **37,** 179.
24. Burg, S. P., and Burg, E. A. (1965). *Science* **148,** 1190.
25. Burg, S. P., and Burg, E. A. (1966). *Proc. Nat. Acad. Sci. U. S.* **55,** 262.
26. Burg, S. P., and Burg, E. A. (1967). *Plant Physiol.* **42,** 144.
27. Burg, S. P., and Burg, E. A. (1968). *Plant Physiol.* **43,** 1069.
28. Burg, S. P., and Burg, E. A. (1968). *In* "Biochemistry and Physiology of Plant Growth Substances" (F. Wightman and G. Setterfield, eds.), p. 1275. Runge Press, Ottawa.
29. Burg, S. P., and Burg, E. A. (1969). *Qual. Plant. Mater. Veg.* **19,** 185.
30. Burton, W. G. (1952). *New Phytol.* **51,** 154.
31. Chadwick, A. V., and Burg, S. P. (1967). *Plant Physiol.* **42,** 415.
32. Chadwick, A. V., and Burg, S. P. (1970). *Plant Physiol.* **45,** 192.
33. Chalutz, E., DeVay, J. E., and Maxie, E. C. (1969). *Plant Physiol.* **44,** 235.
34. Cooper, W. C., and Reece, P. C. (1941). *Proc. Fla. State Hort. Soc.* **54,** 132.
35. Cousins, H. H. (1910). *Annu. Rep. Dep. Agr. Jamaica.*
36. Craker, L. E. (1971). *HortScience* **6,** 137.
37. Craker, L. E., and Abeles, F. B. (1969). *Plant Physiol.* **44,** 1144.
38. Craker, L. E., Standley, L. A., and Starbuck, M. J. (1971). *Plant Physiol.* **48,** 349.
39. Crocker, W., Zimmerman, P. W., and Hitchcock, A. E. (1932). *Contrib. Boyce Thompson Inst.* **4,** 177.
40. Curtis, R. W. (1968). *Plant Physiol.* **43,** 76.
41. D'Auzac, J., and Ribaillier, D. (1969). *C. R. Acad. Sci.* **268,** 3046.
42. Davidson, O. W. (1949). *Proc. Amer. Soc. Hort. Sci.* **53,** 440.
43. Denny, F. E. (1924). *J. Agr. Res.* **27,** 757.
44. Denny, F. E. (1929). *Contrib. Boyce Thompson Inst.* **1,** 59.
45. Denny, F. E. (1930). *Contrib. Boyce Thompson Inst.* **2,** 523.
46. Denny, F. E. (1935), *Contrib. Boyce Thompson Inst.* **7,** 341.
47. Denny, F. E. (1939). *Contrib. Boyce Thompson Inst.* **10,** 191.
48. Deuber, C. G. (1933). *Amer. Gas Ass. Mon.* **15,** 380.
49. Dustman, R. B. (1934). *Plant Physiol.* **9,** 637.
50. Egley, G. H., and Dale, J. E. (1970). *Weed Sci.* **18,** 586.

51. Engelsma, G., and Van Bruggen, J. M. H. (1971). *Plant Physiol.* **48,** 94.
52. Esashi, Y., and Leopold, A. C. (1969). *Plant Physiol.* **44,** 1470.
53. Fischer, C. W. (1949). *N. Y. State Flower Growers, Bull.* **52,** 5.
54. Fischer, C. W. (1950). *Proc. Amer. Soc. Hort. Sci.* **55,** 447.
55. Fujii, K. (1925). *Flora (Jena)* [N. S.] **18–19,** 115.
56. Gahagan, H. E., Holm, R. E., and Abeles, F. B. (1968). *Physiol. Plant.* **21,** 1270.
57. Gane, R. (1936). *New Phytol.* **35,** 383.
58. Goeschl, J. D., and Pratt, H. K. (1968). *In* "Biochemistry and Physiology of Plant Growth Substances" (F. Wightman and G. Setterfield, eds.), p. 1229. Runge Press, Ottawa.
59. Goodspeed, T. H., McGee, J. M., and Hodgson, R. W. (1918). *Univ. Calif., Berkeley, Publ. Bot.* **5,** 439.
60. Gowing, D. P., and Leeper, R. W. (1961). *Bot. Gaz. (Chicago)* **123,** 34.
61. Haber, E. S. (1926). *Proc. Amer. Soc. Hort. Sci.* **23,** 201.
62. Hall, W. C. (1952). *Bot. Gaz. (Chicago)* **113,** 310.
63. Harvey, R. B. (1928). *Minn., Agr. Exp. Sta., Bull.* **247.**
64. Harvey, R. B. (1928). *Science* **67,** 421.
65. Heck, W. W., and Pires, E. G. (1962). *Tex., Agr. Exp. Sta., Misc. Publ.* **MP-603.**
66. Heslop-Harrison, J., and Heslop-Harrison, Y. (1957). *New Phytol.* **56,** 352.
67. Heslop-Harrison, J., and Heslop-Harrison, Y. (1957). *Proc. Roy. Soc. Edinburgh, Sect. B* **66,** 424.
68. Hesselman, C. W., and Freebairn, H. T. (1969). *J. Amer. Soc. Hort. Sci.* **94,** 635.
69. Holm, R. E., and Abeles, F. B. (1967). *Planta* **78,** 293.
70. Holm, R. E., O'Brien, T. J., Key, J. L., and Cherry, J. H. (1970). *Plant Physiol.* **45,** 41.
71. Huelin, F. E. (1933). *Gt. Brit., Dep. Sci. Ind. Res., Food Invest. Bd., Annu. Rep. 1932* p. 53.
72. Huelin, F. E., and Barker, T. (1939). *New Phytol.* **38,** 85.
73. Hyodo, H., and Yang, S. F. (1971). *Archi. Biochem. Biophys.* **143,** 338.
74. Imaseki, H. (1970). *Plant Physiol.* **46,** 170.
75. Imaseki, H., Asahi, T., and Uritani, I. (1968). *In* "Biochemical Regulation in Diseased Plants or Injury," p. 189. Phytopathol. Soc. Jap.
76. Jahn, O. L., Chace, W. G., and Cubbedge, R. H. (1969). *J. Amer. Soc. Hort. Sci.* **94,** 123.
77. Jones, R. L. (1968). *Plant Physiol.* **43,** 442.
78. Kaltaler, R. E. L., and Boodley, J. W. (1970). *HortScience* **5,** Sect. 2, Abstr. 67th Meet., 356.
79. Kang, B. G., and Burg, S. P. (1972). *Plant Physiol.* **49,** 631.
80. Kang, B. G., and Ray, P. M. (1969). *Planta* **87,** 206.
81. Kidd, F., and West, C. (1934). *Gt. Brit., Dep. Sci. Ind. Res., Food Invest. Bd., Annu. Rep. 1933* p. 51.
82. Kidd, F., and West, C. (1938). *Gt. Brit., Dep. Sci. Ind. Res., Food Invest. Bd., Annu. Rep. 1937* p. 108.
83. Kleindorfer, G. B. (1931). *J. Pharmacol. Exp. Ther.* **43,** 445.
84. Knight, L. I., Rose, R. C., and Crocker, W. (1910). *Science* **31,** 635.
85. Kohman, E. F. (1931). *Ind. Eng. Chem.* **23,** 1112.
86. Krassinskii, N. (1936). *Sov. Subtrop.* **7,** 64.
87. Ku, H. S., and Leopold, A. C. (1970). *Plant Physiol.* **46,** 842.
88. Ku, H. S., Suge, H., Rappaport, L., and Pratt, H. K. (1969). *Planta* **90,** 333.
88a. Leather, G. R., Forrence, L. E., and Abeles, F. B. (1972). *Plant Physiol.* **49,** 183.
89. Leblond, C. (1967). *Fruits* **22,** 543.
90. Lieberman, M., Asen, S., and Mapson, L. W. (1964). *Nature (London)* **204,** 756.

91. Lieberman, M., and Mapson, L. W. (1962). *Nature (London)* **196,** 660.
92. Lyon, C. J. (1970). *Plant Physiol.* **45,** 644.
93. Lyons, J. M., and Pratt, H. K. (1964). *Arch. Biochem. Biophys.* **104,** 318.
94. Mack, W. B. (1927). *Plant Physiol.* **2,** 103.
95. McGlasson, W. B., and Pratt, H. K. (1964). *Plant Physiol.* **39,** 120.
96. McMillan, C., and Cope, J. M. (1969). *Amer. J. Bot.* **56,** 600.
97. Miller, E. V. (1946). *Bot. Rev.* **12,** 393.
98. Miller, P. M., Sweet, H. C., and Miller, J. H. (1970). *Amer. J. Bot.* **57,** 212.
99. Minina, E. G. (1938). *Dokl. Akad. Sci. Nauk SSSR* **21,** 298.
100. Minina, E. G. (1952). "Changes in Sex of Plants Produced by Environmental Factors." Acad. Sci. USSR, Moscow.
101. Minina, E. G., and Tylkina, L. G. (1947). *Dokl. Akad. Nauk SSR* **55,** 165.
102. Morgan, P. W., and Powell, R. D. (1970). *Plant Physiol.* **45,** 553.
103. Nakagaki, Y., Hirai, T., and Stahmann, M. A. (1970). *Virology* **40,** 1.
104. Nelson, R. C., and Harvey, R. B. (1935). *Science* **82,** 133.
105. Nichols, R. (1968). *J. Hort. Sci.* **43,** 335.
106. Overholser, E. L. (1927). *Proc. Amer. Soc. Hort. Sci.* **24,** 256.
107. Pratt, H. K., and Goeschl, J. D. (1968). *In* "Biochemistry and Physiology of Plant Growth Substances" (F. Wightman and G. Setterfield, eds.), p. 1295. Runge Press, Ottawa.
108. Pridham, A. M. S., and Hsu, R. (1954). *Proc. Northeast Weed Contr. Conf.* **21,** 221.
109. Radin, J. W., and Loomis, R. S. (1969). *Plant Physiol.* **44,** 1584.
110. Rakitin, Iu. V. (1967). *Sov. Plant Physiol.* **14,** 787.
111. Reinhold, L. (1967). *Science* **158,** 791.
112. Rhodes, M. J. C., and Wooltorton, L. S. C. (1971). *Phytochemistry* **10,** 1989.
113. Richards, H. M., and MacDougal, D. T. (1904). *Bull. Torrey Bot. Club* **3,** 57.
114. Ridge, I., and Osborne, D. J. (1970). *J. Exp. Bot.* **21,** 843.
115. Riov, J., Monselise, S. P., and Kahan, R. S. (1969). *Plant Physiol.* **44,** 631.
116. Rohrbaugh, P. W. (1943). *Plant Physiol.* **18,** 79.
117. Ruge, U. (1962). *Angew. Bot.* **26,** 162.
118. Ryerson, N. (1927). *Calif., Agr. Exp. Sta., Bull* **416.**
119. Schwartz, H. (1927). *Flora (Jena)* [N. S.] **22,** 76.
120. Search, R. W., and Stanley, R. G. (1968). *Plant Physiol.* **43,** Suppl., 52.
121. Seifriz, U., and Urbach, F. (1945). *Growth* **8,** 221.
122. Sfakiotakis, E. M., Dilley, D. R., and Simons, D. H. (1970). *HortScience* **5,** Sect. 2, *Abstr. 67th Meet.,* 341.
123. Smith, W. H. (1939). *Gt. Brit., Dep. Sci. Ind. Res., Food Invest. Bd., Annu. Rep. 1938* p. 165.
124. Smith, W. H., and Parker, J. C. (1966). *Nature (London)* **211,** 100.
125. Smith, W. H., Parker, J. C., and Freeman, W. W. (1966). *Nature (London)* **211,** 99.
126. Soldatenkov, S. V. (1935). *Dokl. Akad. Nauk SSSR* **2,** 313.
127. Sorber, D. G., and Kimball, M. H. (1950). *U. S., Dep. Agr., Tech. Bull.* **996.**
128. Thornton, N. C. (1930). *Contrib. Boyce Thompson Inst.* **2,** 535.
129. Thornton, N. C. (1931). *Contrib. Boyce Thompson Inst.* **3,** 219.
130. Thornton, N. C. (1933). *Contrib. Boyce Thompson Inst.* **5,** 471.
131. Toole, V. K., Bailey, W. K., and Toole, E. H. (1964). *Plant Physiol.* **39,** 822.
132. Traub, H. P., Cooper, W. C., and Reece, P. C. (1940). *Proc. Amer. Soc. Hort. Sci.* **37,** 521.
133. Uota, M. (1969). *J. Amer. Soc. Hort. Sci.* **94,** 598.
134. Vacha, G. A., and Harvey, R. B. (1927). *Plant Physiol.* **2,** 187.
135. Wallace, R. H. (1927). *Bull. Torrey Bot. Club* **54,** 499.
136. Warner, H. L. (1970). Ph.D. Thesis, Purdue University, Lafayette, Indiana.

137. Warner, H. L., and Leopold, A. C. (1967). *BioScience* **17,** 722.
138. Weber, F. (1916). *Sitzungsber. Akad. Wiss. Wien, Math.-Naturwiss. Kl., Abt. 1* **125,** 189.
139. Winston, J. R. (1935). *Citrus Ind.* **16,** 3.
140. Yamaguchi, S. (1954). Ph.D. Thesis, University of California, Los Angeles.
141. Young, R. E., Romani, R. J., and Biale, J. B. (1962). *Plant Physiol.* **37,** 416.
142. Zakharov, N. A. (1936). *Sov. Subtrop.* **7,** 69.
143. Zimmerman, P. W., Crocker, W., and Hitchcock, A. E. (1933). *Contrib. Boyce Thompson Inst.* **5,** 1.
144. Zimmerman, P. W., Crocker, W., and Hitchcock, A. E. (1933). *Contrib. Boyce Thompson Inst.* **5,** 195.
145. Zimmerman, P. W., and Hitchcock, A. E. (1933). *Contrib. Boyce Thompson Inst.* **5,** 351.

Chapter 10

Mechanism of Ethylene Action

I. Attachment Site

The number of ethylene attachment sites in a cell with a volume of 10^4 μm^3 is approximately 500. This estimate is based on the following calculations. A liter of plant material contains 10^{11} cells whose volume is 10^4 μm^3 (1 cell/10^4 μm^3 $\times$ 10^{15} μm^3/liter). At 25°C, 0.1 ppm ethylene will result in a 4.43×10^{-10} *M* solution, which contains 2.67×10^{14} molecules of ethylene per liter of solution (4.43×10^{-10} *M*/liter $\times$ 6.03×10^{23} molecules/mole). The number of ethylene molecules per cell is then 2670 (2.67×10^{14} molecules/liter $\times$ liter/10^{11} cells). Only 267 molecules are in the cytoplasm, assuming that 90% of the cell volume is vacuole. Since 0.1 ppm ethylene is a half-maximal concentration and assuming one ethylene molecule per attachment site, the total number of sites of action is twice this figure or, approximately, 500 ethylene attachment sites per cell. Assuming the active site was a molecule with a molecular weight of 100,000 that could be purified and isolated without any loss during the extraction procedure, a metric ton of plant material would contain approximately 10 mg of site material or 0.000001% of the total biomass,

$$\left(\frac{5 \times 10^2 \text{ sites}}{\text{cell}}\right)\left(\frac{10^{11} \text{ cells}}{\text{liter}}\right)\left(\frac{10^3 \text{ liter}}{\text{metric ton}}\right)\left(\frac{\text{mole}}{6 \times 10^{23} \text{ sites}}\right)\left(\frac{10^8 \text{ mg}}{\text{mole}}\right) = 10 \text{ mg/metric ton}$$

These rough estimates suggest that it is impossible to examine the site of ethylene action directly and that information as to its composition, location, and subsequent action will have to be derived by indirect means. As an initial step it is important to show that the site of ethylene action and attachment is the same for most, if not all, of the physiological processes regulated by the gas. The attachment site is characterized by three types of experiments: dose-response relationships, the action of ethylene analogs, and the competitive inhibition of CO_2.

The dose-response curves for a variety of ethylene effects show a striking similarity, as has been described more fully in Chap. 9. Normal values for threshold effects are 0.01 ppm; for half-maximal responses, 0.1 ppm; and for saturation, 10 ppm. Until oxygen becomes a limiting factor, increasing concentrations of ethylene usually have no further action, and toxic effects that mask or reverse the original observation are not seen.

As discussed in Chap. 9, the relative effectiveness of ethylene analogs, such as CO, acetylene, propylene, and butylene, has been shown to be approximately the same from one ethylene-mediated process to another. While analog experiments are difficult to perform because of potential ethylene contamination of the gases used, a number of investigators have shown that ethylene is the most effective compound, followed by propylene, vinyl chloride, CO, vinyl fluoride, acetylene, allene, methyl acetylene, and 1-butene.

Although CO_2 has long been recognized as having an effect opposite to ethylene, Burg and Burg (22) were the first to clearly state that its action was similar to that observed for competitive inhibitors of enzyme reactions. They pointed out that CO_2 was a close structural analog of allene and CO, compounds that mimic ethylene action. Although CO_2 has some structural resemblance to allene and CO and could occupy the same site in the cell when levels of CO_2 were high (about 10% by volume), it fails to act as an effector probably because of the partial negative charge on each end of the molecule. However, there were some cases where CO_2 did not reverse ethylene action and these exceptions have to be taken into account in the development of ideas concerning the site of ethylene attachment.

Burg and Burg (22) have also suggested that molecular oxygen was required to oxidize the ethylene attachment site, and oxidation was required for full receptiveness or sensitivity to ethylene. While this criterion for characterizing the site of ethylene action may be valid, other investigators have failed to observe similar phenomena (5).

All these experiments can show is that the initial site of ethylene attachment or attack is the same for a wide variety of physiological processes, including growth, development, senescence, and metabolism. They do not supply any information as to the kind of substrate the ethylene is attached to, nor do they indicate the kind of bonding involved. It is axiomatic that ethylene must be bonded to its site in order for it to be effective. While the attachment may be ephemeral, it must occur for some finite period of time in order to alter the structure and function of some subcellular component regulating cellular activity. The question of mode of action can be subdivided into two parts: first, nature of chemical bonding involved, and, second, characterization of the substrate. The kinds of chemical bonding possible are covalent, ionic, hydrogen bonding, coordination, and van der Waals. The kinds of substrate are practically endless, including proteins, nucleic acids, lipids, membranes, and cellular organelles. While the kind of chemical bonding involved in ethylene action is unknown it is possible to eliminate some of the possibilities suggested.

Ionic and hydrogen bonding seem unlikely on the basis of an examination of the structure of ethylene. Ethylene is highly stable and symmetrical and a large amount of energy would be required to dissociate a hydrogen atom resulting in the formation of a charged ethylene molecule. Assuming that dissociation did occur, then it should be possible to observe the exchange of hydrogen between ethylene and plant tissue by exposing plants to deuterated ethylene. An ionized ethylene molecule could pick up hydrogen from the surrounding milieu and in addition undergo *cis-trans* isomerization. Beyer (14) and we (10) have shown that within the error inherent in the methods used, no hydrogen exchange between ethylene and the cell was observable and in addition ethylene did not undergo *cis-trans* isomerism.

$$H^{\oplus} + \begin{array}{c} D \quad D \\ C{=}C \\ D \quad D \end{array} \longleftrightarrow \begin{array}{c} D \quad D \\ C{=}\ddot{C} \\ D \quad \ominus \end{array} + D^{\oplus} + H^{\oplus} \longleftrightarrow \begin{array}{c} D \quad D \\ C{=}C \\ D \quad H \end{array} + D^{\oplus}$$

$$\begin{array}{c} D \quad D \\ C{=}C \\ H \quad H \end{array} \longleftrightarrow \begin{array}{c} D \quad D \\ C{=}\ddot{C} \\ H \quad \ominus \end{array} + H^{\oplus} \longleftrightarrow \begin{array}{c} D \quad H \\ C{=}C \\ H \quad D \end{array}$$

A number of investigators have examined the possibility that ethylene can be incorporated into cellular material by covalent bonds. Behmer (13) exposed apple tissue to 1000 ppm [^{14}C]ethylene for 19 days and reported that no incorporation was observed. Hansen (34) measured levels of ethylene surrounding pear tissue and found they remained constant for long periods of time, which also suggests lack of gas uptake.

I have done similar experiments and have never observed uptake of ethylene by plant tissue. The attachment forces or affinity of tissue for ethylene is extremely weak. In some unpublished research, I have filled a 1.3 meter by 14 mm column with pea embryos from 5-day-old seedlings. Ethylene and other hydrocarbon gases such as methane, ethane, and propylene were injected into the air stream passing slowly over the tissue and the effluent was monitored to determine if retention of ethylene by the column of tissue occurred. In other words, the setup was much like a gas chromatograph except that the solid phase was replaced by living tissue susceptible to ethylene action and hence contained a number of active attachment sites. The effluent of this plant tissue "chromatograph" was monitored with a gas chromatograph. No separation of the mixtures of gases added to the column was observed even though low flow rates and concentrations of ethylene were used. This kind of experiment suggests that the affinity of tissue for ethylene is low and less than that observed had the column been filled with alumina, charcoal, or silica gel.

However, a number of workers have reported incorporation of ^{14}C- and ^{3}H-labeled ethylene into plant tissue. Buhler *et al.* (17) found that [^{14}C]ethylene was incorporated in the order of 0.05% of added label into the organic acid fraction of avocado and pear tissue. On the other hand, experiments with ripe oranges, limes, papayas, green apples, tomatoes, and grapes failed to show any incorporation into fruit tissues.

Hall *et al.* (32) found that cotton and *Coleus* plants incorporated [^{14}C]ethylene into a variety of substances. The rate of incorporation was enhanced by first absorbing ethylene on mercuric perchlorate and then releasing it. Shimokawa and Kasai (93) also studied incorporation of [^{14}C]ethylene regenerated from mercuric perchlorate. They found that the tracer was associated with DNA, RNA, and protein. On the basis of these findings, they suggested that binding of ethylene causes conformational changes in 4 sRNA which, in turn, plays a role in hormonal regulation by ethylene. Further work (94) suggested that regenerated ethylene was incorporated faster in the light than in the dark, and that once it was incorporated it was not transported to other parts of the plant.

Jansen (54) also studied the difference in incorporation between fresh and mercuric perchlorate-regenerated ethylene. Compared with fresh ethylene, regenerated ethylene (*1*) was taken up to a greater extent; (*2*) was not incorporated into volatile products; (*3*) produced radioactive CO_2 equal to 12% of incorporated label compared to 1% for fresh label; and (*4*) labeled succinic, malic, and citric acids which were not tagged when fresh ethylene was used. Obviously, ethylene released from mercuric-ethylene complex was modified to some extent to form some nonethylene ^{14}C-labeled compounds.

Jansen (52) reported that 0.015–0.04% of tritium-labeled ethylene was incorporated into avocados and that 12% of the label wound up in the methyl group of toluene. When he later learned (53) that only 2% of the ^{14}C-labeled ethylene formed toluene, it became obvious that different pathways were being followed for the incorporation of different isotopes.

Radioactive ethylene is not stable. Tolbert and Lemmon (99) have reported that radiation-induced decomposition of ethylene results in the formation of polyethylene, hydrogen, and methane. In view of the known instability of tagged ethylene and the fact that only some workers report incorporation of radioactive ethylene, and then only a small percentage of the starting level, it seems reasonable to assume that the incorporation measured represents an artifact due to impurities in the ethylene used. It seems that this would be especially true in the cases where regenerated ethylene was used.

Beyer (14) and we (10) have examined the physiological effect of deuterated ethylene. If metal-to-ethylene binding were a part of ethylene action, then the deuterated molecule might be more effective than normal ethylene, since Atkinson (12) has shown that the affinity of silver to ethylene through coordination bonding increased with increasing amounts of deuterium. Burg and Burg (21) proposed that a metal may play a central role in the binding site of ethylene. This idea was advanced because of the well known affinity of olefins for metals and the fact that CO, which is an ethylene homolog, binds to a cytochrome which is a metalloprotein. Support for this idea was obtained by Warner (103). He found that low concentrations of EDTA (10^{-5} *M*) reversed ethylene-induced inhibition of growth, ethylene-induced abscission, and ethylene-induced color development in tomato fruit. However, EDTA failed to repress the development of the respiratory climacteric. When chelated EDTA (FeEDTA) was used, it failed to reduce the effectiveness of ethylene.

It seems possible to conclude that, since there is no evidence to the contrary, ethylene may be bound to its site of action by means of weak van der Waals forces.

II. Effect on Isolated Enzymes

A number of investigators have examined the possibility that ethylene has a direct effect on enzyme activity. However, investigations with β-glucosidase, emulsin, salicin (28), α-amylase (29), invertase (29, 91), peroxidase (85), and adenosine triphosphatase (79) have shown that ethylene had no effect on these enzymes. Unpublished observations from my laboratory have also indicated that ethylene had no effect on cellulase and carbonic anhydrase. Carbonic anhydrase was tested because it appeared to be a likely candidate to show some positive response. Carbonic anhydrase contains zinc and has the ability to combine with CO_2—two features that suggest potential sensitivity to ethylene. Nelson (74) reported that ethylene increased the activity of trypsin. However, this effect was thought to be due to the removal of oxygen, since hydrogen had the same effect. Killian and Moritz (58) tested the idea that ethylene could combine with hemoglobin. Hemoglobin is known to combine readily with CO, as evidenced by a change in the absorption spectra of this iron-containing protein. However, no difference between ethylene-treated and untreated hemoglobin was observed. Nord and Weicherz (77) reported that treating cell-free preparations of yeast with ethylene or acetylene resulted in the reduction of the viscosity of these solutions. They interpreted these results as showing a binding of ethylene to protein.

III. Membranes

Since ethylene is more soluble in oil than in water and membranes contain large quantities of lipid, a number of investigators have tested the idea that ethylene regulated physiological phenomena by virtue of some effect on the membrane permeability. However, proponents of this idea have failed to note that CO, an ethylene analog, does not share the lipid solubility characteristics of ethylene, but has the physiological activity of ethylene.

The idea that ethylene has a disruptive effect on membranes which causes a change in permeability and alteration of compartmentalization does not appear to be valid. Ripening fruits exhibited obvious changes in terms of permeability and retention of soluble components. It seemed natural to suggest that ethylene caused changes in membrane permeability which, in turn, led to softening and increased respiration. However, present evidence suggests that changes in membrane characteristics are a result of ripening rather than a cause (15, 19, 23, 89). A similar situation exists in flowers. Nichols (75) pointed out that solute leakage from carnation increased during senescence. Senescence and leakage were promoted by ethylene and reversed by CO_2.

Ethylene has no influence on membrane permeability of potato (74), pea (18), avocado, banana, bean, and *Rhoeo* (89). However, von Guttenberg and Beythien (102) reported that ethylene increased the rate of deplasmolysis of *Rhoeo* leaves. Burg (19), however, failed to confirm their results. Nord and co-workers (76, 77) claimed that ethylene increased permeability of yeast cells. However, the effects were small, and Shaw (91) found no effect of ethylene on yeast cell permeability. Even if ethylene regulated yeast cell permeability, it would tell us little about its action, since, as far as we know, ethylene has no effect on yeast physiology. The same criticism is applicable to studies with red blood cells where ethylene was found to increase permeability (62). In addition, nitrous oxide also increased permeability, which showed that the effect was not specific for ethylene. Similarly, early reports on the response of mitochondria to ethylene suggested an influence of high ethylene levels on conformation (66, 78). While these effects were real, subsequent work pointed out the effects were not typical of normal ethylene action. First of all, high concentrations of ethylene (100 ppm) were required to induce conformational changes and second, saturated gases such as propane and ethane had similar effects (60, 72).

A number of investigators have found that ethylene increased or controlled the rate of enzyme secretion or release from cells. Jones (56) reported that ethylene increased the release of α-amylase from barley half-seeds. However, the effect was not solely on secretion since ethylene inhibited the amount of α-amylase synthesized by the half-seeds when no gibberellic acid was present. Unlike most ethylene-regulated phenomena, Jones reported that high concentrations of ethylene had an inhibitory effect on secretion. Normally, the dose-response curve of ethylene is asymptotic, and no additional effect is seen as the concentration is increased.

Ridge and Osborne (85) studied the regulation of peroxidase activity in pea stem tissue. They reported that ethylene inhibited leakage of peroxidase activity from apical tissue and, to a lesser extent, basal tissue.

Abscission is known to involve the enzymatic dissolution of cell walls with the result that leaves, flowers, fruit, and other parts are cast off from the plant. Horton and Osborne (42) found that the enzyme responsible for the dissolution was cellulase and that its synthesis was increased by ethylene. Later, De la Fuente and Leopold (25) pointed out that, following a short lag, ethylene reduced the breakstrength of abscission-zone explants. Removal of ethylene by flushing the gas phase prevented further reduction in breakstrength. Abeles and Leather (9) examined the possibility that the strict control of breakstrength by ethylene was due to rapid synthesis of cellulase when ethylene was added and the disappearance of cellulase when the gas was removed. They found, however, that while cellulase synthesis followed the addition of ethylene, levels of cellulase remained constant after the gas was removed, and concluded that ethylene must be having another effect in addition to the regulation of enzyme synthesis. Evidence was presented that this additional effect was the control of cellulase secretion and that this accounted for the maintenance of breakstrength once ethylene was removed. Ethylene not only controlled cellulase synthesis, but it also regulated the movement of cellulase from the cytoplasm to the cell wall through the membrane, and once the gas was removed, secretion of cellulase stopped.

In conclusion, the majority of the data suggest that physiological concentrations do not have a disruptive effect on the integrity of cellular membranes.

However, evidence is accumulating that secretory phenomena are affected, and that ethylene can influence transport of materials through membranes.

IV. Regulation of Nucleic Acid and Protein Metabolism

Regulation of RNA and protein synthesis is an important part of ethylene action. Synthesis of new enzymes has been shown to be involved in senescence (8), abscission (1, 9, 24, 64, 82), ripening (44, 65, 71), and swelling of pea stem tissue (20). It is important to emphasize that there is no reason to believe that these processes can be explained solely on the basis of new enzyme synthesis or that the sole function of ethylene is to regulate enzyme synthesis. For example, ethylene has been shown to increase aging (3) and control cellulase secretion (9), in addition to increasing cellulase synthesis during abscission. In the case of other ethylene effects, such as the rapid inhibition of stem elongation and epinasty, there is no evidence that any enzyme synthesis is involved.

A. Enzyme Induction

Regeimbal and Harvey (83) were the first to report that ethylene-treated tissue contained greater quantities of a particular enzyme than controls. They found that ethylene increased the amount of protease and invertase extracted from pineapple fruits. Since that time, reports on the effect of ethylene on other enzymes have appeared. The list includes acid phosphatase (38), ATPase (67, 97), α-amylase (38, 59, 95, 96), catalase (26, 38, 51), cellulase (1, 9, 24, 64, 82), chitinase (2), chlorophyllase (65), cinnamic acid 4-hydroxylase (46), cytochrome c reductase (55), diaphorase (55), β-1,3-glucanase (2, 4), invertase (50, 51, 61, 83, 93, 95), malic enzyme (44, 45, 83, 84), pectin esterase (38), peroxidase (11, 16, 33, 38, 47–51, 69, 70, 73, 80, 85–87, 90, 96), phenylalanine ammonia-lyase (27, 48, 49, 88), polygalacturonase (71), polyphenol oxidase (38, 48, 49, 93, 96), protease (37, 83), and pyruvic carboxylase (44, 45, 83).

Enzyme induction does not always depend solely on the action of ethylene. In some cases, cutting or excising the tissue causes an increase in enzyme activity, and the function of ethylene is to reduce the lag or increase the rate of synthesis. Examples of enzymes whose formation did not strictly depend on ethylene action include β-1,3-glucanase (4), malic enzyme (84), phenylalanine ammonia-lyase (27), and peroxidase (31, 80). Excising tissue can cause wound ethylene, and it is possible that wound ethylene production played some role in enzyme induction in tissue slices. On the other hand, enzyme induction during abscission was dependent on ethylene and, for a reasonable length of time, no cellulase synthesis or abscission occurred following excision unless ethylene was added to the gas phase.

B. Action of Cycloheximide

Cycloheximide, an inhibitor of protein synthesis, has been shown to prevent enzyme synthesis and the action of ethylene. Removing the cycloheximide by washing the tissue with water has demonstrated that its effect was reversible (84). A number of workers (7, 25) have reported that cycloheximide will block abscission. However, cycloheximide also had some secondary effects; notably, the ability to increase ethylene production (83), and this fact had to be considered when studying the action of cycloheximide on ethylene-sensitive tissue such as abscission-zone explants. Fruit ripening was also blocked or delayed with cycloheximide. Frenkel *et al.* (30) reported that cycloheximide blocked ripening and enhanced ethylene synthesis in Bartlett pears. However, it did not have an effect on the respiratory climacteric. Brady *et al.* (15), studying the effect of cycloheximide on bananas, found that both ripening and the respiratory climacteric were inhibited. Burg *et al.* (20) have reported that cycloheximide prevented the swelling of pea stems induced by ethylene.

Cycloheximide blocks the formation of a number of enzymes. The list includes β-1,3-glucanase (4), peroxidase (31, 80), cinnamic acid 4-hydroxylase (46), malate enzyme (84), cellulase (1, 82), and phenylalanine ammonia-lyase (85). It would appear reasonable to conclude that this evidence suggests the de novo synthesis of these enzymes. Critical evidence to establish this point was supplied by Lewis and Varner (64). They found that the deuterium of heavy water was incorporated into cellulase during abscission. This means that existing proteins were hydrolyzed and deuterium was incorporated in the amino acids released. When cellulase was synthesized, some of these deuterium-labeled amino acids were incorporated, giving rise to a cellulase which was denser than preexisting cellulase.

C. Action of Actinomycin D

In a number of cases, the inhibitor of RNA synthesis, actinomycin D, will block ethylene-regulated phenomena as well as enzyme synthesis. Actinomycin D has been shown to block abscission (6, 7, 39) and ethylene-induced swelling of pea stems (20). The synthesis of cellulase (1), malate enzyme (84), peroxidase (80), and cinnamic acid 4-hydroxylase (46) was also blocked by actinomycin D.

However, attention has to be drawn to the fact that some enzymes whose activity was increased by ethylene, and whose synthesis was blocked by cycloheximide, were not greatly affected by actinomycin D. These enzymes include β-1,3-glucanase (4), peroxidase (31, 85), and phenylalanine ammonia-lyase (88). It is interesting to note that these enzymes were also induced by slicing plant tissue. At this point the data suggest that some enzyme systems, induced by slicing tissues, can be increased by ethylene and do not appear to require RNA synthesis. The nature of this preformed or masked RNA, its location in the cell, and mechanism of activation remain unknown.

D. Protein Synthesis

Kidd *et al.* (57) reported an increase in protein content in ripening pears and apples, and similar observations have been made subsequently by others (35, 36, 43). Increased protein synthesis in ripening fruits has been observed in bananas (15), apple (45), pear (30), and figs (68). However, Sacher and Salminen (89) failed to observe an increase in protein synthesis when preclimacteric bananas or avocados were treated with ethylene.

Incorporation of amino acids into protein during abscission has been observed by a number of investigators (6, 7, 63). These observations have been substantiated by data obtained by histochemical staining and radioautography. Webster (104) and Stösser (98) have presented clear and vivid microphotographs showing enhanced protein synthesis in the separation layer of bean and cherry tissue, respectively. Valdovinos *et al.* (101) demonstrated that endoplasmic reticulum of tobacco flower pedicels undergoing abscission was more pronounced than in intact controls.

Ethylene has also been found to increase protein levels in vegetative tissue. Elmer (26) reported that ethylene increased the protein content of potato sprouts. Soybean seedlings fumigated with ethylene were shown to contain greater quantities of protein in the basal and elongating parts, but not in the apical region (40).

In contrast with ripening, abscission, and swelling, ethylene accelerates the loss of protein in aging or senescent tissue (7). However, these changes reflect total protein levels. It is likely that degradative enzymes such as protease and RNase are synthesized during senescence since cycloheximide and, to a lesser extent, actinomycin D will prevent the loss of chlorophyll, RNA, and protein in aging abscission-zone explants (8).

E. RNA Synthesis

The first report on the promotion of RNA synthesis by ethylene was that of Turkova *et al.* (100). They reported an increase in RNA during epinasty of tomato leaves. Whether or not RNA synthesis was required for epinasty was not shown, although the idea is intriguing. Inhibitor studies with actinomycin D suggest that RNA synthesis occurred during abscission (6, 7) and was required for the process to occur. Support for this interpretation stems from the work of a number of investigators (6, 39, 81, 98, 104). It is now known that the increase in RNA synthesis precedes that of protein synthesis (6) and is localized in or near the separation layer (81, 98, 104). The increase in RNA occurred in all fractions, mRNA, rRNA, and sRNA, though the magnitude varied. Differential extraction of the nucleic acids, indicated that the ethylene stimulation was confined to the fraction extracted with sodium lauryl sulfate, with the increase mainly in rRNA and mRNA. 5-Fluorouracil, which blocked 50% of the ethylene-enhanced ^{32}P incorporation, did not inhibit abscission. The greatest inhibition occurred in sRNA and rRNA fractions, indicating that not all fractions were required for abscission. Presumably, as long as sufficient mRNA was being synthesized, enough rRNA and sRNA were already available in the cell to permit abscission. When all RNA synthesis was blocked with actinomycin D, abscission stopped (39).

Ethylene has also been found to increase RNA synthesis in preclimacteric fruit (39). Marei and Romani (68) reported that, as in the case of abscission, the synthesis of all classes of RNA in fig fruit was promoted by ethylene. Hulme *et al.* (45) found that ethylene increased RNA synthesis in apples and that the increase in RNA synthesis was followed by an increase in protein synthesis. However, Sacher and Salminen (89) reported that they failed to observe an increase in RNA synthesis when preclimacteric bananas or avocados were treated with ethylene.

F. Chromatin Activity

Holm *et al.* (40, 41) have reported that ethylene inhibited growth of the apical part of soybean seedlings and increased it in the elongating and basal part. At the same time, RNA levels in the apical portion were reduced while they were increased in the elongating and basal portions. Chromatin from various parts of these seedlings was studied to determine its capacity for RNA synthesis. They found that activity was reduced in the apex and increased in the elongating and basal portions of the seedlings. The rate of response was rapid. The increase in chromatin activity was apparent after 3 hours and the increase in RNA after 6 hours. Nearest neighbor analysis of the RNA synthesized demonstrated that there was a qualitative difference between RNA synthesized by normal tissue and that treated with ethylene. They concluded that ethylene can regulate RNA synthesis so that a change in quantity and kind of RNA formed can occur.

G. DNA Metabolism

Plant growth is either promoted, inhibited, or unaffected by ethylene depending upon the kind of tissue involved. Examples of growth promotion are swelling, epinasty, hook closure, rice seedling elongation, and seed germination. Bud break is probably a special case. Here no growth takes place, as long as ethylene is present. However, after the gas is removed, growth of the buds ensues. Growth inhibition is seen as arrested development of buds, leaves, or apical meristems. Mature tissues, such as stems and leaves, do not undergo any change in size or weight although premature senescence usually occurs.

Since growth, or the lack of it, may be associated with cytokinesis, it is of interest to learn if DNA synthesis is controlled by ethylene. Holm and Abeles (39) reported that DNA synthesis or DNA content in bean leaf tissue was not affected by a 7-hour exposure to ethylene although abscission was promoted. Later (39, 40) they found that soybean seedlings treated with ethylene stopped synthesizing DNA in the apex where growth was inhibited and promoted DNA synthesis in the subapical part where swelling took place. Burg *et al.* (20) found that a similar situation existed in pea seedlings. They found that inhibition of cell division, measured as the reduction in metaphase figures, occurred within 2 hours after ethylene was added to pea seedlings. However, it is not clear if the change in DNA synthesis was the cause or result of inhibited growth. The speed at which ethylene slows growth of pea seedlings is very fast. Warner (103) found pea seedling growth slowed 6 minutes after ethylene was introduced into the gas phase and returned to normal 16 minutes after the ethylene was removed. Kinetic studies on changes in DNA or other postulated sites of action are required to establish the relationship between cause and effect. Burg *et al.* (20) have suggested that ethylene regulated DNA synthesis by some action on microtubule structure essential for spindle fiber formation during mitosis. If the action of ethylene was directed toward microtubules, this might also explain the reorientation of microfibril deposition that occurs during swelling.

V. Conclusions

In conclusion, the data suggest that little is known about the attachment of ethylene to its site of action except that the bonding must be of the weak van der Waals type. There is no indication of what ethylene is attached to except that it must be the same kind of thing for most if not all ethylene-regulated processes. This statement is based on the observation that the dose-response curve, action of analogs, and reversal by CO_2 are similar for many ethylene-regulated processes. In some cases such as ripening, abscission, swelling, and senescence, there is reason to believe that the combination of ethylene with the site results in the regulation of protein or enzyme synthesis which in turn accounts for the physiology observed. In these cases, RNA synthesis, presumably messenger RNA, with the accompanying support of soluble and ribosomal RNA is an initial or early step suggesting regulation of gene action. However, in other cases, especially those associated with the physiology of excised tissue, no RNA synthesis seems to be required and the action of ethylene is directed solely to an increase in protein synthesis. In another class of regulation, such as the inhibition of elongation, the action is directed toward blockage of DNA metabolism. The site of action in epinasty, root initiation, intumescence formation, and floral initiation is even more poorly known if such a thing is possible. The only valid conclusion that appears to arise is that a number of essential features of plant growth and development are susceptible to ethylene action and that in the final analysis, there may be as many mechanisms of ethylene action as there are modes of ethylene operation.

References

1. Abeles, F. B. (1969). *Plant Physiol.* **44,** 447.
2. Abeles, F. B., Bosshart, R. P., Forrence, L. E., and Habig, W. H. (1971). *Plant Physiol.* **47,** 129.
3. Abeles, F. B., Craker, L. E., and Leather, G. R. (1971). *Plant Physiol.* **47,** 7.
4. Abeles, F. B., and Forrence, L. E. (1970). *Plant Physiol.* **45,** 395.
5. Abeles, F. B., and Gahagan, H. E. (1968). *Plant Physiol.* **43,** 1255.
6. Abeles, F. B., and Holm, R. E. (1966). *Plant Physiol.* **41,** 1337.
7. Abeles, F. B., and Holm, R. E. (1967). *Ann. N. Y. Acad. Sci.* **144,** 367.
8. Abeles, F. B., Holm, R. E., and Gahagan, H. E. (1968). *In* "Biochemistry and Physiology of Plant Growth Substances" (F. Wightman and G. Setterfield, eds.(, p. 1515. Runge Press, Ottawa.
9. Abeles, F. B., and Leather, G. R. (1971). *Planta* **97,** 87.
10. Abeles, F. B., Ruth, J. M., Forrence, L. E., and Leather, G. R. (1972). *Plant Physiol.* **49,** 669.
11. Asmaev, P. G. (1937). *Proc. Agr. Inst. Krasnodar* **6,** 49.
12. Atkinson, J. G., Russel, A. A., and Stuart, R. S. (1967). *Can. J. Chem.* **45,** 1963.
13. Behmer, M. (1958). *Klosterneuberg, Aust. Hoehere Bundeslehr Versuchanstalt Wein-, Obst-Garten Bau. Mitt. Ser. B* **8,** 257.
14. Beyer, E. M. (1972). *Plant Physiol.* **49,** 672.
15. Brady, C. J., Palmer, J. K., O'Connell, P. B. H., and Smillie, R. M. (1970). *Phytochemistry* **9,** 1037.
16. Buchanan, D. W., Hall, C. B., Biggs, R. H., and Knapp, F. W. (1969). *HortScience* **4,** 302.
17. Buhler, D. R., Hansen, E., and Wang, C. H. (1957). *Nature* (*London*) **179,** 48.
18. Burg, S. P. (1964). *Colloq. Int. Cent. Nat. Rech. Sci.* **123,** 719.
19. Burg, S. P. (1968). *Plant Physiol.* **43,** 1503.

20. Burg, S. P., Apelbaum, A., Eisinger, W., and Kang, B. G. (1971). *HortScience* **6,** 359.
21. Burg, S. P., and Burg, E. A. (1965). *Science* **148,** 1190.
22. Burg, S. P., and Burg, E. A. (1967). *Plant Physiol.* **42,** 144.
23. Burg, S. P., Burg, E. A., and Marks, R. (1964). *Plant Physiol.* **39,** 185.
24. Craker, L. E., and Abeles, F. B. (1969). *Plant Physiol.* **44,** 1139.
25. De la Fuente, R. K., and Leopold, A. C. (1969). *Plant Physiol.* **44,** 251.
26. Elmer, O. H. (1936). *J. Agr. Res.* **52,** 609.
27. Engelsma, G., and Van Bruggen, J. M. H. (1971). *Plant Physiol.* **48,** 94.
28. Engles, D. T., and Dykins, F. A. (1931). *J. Amer. Chem. Soc.* **53,** 723.
29. Engles, D. T., and Zannis, C. D. (1930). *J. Amer. Chem. Soc.* **52,** 797.
30. Frenkel, C., Klein, I., and Dilley, D. R. (1968). *Plant Physiol.* **43,** 1146.
31. Gahagan, H. E., Holm, R. E., and Abeles, F. B. (1968). *Physiol. Plant.* **21,** 1270.
32. Hall, W. C., Miller, C. S., and Herrero, F. A. (1961). *Plant Growth Regul., Proc. Int. Conf., 4th, 1959* Vol. 4, 751.
33. Hall, W. C., and Morgan, P. W. (1964). *In* "Regulateurs naturels de la croissance végétale" (J. P. Nitsch, ed.), p. 727. CNRS, Paris.
34. Hansen, E. (1942). *Bot. Gaz.* (*Chicago*) **103,** 543.
35. Hansen, E. (1967). *Proc. Amer. Soc. Hort. Sci.* **91,** 863.
36. Hansen, E., and Blanpied, G. D. (1968). *Proc. Amer. Soc. Hort. Sci.* **93,** 807.
37. Harvey, R. B. (1928). *Minn., Agr. Exp. Sta., Bull.* **247,** 1.
38. Herrero, F., and Hall, W. C. (1960). *Physiol. Plant* **13,** 736.
39. Holm, R. E., and Abeles, F. B. (1967). *Plant Physiol.* **42,** 1094.
40. Holm, R. E., and Abeles, F. B. (1967). *Planta* **78,** 293.
41. Holm, R. E., O'Brien, T. J., Key, J. L., and Cherry, J. H. (1970). *Plant Physiol.* **45,** 41.
42. Horton, R. F., and Osborne, D. J. (1967). *Nature* (*London*) **214,** 1086.
43. Hulme, A. C. (1958). *Advan. Food Res.* **8,** 297.
44. Hulme, A. C., Jones, J. D., and Wooltorton, L. S. C. (1963). *Proc. Roy Soc., Ser. B* **158,** 514.
45. Hulme, A. C., Rhodes, M. J. C., and Wooltorton, L. S. C. (1971). *Phytochemistry* **10,** 749.
46. Hyodo, H., and Yang, S. F. (1971). *Arch. Biochem. Biophys.* **143,** 338.
47. Imaseki, H. (1970). *Plant Physiol.* **46,** 170.
48. Imaseki, H., Asahi, T., and Uritani, I. (1968). *In* "Biochemical Regulation in Diseased Plants or Injury," p. 189. Phytopathol. Soc. Jap.
49. Imaseki, H., Uchiyama, M., and Uritani, I. (1968). *Agr. Biol. Chem.* **32,** 387.
50. Ivanoff, N. N. (1932). *Biochem. Z.* **254,** 71.
51. Ivanoff, N. N., Prokoshev, S. M., and Gabunya, M. K. (1930–1931). *Bull. Appl. Bot., Genet. Plant Breed.* **25,** 262.
52. Jansen, E. F. (1963). *J. Biol. Chem.* **238,** 1552.
53. Jansen, E. F. (1964). *J. Biol. Chem.* **239,** 1664.
54. Jansen, E. F. (1969). *Food Sci. Technol., Proc. Int. Congr., 1st, 1962* Vol. 1, p. 475.
55. Jones, J. D., Hulme, A. C., and Wooltorton, L. S. (1965). *New Phytol.* **64,** 158.
56. Jones, R. L. (1968). *Plant Physiol.* **43,** 442.
57. Kidd, F., West, C., and Hulme, A. C. (1939). *Gt. Brit., Dep. Sci. Ind. Res., Food Invest. Bd., Annu. Rep. 1938* p. 119.
58. Killian, H., and Moritz, H. (1931). *Z. Gesamte Exp. Med.* **79,** 173.
59. Kraynev, S. I. (1937). *Proc. Agr. Inst. Krasnodar* **6,** 101.
60. Ku, H. S., and Leopold, A. C. (1970). *Plant Physiol.* **46,** 842.
61. Kursanov, A., and Krukova, N. (1938). *Biokhimiya* **3,** 202.
62. Leake, C. N., Lapp, H., Tenney, J., and Waters, R. M. (1927). *Proc. Soc. Biol. Med.* **25,** 93.

63. Leopold, A. C. (1967). *Symp. Soc. Exp. Biol.* **21,** 507.
64. Lewis, L. N., and Varner, J. E. (1970). *Plant Physiol.* **46,** 194.
65. Looney, N. E., Patterson, M. E. (1967). *Nature* (*London*) **214,** 1245.
66. Lyons, J. M., and Pratt, H. K. (1964). *Arch. Biochem. Biophys.* **104,** 318.
67. Madeikyte, E., and Turkova, N. S. (1965). *Liet. TSR Mokslu Akad. Darb., Ser. C* No. 2, p. 37.
68. Marei, N., and Romani, R. (1971). *Plant Physiol.* **48,** 806.
69. Matoo, A. K., and Modi, V. V. (1969). *Plant Physiol.* **44,** 308.
70. Matoo, A. K., Modi, V. V., and Reddy, V. V. R. (1968). *Indian J. Biochem.* **5,** 111.
71. McCready, R. M., and McComb, E. A. (1954). *Food Res.* **19**, 530.
72. Mehard, C. W., and Lyons, J. M. (1970). *Plant Physiol.* **46,** 36.
73. Nakagaki, Y., Hirai, T., and Stahmann, M. A. (1970). *Virology* **40,** 1.
74. Nelson, R. C. (1939). *Food Res.* **4,** 173.
75. Nichols, R. (1968). *J. Hort. Sci.* **43,** 335.
76. Nord, F. F., and Franke, K. W. (1928). *J. Biol. Chem.* **79,** 27.
77. Nord, F. F., and Weichherz, J. (1929). *Hoppe-Seyler's Z. Physiol. Chem.* **183,** 191.
78. Olson, A. O., and Spencer, M. (1968). *Can. J. Biochem.* **46,** 277.
79. Olson, A. O., and Spencer, M. (1968). *Can. J. Biochem.* **46,** 283.
80. Osborne, D. J. (1968). *Sci.* (*Soc. Chem. Ind., London*) *Monogr.* **31,** 236.
81. Osborne, D. J. (1968). *In* "Biochemistry and Physiology of Plant Growth Substances" (F. Wightman and G. Setterfield, eds.), p. 815. Runge Press, Ottawa.
82. Ratner, A., Goren, R., and Monselise, S. P. (1969). *Plant Physiol.* **44,** 1717.
83. Regeimbal, L. O., and Harvey, R. B. (1927). *J. Amer. Chem. Soc.* **49,** 1117.
84. Rhodes, M. J. C., Wooltorton, L. S. C., Galliard, T., and Hulme, A. C. (1968). *Phytochemistry* **7,** 1439.
85. Ridge, I., and Osborne, D. J. (1970). *J. Exp. Bot.* **21,** 720.
86. Ridge, I., and Osborne, D. J. (1970). *J. Exp. Bot.* **21,** 843.
87. Ridge, I., and Osborne, D. J. (1971). *Nature* (*London*) **229,** 205.
88. Riov, J., Monselise, S. P., and Kahan, R. S. (1969). *Plant Physiol.* **44,** 631.
89. Sacher, J. A., and Salminen, S. O. (1969). *Plant Physiol.* **44,** 1371.
90. Shannon, L. M., Uritani, I., and Imaseki, H. (1971). *Plant Physiol.* **47,** 493.
91. Shaw, F. H. (1935). *Aust. J. Exp. Biol.* **13,** 95.
92. Shcherbakov, A. P. (1939). *Izv. Akad. Nauk SSSR, Ser. Biol.* No. 6, p. 975.
93. Shimokawa, K., and Kasai, Z. (1968). *Agr. Biol. Chem.* **32,** 680.
94. Shimokawa, K., Yokoyama, K., and Kasai, L. (1969). *Mem. Res. Inst. Food Sci., Kyoto Univ.* **30,** 1.
95. Smirnov, A. I., and Krainev, S. I. (1940). *Izv. Akad. Nauk SSSR, Ser. Biol.* No. 4, p. 577.
96. Stahmann, M. A., Clare, B. G., and Woodbury, W. (1966). *Plant Physiol.* **41,** 1505.
97. Stewart, E. R., and Freebairn, H. T. (1969). *Plant Physiol.* **44,** 955.
98. Stösser, R. (1971). *Z. Pflanzenphysiol.* **64,** 328.
99. Tolbert, B. M., and Lemmon, R. M. (1955). *Radiat. Res.* **3,** 52.
100. Turkova, N. S., Vasileva, L. N. and Cheremukhina, L. F. (1965). *Sov. Plant Physiol.* **12,** 721.
101. Valdovinos, J. G., Jensen, T. E., and Sicko, L. M. (1971). *Plant Physiol.* **47,** 162.
102. von Guttenberg, H., and Beythien, A. (1951). *Planta* **40,** 36.
103. Warner, H. L. (1970). Ph.D. Dissertation, Purdue University, Lafayette, Indiana.
104. Webster, B. D. (1968). *Plant Physiol.* **43,** 1512.

Chapter 11

Air Pollution and Ethylene Cycle

I. Background

The role of ethylene as an air pollutant has been reviewed earlier by a number of workers (15, 25, 77). However, most of the contemporary research and concern has centered around other components of polluted air, such as SO_2, ozone, peroxyacetyl nitrate, CO, fluorides, and particulates, and has either ignored ethylene entirely or lumped it together with other gases called hydrocarbons. When reference is made to ethylene the reader is left with the impression that it is of limited consequence and its effects are of concern primarily to orchid growers. More attention has been focused on the other gases because, unlike ethylene, they have a direct effect on human health and comfort and cause visible damage in the form of lesions and discolored plant material.

This chapter reviews the question of whether or not ethylene air pollution is a serious (i.e., costly) problem for growers and others dependent on normal vegetation for livelihood or an acceptable way of life. A number of ideas should be kept in mind at this point. Ethylene, unlike other pollutants, is a plant hormone normally made and used by the plant to control a variety of phenomena, including growth, flowering, organ initiation, tropistic responses, and aging.

Acceleration of aging causes the greatest damage to plants. As an aging hormone, ethylene speeds up fruit ripening, floral senescence, and leaf abscission. Most of the damage attributed to ethylene is associated with these processes. The losses due to overripe fruit shipments, misshapen or senescent flowers, and defoliated vegetation are striking and expensive. As a hormone, ethylene regulates these processes with small quantities of gas. Most ethylene effects have the same dose-response curve: no effect at 0.001–0.01 ppm, discernible effects between 0.01 and 0.1 ppm, half-maximal responses at 0.1–1 ppm, and saturation from 1 to 10 ppm.

II. Sources of Ethylene

A. Natural Sources

1. Plants

Atmospheric ethylene arises from natural and man-made sources. Plants are probably the largest natural source. In the United States, which has a surface area of 3 million square miles, it is estimated that vegetation produces 2×10^4 tons (1.8×10^4 metric tons) of ethylene each year (1). This value was derived from the assumption that each acre was covered with a ton of vegetation and produced 0.5 nl ethylene per gram fresh weight per hour. It is reasonable to assume that true rates of production will have to take into account differences in vegetational cover and seasonal conditions. Foster (34) reported that flowering in bromeliads in Costa Rica was induced by volcanic gases. Whether the gases themselves contained ethylene or ethylene analogs or caused fires which, in turn, produced ethylene was not discussed.

2. Soil

Anaerobic soils may also be a source of ethylene. Smith and Russell (73) reported that the gas phase of anaerobic soils contained up to 10 ppm ethylene in addition to other hydrocarbons such as methane, ethane, and propane. In aerated soils, levels were 0.07 ppm at the surface and increased up to 0.14 ppm at a depth of 60 cm.

3. Natural Gas

Leaking natural gas may also introduce ethylene into the atmosphere. Crocker *et al.* (21) reported that a sample of West Virginia natural gas they examined contained 10 ppm ethylene. However, this is the exception since most natural gases contain only saturated hydrocarbon gases. Crocker *et al.* (21) also pointed out that the decomposition of stored coal also gave rise to detectable amounts of ethylene.

4. Burning Vegetation

Most smoke arising from burning vegetation is the result of human activity, although some forest and prairie fires can be considered natural to the extent that they were started by lightning. Hydrocarbon emissions from burning straw were studied by Boubel *et al.* (11). They found that 6.2 pounds/ton of fuel were converted into hydrocarbons, of which 1.7 pounds were ethylene. Darley *et al.* (23) performed similar studies on burning wood chips and brush. The yield of ethylene per metric ton of fuel from these fires varied from 0.1 kg for the wood chip fires to 2.7 kg for green brush. Higher rates of conversion were reported by Feldstein *et al.* (30). Measuring ethylene production by fires used to clear land, they found that a single burn of 6700 metric tons of fuel resulted in the production of 101 metric tons of ethylene, or a yield of 1.5‰. Darley *et al.* (23) estimated that 136,000 metric tons of agricultural wastes were burned in California in 1966, giving rise to the emission of 136 metric tons of ethylene. On the basis of amounts of burning occurring in the United States yearly, we estimated that open fires contributed 54,000 metric tons of ethylene to the atmosphere yearly. Compared with other sources, this is not a high figure, but the fact that fires are point sources results in localized concentrations of significantly high levels of ethylene. This fact has been taken advantage of by both pineapple and mango growers. Rodriguez (67) reported that fires alongside of pineapple fields and mango plantations were used to accelerate the formation of flowers. McElroy (57) recorded 4 ppm ethylene in the smoke of rice stubble burns. At a height of 30 meters above the burn, ethylene levels dropped to 300 to 500 ppb, and at 914 meters to preburn levels of less than 50 ppb. McElroy also reported that ethylene levels dropped to preburn levels 1.6 km downwind from the fire.

B. Man-Made Sources

The major source of atmospheric ethylene is probably man-made. Ethylene is produced by industrial sources, combustion of coal and oil, refuse burning, operation of motor vehicles, and, prior to the introduction of natural gas as a fuel, leaking gas mains.

1. Industrial

Ranjan and Jha (64) observed damage to mangos called black tip disease in the vicinity of brick kilns. While SO_2 was also considered as a source of damage, the authors concluded that ethylene was the causal agent in this case. According to their report, 4% of the fumes from the kiln was ethylene. Hall *et al.* (40) reported that cotton grown in the vicinity of a polyethylene plant in Texas showed symptoms similar to those gassed with ethylene. The ethylene content of the air near the plant was 3 ppm and dropped to ambient levels 6.5 km away. Most of the vegetation damage they recorded occurred within a 3.2-km radius of the plant. Damage included loss of leaves and other lateral appendages, loss of apical dominance, prostrate and vinelike growth, forcing of lateral buds, shortened internodes, and reddening and chlorosis of the leaves. Heck *et al.* (44) demonstrated that growth of cotton is sensitive to low ethylene concentrations. A reduction in growth from 25 to 50% followed a 27-day fumigation of between 40 to 100 ppb.

Fukuchi and Yamamoto (35) have reported premature abscission of mandarin fruit in the vicinity of gas works. The problem was controlled by passing effluents from the gas works over catalyst beds heated with steam.

2. Greenhouses

Operators of greenhouses are especially vulnerable to loss by ethylene since their crop is produced within a confined air space. However, not all of the ethylene is derived from sources outside these buildings. Tija *et al.* (83) reported that chrysanthemum growers suffered losses when the exhaust gases of oil heaters or open flame burners were used for heating during cold weather. Chrysanthemums under these conditions showed abnormalities in growth and failed to enter the reproductive stage even though subjected to photoinductive conditions favorable for flowering. The air in these greenhouses contained small amounts of ethylene, and chrysanthemum plants subjected to 1–4 ppm ethylene failed to flower under short days and showed typical epinastic symptoms shortening of internodes, loss of apical dominance, and thickening of stems. Hasek *et al.* (42) pointed out that proper adjustment and venting of heaters prevented damage to greenhouse crops. They also reported that weed oils and surfacing oils produced ethylene, and that these compounds could cause damage if used in closed greenhouses.

Ethylene damage from the use of dithio smoke generators used to control insects in greenhouses was reported to cause ethylene damage to orchids (43). The manufacturer of this chemical warns users that defoliation of roses can occur if inadequate ventilation is provided after fumigation. Eaves and Forsyth (28) reported that carnations located near a greenhouse in which tomatoes were being harvested were reported to wilt earlier than controls. The ethylene content of the air in the house in which the tomatoes were being harvested was several parts per million, enough to cause sleepiness of the carnations.

3. Laboratories

Sources of ethylene that concern experimental scientists are rubber stoppers, fluorescent light, and agar media. Jacobsen and McGlasson (46) observed that autoclaved rubber stoppers produced ethylene. Wills and Patterson (90) reported that ballast chokes from fluorescent lighting fixtures also produced ethylene from the breakdown of lacquered copper wire. Personal experience has indicated that autoclaved agar, especially when organic growth media were included, produced significant quantities of ethylene. Unless care is taken to monitor the gas phase surrounding plant tissue under certain experimental conditions, it is possible that these sources of ethylene can cause unexpected results.

4. Automobiles

Automobile exhaust consists of 18% CO_2 (50), 6% CO, 0.4% nitric oxide, 0.6% hydrocarbons (70), and a balance of nitrogen. The major hydrocarbon fractions are acetylene, 27%; ethylene, 20%; propene, 8%; butane, 7%; and toluene, 5%. The remaining 33% is made up of a large number of other compounds of which forty-four have been identified (9). Effluents from transportation account for 70% of the hydrocarbons produced by man, and various estimates place this annual production in the order of 10^6 (3) to 3×10^7 (9) metric tons per year for the United States. According to our estimates, 12 million metric tons of ethylene were produced by cars in the United States in 1966 (1). This is close to the amount produced for industrial purposes (10 million metric tons) and represents 14 billion dollars, if the ethylene from exhaust could have been separated and stored for subsequent use. Since the United States is about 3 million square miles in area, the rate of production per unit surface equals 4.5 metric tons/mile^2/year. However, cars are localized in urban areas, which results in a high rate of production for cities like Detroit, Washington, D.C., and San Francisco, which have rates of 32, 24, and 20 metric tons/mile^2/year, respectively (9). It is reasonable to assume that rural areas contribute little to atmospheric ethylene.

Ethylene is only one component of the hydrocarbons found in air. The major fraction is methane; levels between 1.1 and 1.5 ppm have been recorded. Unlike the other hydrocarbons, it is produced mainly by decomposition of organic matter by bacteria. It is estimated that 270 million metric tons of methane are produced world-wide annually to maintain these levels (3, 9). Another class of hydrocarbons produced by living matter is the terpenes. Estimates of production vary from 1 million metric tons per year (3) to 440 million metric tons (9). It is difficult to have any perspective on the size of these numbers, but it may be helpful to note that CO_2 production in the United States by human activity amounts to 2 billion metric tons annually (3).

Rates of ethylene production from autos vary little according to operating conditions. Ethylene content of auto exhaust ranges from 100 to 400 ppm under conditions of idle, acceleration, cruise, or deceleration. For any given length of time, the rate of exhaust production is a more important factor. An idling car produces 1.7 cubic meters per minute, accelerating 27, cruising 7.7, and decelerating 3.2 (50). Research has also shown that it makes little difference if regular or premium fuel is used, if the observations are made in winter or summer, or if hot versus cold starts are studied (50, 58, 70, 82). While fork lifts using liquified petroleum gas produced less ethylene, the rates were still significant and in the order of 60–200 ppm (82). The management of a large indoor shopping mall in Ohio has traced the abscission of ornamental rubber (*Ficus*) plants to the operation of propane-powered machines used to clean floors.*

5. *Levels in the Air and Variation*

The amount of ethylene in the air is influenced by location, time of day, and weather. Low levels are associated with rural areas or small towns. For example, values in the order of 3–5 ppb have been recorded for desert areas and Davis, California (9, 63). In contrast, San Francisco air contained 100 ppb for an hour or more on various occasions (4, 9, 16, 37). Average values for this city are in the order of 50 ppb (4, 9, 79). An unusually high value of 500 ppb has been reported for Pasadena (71). Surrounding areas, such as San Gabriel Valley, have lower maximum values, i.e., 30 ppb (9, 37). A similar situation has been reported for the East coast. Average values of ethylene in the Washington, D.C., area are in the order of 10–20 ppb, with maximum values near 60 ppb. In Frederick, Maryland, 50 miles west, ethylene levels are in the order of 1–5 ppb (2). Ethylene levels remain essentially constant for some distance above street levels. The value of 45 ppb was recorded at the top of a five-story building and 51 ppb at the second floor (78).

Diurnal variation in ethylene levels has been demonstrated as early as 1957 (71). Figure 11-1 shows that levels of ethylene coincide closely with traffic patterns in Pasadena. In addition, other common compounds of auto exhaust, such as CO and acetylene, closely follow the pattern demonstrated for ethylene. Scott *et al.* (71) also studied changes in NO_x (NO_x represents an unknown proportion of NO and NO_2), ozone, and peroxyacetyl nitrate. Levels of NO_x closely followed those for other auto emissions, and ozone and peroxyacetyl nitrate were low in the morning, increased

* Unpublished results (Abeles, 1970).

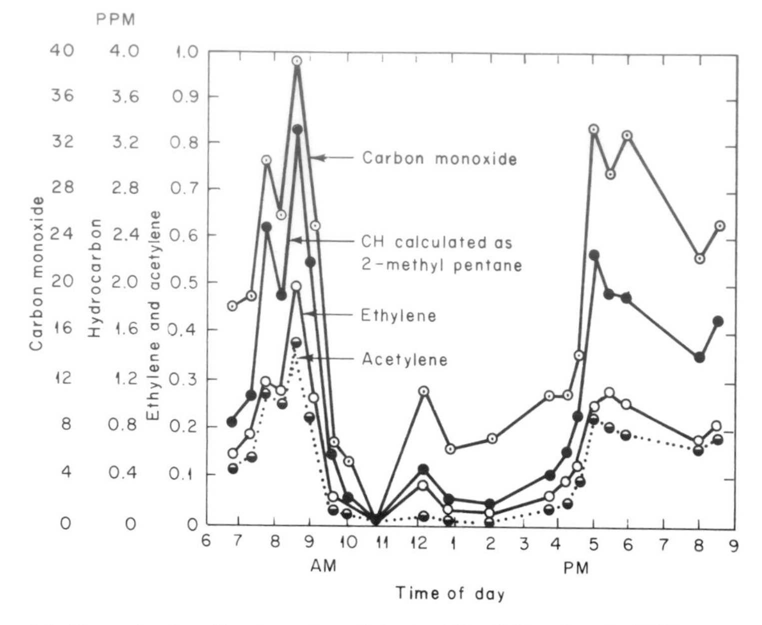

Fig. 11-1. Concentration of various air pollutants at South Pasadena in 1956 on a smogless day. [Data modified from Scott *et al.* (71).]

after noon, and decreased again after 4:00 PM. They also compared ambient air with auto exhaust and incinerator effluent and concluded that incinerators contributed little to the hydrocarbons of ambient air. Ambient air contained little methanol, which is an important component of incinerator effluent, but did contain propylene, which is released by cars and is not formed by incinerators. Essentially similar data showing diurnal variation in ethylene levels and other components of auto exhaust have been presented subsequently by others (4, 37).

According to Scott *et al.* (71), accumulation of ethylene during the night occurs as a result of layering of cooler air just at ground level in which pollutants can be trapped. In the morning, the inversion is destroyed by sunlight permitting rapid exchange of the air. This inversion can play an important role in ethylene levels. Hasek *et al.* (43) reported that on clear, cold nights in San Francisco, ethylene levels exceeded daytime levels, while the opposite was true when the weather was windy or rainy. In winter months, when radiation inversions are more common, ethylene accumulation was more pronounced than during the summer months.

An interesting example of this effect of winter weather and increasing accumulation of ethylene over the years has been presented by James (47).

As Davidson (26) originally pointed out in 1949, "dry sepal" injury to orchids was associated with calm climacteric conditions in New York during the winter months. A Chicago florist also reported that the greatest damage occurred during the winter months, especially when the air moved slowly in a southeastern direction (18). Concentrations required to cause damage varied according to the duration of exposure. Premature senescence of sepals and prevention of normal blossom opening was caused by 40–100 ppb for 8 hours, 2–20 ppb for 24 hours, and no effect at 1 ppb for prolonged periods of time. Similarly, low levels of ethylene (50 ppb) prevented normal development of marigold leaves (59). Ethylene levels in urban California are high enough to cause damage to orchids. James (47) and co-workers have accumulated crop loss data from a number of orchid growers in the San Francisco area and found that the greatest losses occurred in the winter (see Fig. 11-2). Yearly trends of November losses are shown in Fig. 11-3. The data collected show that, in a 10-year period, crop loss has steadily increased until a number of growers stopped producing plants or went into another form of business. These data, perhaps as well as any other, demonstrate that ethylene air pollution became an increasingly important problem in San Francisco over the years until certain types of agricultural pursuits were terminated. Additional examples of plant responses caused by ethylene air pollution are smog-induced flowering bromeliads (14) and changes in the levels of certain enzymes such as glucanase (2). Glucanase levels in bean leaves harvested in Frederick, Maryland, are low but can be readily increased by ethylene fumigation. In Washington, D.C., glucanase levels are higher, and leaves from these plants are no longer sensitive to additional ethylene exposures.

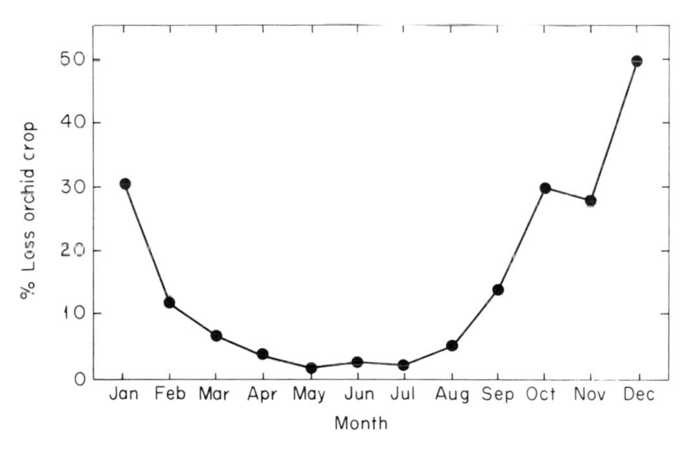

Fig. 11-2. Percent losses in *Cattleya* orchids attributed to ethylene at San Leandro, California, in 1958. [Data modified from James (48).]

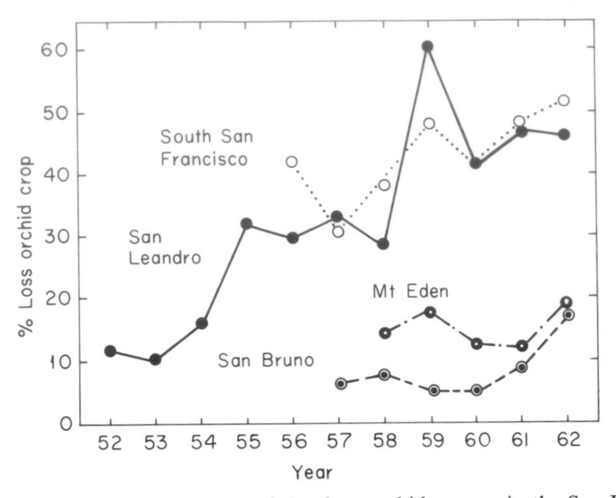

Fig. 11-3. Trends in November losses of *Cattleya* orchids grown in the San Francisco area from 1952 to 1962. [Data of James (48).]

6. *Illuminating Gas*

With the introduction of the automobile, petroleum production became an important industry. An indirect result of drilling for oil was the production of natural gas, and, since this fuel was cheap, it readily replaced manufactured gas in the United States and elsewhere. As an indirect result, reports of damage to trees and other plants by illuminating gas became less frequent even though leaking gas mains are still a problem in gas distribution. Crocker (20) estimated that 10% of the gas produced by gas plants was lost because of leaks in the distribution system. Illuminating gas became an important source of light and heat in the middle of the nineteenth century. The three major types were coal gas, water gas, and oil or Pintsch's gas. The first is formed by the destructive distillation of coal and was freed of ammonia, tar, H_2S, and SO_2 by washing and passing through slaked lime. Water gas was made by passing steam over burning coal to form CO and H_2. Since it contained no hydrocarbon gases which are required for a luminous flame, it was "carburetted" by passing through hot firebrick upon which a stream of high boiling petroleum was sprayed. The petroleum was cracked, giving rise to ethylene and other hydrocarbon gases. Oil gas was made directly by cracking oil.

Damage to shade trees by leaking illuminating gas was first reported by Girardin in 1864 (36). Since that time a number of other workers reported the same phenomenon (21, 27, 29, 45, 52, 53, 62, 72, 75, 76, 80, 81, 84, 85). In one city in Massachusetts, 500 trees were injured along 4 miles of newly laid gas pipes (80). It became standard to settle these claims (5 to 150 dollars per tree) out of court. In a suit against the Illinois Public Service Company in Chicago, a rose grower was awarded 90,000 dollars for the loss of a rose crop (87). Early investigators suspected CO, acetylene, and HCN as the primary cause of damage.

While these gases can cause damage, the most important constituent was probably ethylene, and work by Neljubov (61) in Russia and Crocker (20) in this country demonstrated the central role ethylene played in losses due to illuminating gas.

Figure 11-4 shows the effect of a leaking gas main on shade trees around 1900. The trees in the immediate vicinity of the gas main (*Gasrohr*) are defoliated while trees more distant are free of injury. One report indicates that defoliation occurred in the presence of leaking natural gas (69). Since natural gas normally contains little or no ethylene, the author suggested that the damage was caused by anaerobiosis of the soil caused by displacement of air by the leaking gas.

The effect of a leaking gas main can occur some distance from the main itself if the ground is frozen or paved. An example of this effect is shown in Fig. 11-5. In this case, gas from a leaking main moved horizontally under the frozen surface and into a greenhouse, causing complete defoliation of a rose crop. The prevalent use of illuminating gas in laboratories caused

Fig. 11-4. Effect of leaking illuminating gas on shade trees in Germany around 1900. "Gasrohr" indicates the location of the gas line. [From Wehmer (84).]

Fig. 11-5. Defoliation of a rose crop in a Nebraska greenhouse in 1911. Frame (A) shows the location of the main supplying street lights with gas. Man standing over the disturbed area indicates the location of the leak. The illuminating gas moved horizontally under the frozen ground into the greenhouse on the left where it defoliated the rose crop shown in (B). [From Wilcox (89).]

problems for a number of early plant physiologists. The early literature contains numerous references to physiological effects of "room air" on the growth of seedlings and other plants (35, 54, 65, 66). The fact that most laboratories at the turn of the century were equipped with illuminating gas is the probable cause for the effects observed. Wiesner (88) performed a series of phototropic experiments utilizing a gas lamp and a window as light sources. He concluded that the differences in growth he observed were due to the fact that natural light was composed of parallel light rays while those from the gas lamp were not. A more probable explanation is that plants in the vicinity of the gas jet were responding to traces of ethylene leaking from the lamp.

III. Economic Losses Due to Ethylene Air Pollution

Table 11-1 summarizes reports in the literature on ethylene damage to agronomic crops. This ethylene is from three sources: illuminating gas, poorly vented storage facilities, and air polluted with automobile exhaust. The earlier reports reflect losses to trees and greenhouse crops due to escaping illuminating gas. With the change to natural gas, these losses have ceased to be important. Accumulation of ethylene in storage areas has been, and continues to be, a problem for handlers of fruit, flowers, and nursery stock. A number of sources of ethylene have been identified, but the major ones appear to be the plants themselves, diseased material, or

Table 11-1

Reports of Ethylene Damage to Plant Material

Date of damage	Crop	Amount	Cause	Reference number
1906	Trees	400 trees	Illuminating gas	80
1913	Carnations	$20,000	Illuminating gas	20
1931	Roses	$90,000	Illuminating gas	87
1940	Flowers	—	Apple volatiles	56
1942	Carnations	—	Pintsch's gas	31
1942	Nursery stock		Unventilated storage	60
1949	Orchids		Air pollution	26
1949	Flowers	—	Unventilated storage	32
1952	Nursery stock	—	Apple volatiles	22
1956	Lettuce	25% of shipment	Unventilated storage	68
1959	Flowers	$70,000	Air pollution	16, 24
1963	Horticultural crops	—	Air pollution	49
1963	Carnations	$700,000	Air pollution	49
1964	Carnations	—	Diseased plants	74
1964	Orchids	$100,000	Air pollution	9, 48
1968	Orchids	$150,000	Air pollution	18
1970	Flowers	$9200	Air pollution	86

the inclusion of climacteric fruit in the same area. Various controls for losses such as ventilation, ethylene absorbants ($KMnO_4$ or bromine on adsorbants), and segregation of material are available. The third major source of damage, ethylene air pollution, has increased since it was first observed in 1949. Most damage has occurred in greenhouses located in or near major urban areas in the United States. The Los Angeles Board of Supervisors recently "planted" 74,500 dollars worth of plastic plants along a 1½-mile stretch of Jefferson Boulevard in Los Angeles because normal vegetation was sensitive to auto exhaust.* Increasing use of plastic plants is also being made by fast food franchises and service stations. A compilation of damage such as that presented in Table 11-1 must be incomplete since most growers affected by ethylene pollution do not report their losses in scientific or technical literature or are not aware of what is causing the damage they observe.

IV. Control of Air Pollution

Since cars produce 90% of the ethylene in the air, it seems reasonable that the easiest way to control ethylene air pollution is to limit hydrocarbon production by emission control devices or by selecting power plants that completely oxidize their fuel to CO_2 and water. In the meantime, a number of alternatives have been examined to reduce the ethylene levels in greenhouses and storage areas. They are brominated charcoal, ozone, and $KMnO_4$.

Brominated charcoal is an extremely effective trap for ethylene and has been extensively studied as a purifying system both for greenhouses and fruit storage areas (22, 49). However, it appears that use of brominated charcoal is not a practical means of purifying air. Brominated charcoal slowly releases bromine, which accelerates corrosion of the air-handling system. In addition, brominated charcoal has an affinity for moisture and becomes soaked and waterlogged at high relative humidities. Because of the waterlogging problems, further research with these systems has been discontinued (48).† Ordinary charcoal, which has been successfully used to free air of oxidants in greenhouses, has little affinity for hydrocarbons with a molecular weight below 50. Because of this, it is valueless as an ethylene absorbent.

Ozone reacts rapidly with ethylene to form formaldehyde and formic acid. Colbert (17) studied extensively the feasibility of using ozone to oxidize

* *The Washington Post*, Fl, March 5, 1972.

† O. W. Davidson, personal communication (1970).

ethylene and trapping the unreacted ozone and reaction products on activated carbon. He estimated that 89 metric tons of apples produced 15 mg of ethylene per minute and required the production of 0.25 g ozone per minute for complete oxidation. Even though the feasibility of such a system was demonstrated, no further reports on its adaptation to commercial situations have been made. To make such a system feasible, care would have to be taken to prevent ozone from coming into contact with the stored material. Plants are extremely sensitive to minute quantities of ozone. For example, 10 ppb ozone for a few hours can cause damage to sensitive plants. Craker (19) has pointed out that in some cases ozone causes symptoms typical of ethylene damage to plants. He demonstrated that ethylene production by plants was increased by ozone and that the ethylene was responsible for the responses of the plant tissue to ozone.

The third effective absorbent for ethylene is Celite or silica gel treated with $KMnO_4$. This material has been tested as a means of reducing ethylene levels in storage areas (28) and growth chambers (2). The $KMnO_4$ oxidizes the ethylene to ethylene glycol and, in addition, is an effective trap for other pollutants such as ozone and SO_2. While ethylene absorbents may have some practical value for areas in which plant matter is stored or shipped, it has not been used successfully in greenhouses. The primary problem is the high cost of installing and operating the air-filtering system.

The effect of growing plants in air purified with Purafil is shown in Fig. 11-6. Purafil is a commercially available product (see Chap. 2) consisting of alumina pellets impregnated with $KMnO_4$. In addition to removing ethylene, the material also removes O_3, SO_2, NO_2, H_2S, and NH_3 from air.

V. Degradation of Atmospheric Ethylene

As early as 1913 Knight and Crocker (51) pointed out that ethylene was a hazard and a potential air pollutant. They indicated that fires were a source of the gas and wondered why levels of ethylene did not eventually accumulate to the point that permanent damage to vegetation occurred. Levels of atmospheric ethylene are low (<5 ppb), and there are no data to suggest that the concentration of the gas is increasing with time. Because of this, mechanisms must exist to consume and degrade ethylene, otherwise atmospheric levels would have increased as a result of long-term production by vegetation and fires and, more recently, as a result of human activity.

At the present time, three mechanisms for ethylene degradation are known: ozone oxidation, a light-dependent reaction with NO, and uptake by soil microflora.

Fig. 11-6. Effect of ethylene air pollution on plant growth in Beltsville, Maryland. Plants were grown under the conditions indicated in the figures for varying periods of time. Ambient levels of ethylene varied from 1 to 60 ppb depending on traffic and weather conditions. $KMnO_4$ filter removed 75% of ambient ethylene levels. Other plants shown were exposed to levels of ethylene indicated superimposed on ambient air. (Unpublished work of F. B. Abeles and H. E. Heggestad, 1972.) (A), Red kidney beans, 16 days old. (B), Pink Cascade petunia after 45-day exposure to conditions indicated. (C), A, and C, cucumbers after 65-day exposure to conditions indicated.

A. Ozone

Ethylene reacts with ozone to form H_2O, CH_2O, CO_2, CO, and formic acid (55, 71). When 32 ppm each of ethylene and ozone were mixed in a chamber the following products were formed. The number above each product represents parts per million formed (71).

$$\overset{32\text{ ppm each}}{C_2H_4 + O_3} \longrightarrow \overset{28}{H_2O} + \overset{17}{CH_2O} + \overset{4}{CO_2} + \overset{28}{CO} + \overset{1}{HCOOH} \qquad (11\text{-}1)$$

The lack of stoichiometry suggests that additional unknown compounds were formed. According to Colbert (17), the reaction is second-order with a rate constant at 0°C of about 4×10^{-5} moles/second/ppm for ozone concentrations of 300–400 ppm.

Atomic oxygen (·O) can also react with ethylene to form CO, methane, ethane, propylene, acetaldehyde, propanol, butanol, H_2O, ethylene oxide, and possibly ketene. Atomic oxygen arises from:

$$NO_2 + h\nu \longrightarrow NO + \cdot O \qquad (11\text{-}2)$$

$$O_3 + h\nu \longrightarrow O_3 + \cdot O \qquad (11\text{-}3)$$

However, these reactions are of little significance since concentrations of atomic oxygen were estimated at 1×10^{-5} ppb and 99.9% of it reacts with molecular oxygen to form ozone.

B. Photolysis with NO_x

Haagen-Smit and Fox (38) and others (5–7, 10, 70, 78, 79) have studied the production of ozone and peroxyacetyl nitrate by mixtures of hydrocarbons with nitrogen oxides. Table 11-2 shows the relative effectiveness

Table 11-2

Production of Ozone and Peroxyacetyl Nitrate from Auto Exhaust

Exhaust hydrocarbon	% Total emission	Ozone formation: Crack depth[a]	Ozone formation: ppm[b]	Peroxyacetyl nitrate formation[c]
Acetylene	27	0.5	—[d]	—
Ethylene	20	2	—	0
Propene	8	5	0.68/75	0.25
Butane	7	0	—	—
Toluene	5	—	—	—
1-*n*-Butene	3	5	0.58/45	—
1-*n*-Pentene	1.4	5	0.62/45	—
2,3-Dimethyl-2-butene	1.4	—	0.45/38	0.69
Butadiene	1.1	12	0.20/60	—
trans-2-Butene	0.6	—	0.73/35	0.51
Hexene	0.3	4	0.45/35	0.20

[a] Total crack depth in rubber, mm (1 mm = 3 ppm O_3) (38).

[b] Initial hydrocarbon concentration = 3 ppm; NO or NO_2 concentration = 1 ppm. Value shown represents ozone formation after irradiation for indicated time in minutes (9).

[c] Relative yield (9).

[d] No data available.

of ethylene and other exhaust constituents in forming these oxidants. The lack of reliable data makes it difficult to compare various exhaust fractions and their relative contributions to ozone and peroxyacetyl nitrate production. However, since ethylene is the second largest fraction of hydrocarbon emissions it represents an important contributor to total ozone production. Unlike other hydrocarbons, however, photolysis with NO_x does not produce peroxyacetyl nitrate. The chemical reactions occurring during photolysis are only partially known. Work by Altshuller and co-workers (5, 6) has established the following reactions. In the presence of UV light, ethylene reacts with either NO or NO_2 to form ozone and ⅓ mole of formaldehyde. Maximum rate of production occurs when the ratio of ethylene to NO_x is 1:1. As a result of secondary reactions between oxide and formaldehyde, CO is formed. It is reasonable to assume that ozone formed can also back-react with undergraded ethylene, giving rise to a series of products indicated in Eq. (11-1).

The fact that these reactions occur at low ethylene concentrations adds to the difficulty of characterizing other reaction products. Estimates of half-times at which these reactions occur include 6 hours for 45 ppm ethylene, 70 minutes for 1 ppm, and 210 minutes for 0.2 ppm. It is useful to keep in mind that these concentrations of ethylene are greater than those that normally occur in polluted air (0.01–0.1 ppm).

C. Soil Microflora

A third means of removing ethylene from the atmosphere may be by soil microflora. It is known that bacteria are able to consume hydrocarbons, including ethylene (91). We (1) have found that when soil samples were placed in an atmosphere containing auto exhaust, ethylene and other hydrocarbon gases were removed. Half-times for removal were in the order of 3–4 days for an initial concentration of 30 ppm. Heating, ethylene oxide, anaerobiosis, and low temperatures prevented hydrocarbon uptake. Extrapolating from our laboratory experiments, we calculated that 7×10^6 tons of ethylene may be removed from the air by microbial degradation. Since the rate of production exceeds uptake by soil by a factor of two, oxidation by ozone and photolysis with NO_x undoubtedly play an important role in regulating atmospheric levels of ethylene.

D. Plants

Can vegetation take up hydrocarbons from auto exhaust? Except for ethylene, no data are available on the interaction of vegetation with hydrocarbons. A number of investigators have exposed plants to [^{14}C]ethylene. Some have reported uptake in some fruits but not in others (12). Reports of uptake can probably be discounted as due to artifacts of the experimental system studied. First of all, only a small fraction (<0.05%) of the added ethylene becomes incorporated (12, 15). Second, aged ethylene and ethylene regenerated from mercuric perchlorate were taken up more rapidly than fresh ethylene (39). Finally, if ethylene levels surrounding plant material are monitored, they are never seen to decrease (41). Since other components of auto exhaust such as CO, acetylene, and propylene act physiologically as ethylene analogs, there is reason to believe that they, too, are not consumed by plants.

VI. Air Quality Standards

Various standards for acceptable levels of ethylene have been proposed or adopted. On March 1962, standards for ethylene were proposed by the California State Board of Health. According to these standards, maximum levels of ethylene for 1 hour are 0.5 ppm and for 8 hours 0.1 ppm (13). The American Industrial Hygiene Association has also proposed air quality guides for ethylene. These are shown in Table 11-3. According to the available data, present air quality standards are too high. A more reasonable level would be one-tenth that shown in Table 11-3. As discussed above, significant losses of greenhouse crops already occur in areas that have maximum ethylene levels of 0.1 ppm and an 8-hour exposure of 0.05 ppm. Under the conditions capable of generating 0.1 ppm ethylene, other aspects of air quality have also degraded to the point where ethylene values represent only a part of the total problem.

Table 11-3

AIR QUALITY STANDARDS

	Recommended levels of ethylene in air (ppm)[a]	
Area	1-hour maximum	8-hour maximum
Rural	0.25	0.06
Residential	0.50	0.1
Commercial	0.75	0.15
Industrial	1.00	0.20

[a] Anonymous (8).

References

1. Abeles, F. B., Craker, L. E., Forrence, L. E., and Leather, G. R. (1971). *Science* **173,** 914.
2. Abeles, F. B., Forrence, L. E., and Leather, G. R. (1971). *Plant Physiol.* **48,** 504.
3. Altshuller, A. P. (1968). *Advan. Chromatogr.* **5,** 229.
4. Altshuller, A. P., and Bellar, T. A. (1963). *J. Air Pollut. Contr. Ass.* **13,** 81.
5. Altshuller, A. P., and Cohen, I. R. (1963). *Int. J. Air Water Pollut.* **7,** 787.
6. Altshuller, A. P., and Cohen, I. R. (1964). *Int. J. Air Water Pollut.* **8,** 611.
7. Altshuller, A. P., Klosterman, D. L., Leach, P. W., Hindawi, I. J., and Sigsby, J. E. (1966). *Int. J. Air Water Pollut.* **10,** 81.
8. Anonymous. (1968). *Amer. Ind. Hyg. Ass., J.* **29,** 627.
9. Barth, D. S. (1970). "Air Quality Criteria for Hydrocarbons," Nat. Air Pollut. Contr. Admin. Publ. No. AP-64. U. S. Dept. of Health, Education and Welfare, Washington, D. C.
10. Bellar, T., Sigsby, J. E., Clemons, C. A., and Altshuller, A. P. (1962). *Anal. Chem.* **34,** 763.
11. Boubel, R. W., Darley, E. F., and Schuck, E. A. (1969). *J. Air Pollut. Contr. Ass.* **19,** 497.
12. Buhler, D. R., Hansen, E., and Wang, C. H. (1957). *Nature* (*London*) **179,** 48.
13. California Standards for Ambient Air Quality and Motor Vehicle Exhaust. (1962). Tech. Rep., Suppl. No. 2. State Board of Public Health.
14. Cathey, H. M., and Downs, R. J. (1965). *Exch. Flower, Nursery Gard. Cent. Trade* **143,** 27.
15. Clayton, G. D. (1966). G. D. Clayton & Assoc., Inc. Prepared for Auto. Mfg. Ass., Detroit, Michigan.
16. Clayton, G. D., and Platt, T. S. (1967). *Amer. Ind. Hyg. Ass., J.* **28,** 151.
17. Colbert, J. W. (1952). *Refrig. Eng.* **60,** 265.
18. Cottrell, G. G. (1968). *Florist Nursery Exch.* **148,** 5.
19. Craker, L. (1971). *Environ. Pollut.* **1,** 299.
20. Crocker, W. (1913). *Sch. Sci. Math.* **13,** 277.
21. Crocker, W., Zimmerman, P. W., and Hitchcock, A. E. (1932). *Contrib. Boyce Thompson Inst.* **4,** 177.
22. Curtis, O. F., and Rodney, D. R. (1952). *Proc. Amer. Soc. Hort. Sci.* **60,** 104.
23. Darley, E. F., Burleson, F. R., Mateer, E. H., Middleton, J. T., and Osterli, V. P. (1966). *J. Air Pollut. Contr. Ass.* **16,** 685.
24. Darley, E. F., Dugger, W. M., Mudd, J. B., Ordin, L., Taylor, O. C., and Stephens, E. R. (1963). *Arch. Environ. Health* **6,** 761.
25. Das Gupta, S. N. (1957). *Proc. Indian Sci. Congr., 1957* No. 2, p. 88.
26. Davidson, O. W. (1949). *Proc. Amer. Soc. Hort. Sci.* **53,** 440.
27. Deuber, C. G. (1933). *Amer. Gas Ass. Mon.* **15,** 465.
28. Eaves, C. A., and Forsyth, F. R. (1969). *Florists' Rev.* **145,** 61.
29. Ehrenburg, P., and Schultze, K. (1916). *Z. Pflanzenkr. Pflanzenschutz* **26,** 65.
30. Feldstein, M., Duckworth, S., Wohlers, H. C., and Linsky, B. (1963). *J. Air Pollut. Contr. Ass.* **13,** 542.
31. Ferguson, W. (1942). *Sci. Agr.* **22,** 509.
32. Fischer, C. W. (1949). *N. Y. State Flower Growers, Bull.* **49,** 9.
33. Fitting, H. (1911). *Jahrb. Wiss. Bot.* **49,** 187.
34. Foster, N. B. (1953). "Bromeliads." Bromeliad Soc., Inc.
35. Fukuchi, T., and Yamamoto, T. (1969). *Kogai To Taisaku* **5,** 17.
36. Girardin, J. P. L. (1864). *Jahresber. Agrikulturchem.* **7,** 199.
37. Gordon, R. J., Mayrsohn, H., and Ingels, R. M. (1968). *Environ. Sci. Technol.* **2,** 1117.
38. Haagen-Smit, A. J., and Fox, M. M. (1956). *Ind. Eng. Chem.* **48,** 1484.

39. Hall, W. C., Miller, C. S., and Herrero, F. A. (1961). *Plant Growth Regul., Proc. Int. Conf., 4th, 1959* Vol. 4, 751.
40. Hall, W. C., Truchelut, G. B., Leinweber, C. L., and Herrero, F. A. (1957). *Physiol. Plant.* **10,** 306.
41. Hansen, E. (1942). *Bot. Gaz.* **103,** 543.
42. Hasek, R. F., James, H. A., and Sciaroni, R. H. (1969). *Florists' Rev.* **144,** 16.
43. Hasek, R. F., James, H. A., and Sciaroni, R. H. (1969). *Florists' Rev.* **144,** 21.
44. Heck, W. W., Pires, E. G., and Hall, W. C. (1961). *J. Air Pollut. Contr. Ass.* **11,** 549.
45. Hitchcock, A. E., Crocker, W., and Zimmerman, P. W. (1934). *Contrib. Boyce Thompson Inst.* **6,** 1.
46. Jacobsen, J. V., and McGlasson, W. B. (1970). *Plant Physiol.* **45,** 631.
47. James, H. A. (1959). "Report of a District-wide Survey of Air Pollution Damage to flora. Fall and Winter 1958." Bay Area Air Pollut. Contr. Dist., 1480 Mission St., San Francisco, California.
48. James, H. A. (1963). *Inform. Bull.* **8;** see James (47).
49. James, H. A. (1964). Commercial Crop Losses in the Bay Area Attributed to Air Pollution; see James (47).
50. Jones, J. L., Schuck, E. A., Eldridge, R. W., Endow, N., and Cranz, F. W. (1963). *J. Air Pollut. Contr. Ass.* **13,** 73.
51. Knight, L. I., and Crocker, W. (1913). *Bot. Gaz.* (*Chicago*) **55,** 337.
52. Kny, L. (1871). *Bot. Ztg.* **29,** 852.
53. Krone, P. R. (1937). *Mich., Agr. Exp. Sta., Spec. Bull.* **285.**
54. Lackner, C. (1873). *Gertnerishe Plaudereien Monatsschr. Ver Beförd. Gartenbaues Kgl. Preuss. Staaten* **16,** 16.
55. Leighton, P. A. (1961). "Photochemistry of Air Pollution." Academic Press, New York.
56. Lumsden, D. V., Wright, R. C., Whiteman, T. M., and Byrnes, T. W. (1940). *Science* **92,** 243.
57. McElroy, J. J. (1960). *Calif. Agr.* **14,** 3.
58. McMichael, W. F., and Sigsby, J. E. (1966). *J. Air Pollut. Contr. Ass.* **16,** 474.
59. Middleton, J. T., Darley, E. F., and Brewer, R. F. (1958). *J. Air Pollut. Contr. Ass.* **8,** 9.
60. Milbrath, J. A., and Hartman, H. (1942). *Oreg., Agr. Exp. Sta., Bull.* **413.**
61. Neljubov, D. (1901). *Beih. Bot. Zentralbl.* **10,** 128.
62. Pfeiffer, O. (1898). *Schillings J. Gasbeleucht.* **41,** 137.
63. Pratt, H. K., and Goeschl, J. D. (1969). *Annu. Rev. Plant Physiol.* **20,** 541.
64. Ranjan, S., and Jha, V. R. (1940). *Proc. Indian Acad. Sci., Sect. B* **11,** 267.
65. Richter, O. (1908). *Verh. Ges. Deut. Naturf. Arzte* **80,** 189.
66. Richter, O. (1913). *Verh. Ges. Deut. Naturf. Arzte* **85,** 649.
67. Rodriguez, A. B. (1932). *J. Dep. Agr. P. R.* **26,** 5.
68. Rood, P. (1956). *Proc. Amer. Soc. Hort. Sci.* **68,** 296.
69. Schollenberger, C. J. (1930). *Soil Sci.* **29,** 261.
70. Schuck, E. A., and Ford, H. W. (1958). *Air Pollut. Found., Rep.* **26.**
71. Scott, W. E., Stephens, E. R., Hanst, P. C., and Doerr, R. C. (1957). *Proc. Amer. Petrol. Inst.* **37,** 171.
72. Shonnard, F. (1903). Dept. Public Works, Yonkers, New York.
73. Smith, K. A., and Russell, R. S. (1969). *Nature* (*London*) **222,** 769.
74. Smith, W. H., Meigh, D. F., and Parker, J. C. (1964). *Nature* (*London*) **204,** 92.
75. Sorauer, P. (1916). *Z. Pflanzenkr. Pflanzenschutz* **26,** 129.
76. Späth, J. L., and Meyer, K. (1873). *Landwirt. Vers. Sta.* **16,** 336.
77. Stahl, Q. R. (1969). *Nat. Air Pollut. Contr. Admin. Publ.* No. APTD 69-35.
78. Stephens, E. R., and Burleson, F. R. (1967). *J. Air Pollut. Contr. Ass.* **17,** 147.

79. Stephens, E. R., Darley, E. F., and Burleson, F. R. (1967). *Amer. Petrol. Inst. Div. Refin. Meet.*
80. Stone, G. E. (1906). *Mass., Agr. Exp. Sta., Rep.* p. 180.
81. Stone, G. E. (1913). *Mass., Agr. Exp. Sta., Bull.* **31,** 45.
82. Swartz, D. J., Wilson, K. W., and King, W. J. (1963). *J. Air Pollut. Contr. Ass.* **13,** 154.
83. Tjia, B. O. S., Rogers, M. N., and Hartley, D. E. (1969). *J. Amer. Soc. Hort. Sci.* **94,** 35.
84. Wehmer, C. (1900). *Z. Pflanzenkr. Pflanzenschutz* **10,** 267.
85. Wehmer, C. (1918). *Ber. Deut. Bot. Ges.* **36,** 460.
86. Weideusaul, T. C., and Lacasse, N. L. (1970). "Statewide Survey of Air Pollution Damage." Center for Air Environment Studies, Pennsylvania State University, University Park, Pennsylvania.
87. Weinard, F. F. (1931). *Florists' Rev.* **67,** 19.
88. Wiesner, J. (1878). *Sitzungsber. Kaiserl. Akad. Wiss. Wien, Math.-Naturwiss. Kl.* **77,** 15.
89. Wilcox, E. M. (1911). *Nebr. State Hort. Soc. Annu. Rep.* p. 278.
90. Wills, R. B., and Patterson, B. D. (1970). *Nature* (*London*) **225,** 199.
91. Zobell, C. E. (1950). *Advan. Enzymol.* **10,** 443.

Author Index

Numbers in parentheses are reference numbers and indicate that an author's work is referred to although his name is not cited in the text. Numbers in bold show the page on which the complete reference is listed.

H

I

J

L

M

R

T

Subject Index

D

H

I

Q

R